High-Frequency Analog Integrated-Circuit Design

WILEY SERIES IN MICROWAVE AND OPTICAL ENGINEERING

KAI CHANG, Editor
Texas A&M University

INTRODUCTION TO ELECTROMAGNETIC COMPATIBILITY
Clayton R. Paul

OPTICAL COMPUTING: AN INTRODUCTION
Mohammad A. Karim and Abdul Abad S. Awwal

COMPUTATIONAL METHODS FOR ELECTROMAGNETICS AND MICROWAVES
Richard C. Booton, Jr.

FIBER-OPTIC COMMUNICATION SYSTEMS
Govind P. Agrawal

ANTENNAS FOR RADAR AND COMMUNICATIONS: A POLARIMETRIC APPROACH
Harold Mott

OPTICAL SIGNAL PROCESSING, COMPUTING, AND NEURAL NETWORKS
Francis T. S. Yu and Suganda Jutamulia

MULTICONDUCTOR TRANSMISSION LINE STRUCTURES
J. A. Brandão Faria

MICROWAVE DEVICES, CIRCUITS AND THEIR INTERACTIONS
Charles A. Lee and G. Conrad Dalman

MICROSTRIP CIRCUITS
Fred Gardiol

HIGH-FREQUENCY ANALOG INTEGRATED-CIRCUIT DESIGN
Ravender Goyal

High-Frequency Analog Integrated Circuit Design

RAVENDER GOYAL
Mentor Graphics Corporation
Wilsonville, Oregon

A WILEY-INTERSCIENCE PUBLICATION
JOHN WILEY & SONS, INC.
NEW YORK / CHICHESTER / BRISBANE / TORONTO / SINGAPORE

This text is printed on acid-free paper.

Library of Congress Cataloging in Publication Data:

Goyal, Ravender.
High frequency analog integrated circuit design / Ravender Goyal.
p. cm. -- (Wiley series in microwave and optical engineering)
"A Wiley-Interscience publication."
Includes index.
ISBN 0-471-53043-3 (alk. paper)
1. Linear integrated circuits--Design and construction. 2. Very high speed integrated circuits--Design and construction. I. Title.
II. Series
TK7874.G67 1994
621.381'32--dc20 93-46163

Printed in the United States of America

10 9 8 7 6 5 4 3 2 1

In memory of
my loving and caring mother

Mrs. Brij Bala Goyal

who left us for her heavenly abode on
January 9, 1995

Contributors

P. E. Allen, Georgia Institute of Technology, School of Electrical and Computer Engineering, Atlanta, GA 30332-0250

C. M. Breevoort, Consultant, 682 Day Lily Court, Acworth, GA 30102-8120

Tzu-Hung Chen, Hexawave Inc., 2 Prosperity Road II, 1F, Science-Based Industrial Park, Hsinchu, Taiwan, ROC

Donald Estreich, Microwave Technology Division, Hewlett Packard Company, Building 1US-E, 1412 Fountaingrove Parkway, Santa Rosa, CA 95403

Ravender Goyal, Mentor Graphics Corporation, 8005 S. W. Boeckman Road, Wilsonville, OR 97070-7777

V. S. Rao Gudimetla, Department of Electrical Engineering and Applied Physics, Oregon Graduate Institute of Science and Technology, 19600 N. W. von Neumann Drive, Beaverton, OR 97006-1999

David Haigh, Department of Electronic and Electrical Engineering, University College, Torrington Place, London WC1E 7JE, UK

Larry E. Larson, Hughes Research Laboratories, MS RL55, 3011 Malibu Canyon Road, Malibu, CA 90265

Pascal Philippe, Laboratoires D'Electronique Philips, 94453 Limeil-Brévannes Cédex, France

Chris Toumazou, Department of Electrical and Electronic Engineering, Imperial College of Science Technology and Medicine, London, UK

Preface

Analog circuit design using silicon bipolar junction transistor (BJT), n-type, complementary, and bipolar complementary metal oxide semiconductor (NMOS, CMOS, and biCMOS) devices, is a well-established technology. Silicon-technology-based analog circuits perform reasonably well up to the frequencies at the lower end of the gigahertz range. Gallium arsenide (GaAs) integrated-circuit (IC) technology became commercially available abundantly as a foundry process since the early 1980s. Most of the ICs using GaAs technology, were previously designed at microwave frequencies, using traditional impedance-matching design techniques. Recent rapid growth in mobile and wireless communications has led to expanded GaAs technology development and its new applications in the lower microwave frequency range, around the subgigahertz to a few gigahertz. As a result there is significant interest in the commercial applications of GaAs analog and digital ICs in systems at L- and C-bands, such as in cellular telephone transmitter and receiver systems at 900 MHz–1.1 GHz, wireless local area network (WLAN), personal communication systems (PCS), personal computer networking (PCN), global positioning satellite (GPS) systems, set-top TV signal converter systems, high-definition television (HDTV) signal processing, and fiber optic transreceivers. Advances in digital circuits based on GaAs technology have proceeded faster than their analog counterparts because of the lack of precise control of electrical properties of GaAs devices fabricated across the wafer and/or from wafer to wafer. However, lately, there have been breakthrough improvements in the manufacturing technology of GaAs metal semiconductor field-effect transistor (MESFET)-based IC processing. This has provided analog designers with a technology with high-performance and high-gain devices with sufficient control of their electrical properties, to design useful high-precision and high-performance analog circuits.

There is an abundance of excellent texts on the topic of silicon-based analog IC design and fabrication design techniques. Also, several texts address design techniques for monolithic microwave integrated circuits (MMICs). The main difference between microwave (including the analog circuits) and traditional analog circuit design techniques is the matching techniques. Microwave circuit designs typically use passive components such as resistors, capacitors, transmission lines, and spiral inductors for matching networks over a desired frequency band. Analog circuits, on the other hand, mostly use active devices for realizing a circuit function such as a differential amplifier, current source, buffer amplifier, mixer, oscillator, and analog-to-digital and digital-to-analog (A/D and D/A) converters. Analog circuit design techniques using high-performance GaAs MESFET devices are similar, in principle, to the traditional analog IC design techniques used in silicon BJT, NMOS, CMOS, and similar technologies. However, because of differences between GaAs MESFET and silicon BJT or MOS device behavior, some of the design techniques are different. This text addresses design techniques for analog ICs based on GaAs MESFETs. This book will prove useful for the designers who are experienced in analog circuit design using silicon devices and who plan to extend their circuit design knowledge to the circuits operating at radiofrequency (RF) and lower microwave frequencies. At the same time designers trained in microwave IC design and familiar with GaAs MESFET will find this text very useful to extend their knowledge to analog design techniques. The topics are covered in a relatively basic manner, to address the needs of new engineer readers. At the same time extensive practical design examples are provided in appropriate circuit design chapters, which makes this text valuable for expert professionals in the field.

Chapter 1 provides an overview of GaAs based analog IC design, technology, and device tradeoffs between MESFET and other GaAs-based three-terminal devices. An overview of silicon-based analog ICs versus GaAs-based analog ICs and high-frequency analog IC applications in general are also discussed in this chapter.

Chapter 2 first describes growth techniques for the substrate material and then details basic steps in fabrication technology. Next, MESFET device fabrication steps are described in detail. The chapter also gives an example of a typical analog IC production process with detailed illustrations.

For successful analog circuit design, it is very important to understand the basic operation of the devices used in circuit design. Chapters 3 and 4 describe active and passive device design considerations and then a variety of device models. Both process-based and empirical models are described for MESFETs. Device characterization and parameter extraction for device modeling are described in detail. Nonlinear, linear, and noise models for MESFETs and Schottky diodes are presented.

Most of the published work in the area of GaAs MESFET-based analog IC design does not discuss the step-by-step details of the circuit design. This is one of the first texts to discuss IC design considerations and actual designs from the very basic concepts such as biasing network, current source, current mirrors, and differential-circuit designs to intermediate-complexity circuit designs such as amplifiers, mixers, oscillators, and operational-amplifier designs, which are presented in detail. Finally higher-level functions such as A/D and D/A converters and their implementation in GaAs technology are discussed in detail.

Chapter 5 describes basic building-block circuits for analog IC design. Over the years a number of basic circuit configurations, or subcircuits, that make efficient use of the advantages of monolithic technology while avoiding most of its shortcomings have been developed. Chapter 5 outlines biasing techniques suitable for GaAs MESFETs, along with the basic design philosophy and the guidelines associated with them. The basic subcircuits covered in this chapter mostly relate to the direct-current (dc) design of a monolithic circuit, with regard to the current and voltage bias levels within the circuit and their drift with temperature and power supply changes. These basic subcircuits, such as active and passive load circuits, current sources and current mirrors, and voltage-shift networks, form a useful set of building blocks, which, along with the basic gain stages, serve as the starting point for designing more complex analog functions and subsystems.

Chapter 6 describes various amplifier design techniques. Important design considerations for amplifier design are discussed and then single-stage, multistage, gain-control, phase-splitting, and transimpedance amplifier circuit designs are detailed.

The operational amplifier (op amp) is by far the most widely known and used class of analog ICs. Since its introduction in the 1960s, the monolithic op amp has proliferated into numerous designs and has found its way into countless applications. Chapter 7 describes op-amp design techniques. The chapter starts with the design techniques for the basic building blocks necessary for designing op amps. It then describes in detail the design of several practical op amps which have been successfully designed in industry as well as in research laboratories.

Chapter 8 describes design techniques for mixers and oscillators, which are important parts of any wireless communication system. This chapter starts with the basic theory and principle of operation of mixers and oscillators. Circuit topologies for a variety of mixers—active and passive, single- and double-balanced—are described, and then practical mixer circuits are provided as design examples. In Section 8.2, different oscillator circuit topologies are described and then practical design examples are given.

Chapter 9 describes A/D and D/A conversion techniques compatible with GaAs MESFET technology. The analog signal processing marketplace is currently dominated by Si bipolar and Si MOSFET technologies. In these components accuracy and speed are the two main driving forces. Most applications for data converters and other signal processing functions require a minimum of 6–8 bits of resolution.

Chapter 9 demonstrates that GaAs is a viable technology for analog signal processing applications including high-speed A/D and D/A converters.

Although silicon technology is clearly leading in innovative approaches toward high-speed architectures, GaAs MESFET technology has no apparent limitations that cannot be overcome by architectural or circuit design solutions. A 12-bit D/A converter operating above 1 GHz has been demonstrated using a current-steering architecture with 50-Ω external load resistors. The inclusion of NiCr resistors on the die greatly enhances the achievable accuracy. GaAs MESFET A/D converters are limited in resolution because of imperfections in the material. Although the performance of A/D converters should steadily improve with improvements in technology, MESFET technology is best suited for architectures that make trade-offs between the performance of analog components and speed. In particular, the pipelined and the ΔΣ architectures are the most promising using present-day GaAs MESFET technology. The former architecture allows for performance trade-offs among several analog components, half the speed of parallel converters, and at least an 8-bit resolution with the use of digital error correction, while the latter is suitable because of its low analog components contents.

Chapter 10 covers the advanced topic of synthesizing linearized circuits with high efficiency and large-signal swings, which can play a major role in the emerging new generation of high-frequency ICs in communication applications. This chapter describes the basic theory for the synthesis of these linearized conductance functions, which use the inherent square-law characteristics of the GaAs MESFET. These linearized circuits are synthesized with linearized conductance functions that can be designed using basic building blocks such as high-gain, high-frequency op amps and sum-and-difference circuits. These circuits can also be easily tuned. Such techniques may eventually lead to important engineering benefits such as an efficient engineering design cycle, reduced time to manufacturing, and lower cost of material and resources.

In summary, this book provides detailed information on GaAs MESFET-based high-frequency analog IC technology and design. It contains appropriate (background and intermediate-level) material for a two-quarter circuit design course at the senior or graduate-level. With the appropriate selection of topics, one can use this book for a one-semester, a one-quarter, or a two-quarter course. A one-semester course may include Chapters 1 to 7, which will equip the students with a good background in high-performance analog circuit design techniques. In a two-quarter course, the first quarter may include Chapters 1 to 6, and the second quarter may include Chapters 7 to 10. This will equip the students with a solid background in designing basic circuits, such as biasing techniques, gain blocks, and advanced circuits such as op amps, mixers, oscillators, signal processing circuits, and synthesis techniques.

I want to most warmly thank all the authors who have contributed to this text and spent their valuable time to deliver the manuscript under tight schedules. I would also like to thank 400–500 of those designers, engineers, experts, and managers who have attended my seminars on similar topics, whose curiosity, questions, and feedback have really helped me focus on developing the framework of this book. Additionally, I would like to thank numerous experts in the universities and industry—practically all over the world, who have provided valuable critical reviews and suggestions in the preparation of this text. I wish to thank all my present and former colleagues at Mentor Graphics Corporation, Anadigics Inc., Avantek, and Honeywell. Many thanks to George Telecki and Cynthia Hess of John Wiley & Sons, who have been very patient, cooperative, and supportive all along the development of the manuscript of this book. I thank my parents, whose hard-working nature has always inspired me in my life; my sister Dr. Anita Bansal and my brother Dr. Rakesh Goyal, who always supported my decisions; and last but not least, my wife Poonam, who has been patient and supportive throughout the preparation of this text.

RAVENDER GOYAL

Lake Oswego, Oregon
September 1994

Contents

High-Frequency Analog Integrated-Circuit Design

CHAPTER ONE

Overview

RAVENDER GOYAL
Mentor Graphics Corporation
Wilsonville, Oregon

Significant progress has been made in the advancement of GaAs substrate material and GaAs-based device manufacturing technologies during the past several decades. Presently, GaAs MESFET technology has established a secure position as one of the best solutions for high-quality microwave and millimeter-wave discrete devices. Research activities began on the design and manufacturing of MMICs based on already matured MESFET technologies, sometime in the late 1960s at Texas Instruments [1,2]. Further, the level of interest and the activities in MMIC development increased dramatically after Plessey reported the first GaAs X-band amplifier with a single MESFET and lumped matching elements on a single chip [3]. MMICs have established themselves now as a major component in microwave and RF subsystems and systems for commercial applications. The increased enthusiasm for GaAs MMICs is due to significant improvements in a number of diverse technical areas including inherent superiority of the material, lower cost, small size, light weight, and improved reliability and reproducibility, as well as better tolerance to heat and radiation.

Successful application of GaAs substrate to the fabrication of MMICs encouraged developers to use silicon substrate to design high-frequency circuits. Because of the lossy nature of silicon substrate, it was not possible to produce ICs high-frequency based on silicon substrate. This led to the use of oxide-isolated, complex device structures using bipolar devices, which improved the performance of silicon-based MMICs at the expense of

High-Frequency Analog Integrated-Circuit Design, Edited by Ravender Goyal
ISBN 0-471-53043-3

complex processing technology. However, GaAs still provides a competitive edge over silicon for high-frequency circuit design and manufacturing. While currently it is well established that GaAs is the only common material useful for commercial MMICs at frequencies above 1–2 GHz, silicon still leads the market in the subgigahertz (<1 GHz) region with some exceptions where noise is the major consideration. Furthermore, manufacturing process technologies for GaAs are still in the early stages of development compared to silicon technology. Today, several million transistors can be integrated on a single chip (state-of-the-art microprocessor chips are reported to have approximately 3 million transistors on a single chip) in silicon, in production quantities. However, intrinsic GaAs properties permit the production of amplifiers with lower noise figure, wider bandwidth, higher power, and better isolation than can be achieved in silicon, which are all the major considerations while designing and fabricating high-frequency circuits.

GaAs technology, as yet, is widely used for monolithic microwave and millimeter-wave devices and high-speed digital circuits. However, there is an emerging and fast-growing area of high-performance and high-precision analog and linear devices, using GaAs substrate. These include operational amplifiers, intermediate-frequency (IF) amplifiers, analog and digital converters, switched capacitor filters, sample-and-hold circuits, and transimpedance amplifiers.

1.1 WHY GALLIUM ARSENIDE?

The first article published on the electronic properties of III–V compound semiconductors appeared in 1952. Since that time, the importance of GaAs as III–V semiconductor has been studied for its application in RF and microwave circuit application. GaAs as a semiconductor material has two basic advantages over silicon. First, the ability to produce semi-insulating GaAs substrate allows greatly reduced parasitic capacitance, which allows true monolithic IC implementation, operating at high frequencies from 1–40 GHz or more. The bulk resistivity of semi-insulating GaAs material typically used in manufacturing ICs is above $10^8 \Omega \cdot cm$, whereas typical bulk resistivity of silicon used in manufacturing ICs is approximately $100 \Omega \cdot cm$. This resistivity difference of above six orders of magnitude results in reduced parasitic effects when ICs are implemented on a GaAs semi-insulating substrate. This also simplifies the integration of many devices on the same IC, as the electrical isolation between the devices is obtained inherently compared to a complex oxide isolation process typically used in silicon devices. Second, there is higher electron mobility and higher saturated electron drift velocity in GaAs compared to that in silicon, as shown in Figure 1.1. This means that electrons move faster in GaAs material than in silicon, which makes GaAs a more attractive material of choice for fabricating devices that could operate at higher frequencies in the

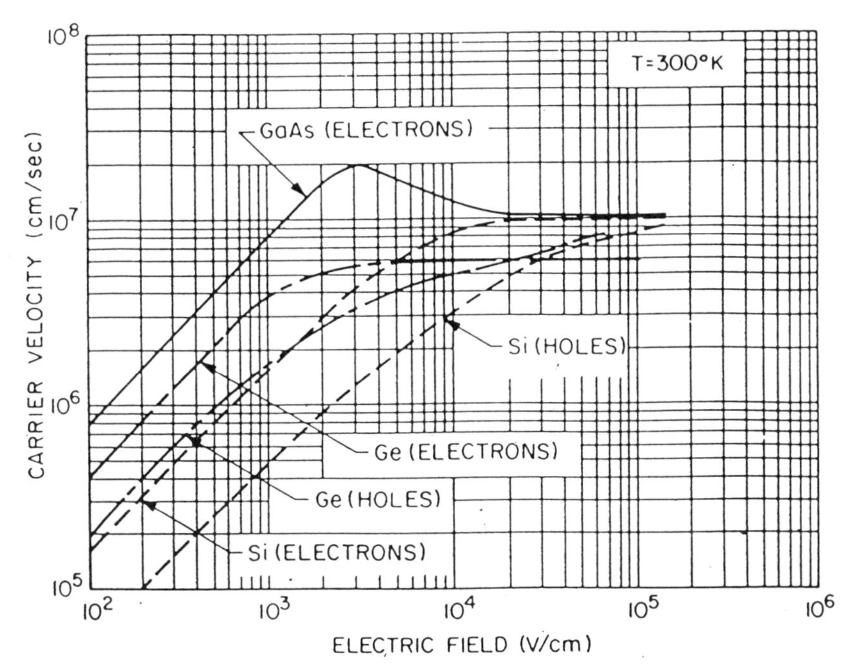

FIGURE 1.1 Measured carrier velocity versus electric field for high-purity Ge, Si, and GaAs. *Source*: S. M. Sze, *Physics of Semiconductor Devices*, Wiley, New York, 1985.

RF and microwave regions. Table 1.1 provides a comparison of electrical properties of semi-insulating silicon and GaAs materials. As evident from this table, energy bandgap in GaAs (1.43 eV) is larger compared to that in silicon (1.12 eV), which makes GaAs devices more radiation-tolerant than silicon MOS and bipolar devices.

Semi-insulating GaAs substrate, used in IC manufacturing, must meet the following requirements to achieve high-performance and high-yielding ICs:

The material must have as high a resistivity as possible, better than $10^8\ \Omega \cdot \text{cm}$.

There must be minimum density of crystalline defects such as dislocations, stacking faults, and precipitates.

There must be minimal or no undesirable substate-to-active-layer (channel) interface effects, which may result in adverse electrical effects such as backgating and light sensitivity.

High resistivity of the semi-insulating substrate must be maintained during the high-temperature processing steps such as epitaxial growth or annealing of ion-implanted active layer.

There should be no degradation of active-layer properties by outdiffusion of impurities from substrate during high-temperature processing steps.

TABLE 1.1 Properties of Si and GaAs (at 300 K)

Properties	Si	GaAs
Atoms/cm^3	5.0×10^{22}	2.21×10^{22}
Atomic weight	28.02	144.63
Breakdown field (V/cm)	$\sim 3 \times 10^5$	$\sim 4 \times 10^5$
Dielectric constant	11.8	10.9
Effective density of states in conduction band, N_c (cm^{-3})	2.8×10^{19}	4.7×10^{17}
Effective density of states in valence band, N_v (cm^{-3})	1.02×10^{19}	7.0×10^{18}
Effective mass m^*/m_0		
Electrons	$m_l^* = 0.97$ $m_t^* = 0.19$	0.068
Holes	$m_{lk}^* = 0.16$ $m_{hh}^* = 0.5$	0.12, 0.5
Intrinsic carrier concentration (cm^{-3})	1.6×10^{10}	1.1×10^7
Melting point (°C)	1420	1238
Minority carrier lifetime (s)	2.5×10^{-3}	$\sim 10^{-8}$
Mobility (drift) (cm^2/V-s)		
μ_n (electrons)	1500	8500
μ_p (holes)	600	400
Work function (V)	4.8	4.7
Thermal conductivity at 300 K (W/cm °C)	1.45	0.46

The difficulty in achieving these requirements for semi-insulating bulk material has led to the development of buffer-layer technology, in which a relatively thick, high-resistivity epitaxial layer is grown on the semi-insulating substrate. The active layer is then produced by either a second epitaxial-layer growth on top of the high-resistivity epitaxial layer or by ion implantation into the buffer layer. This procedure is successfully used in masking undesirable substrate properties, but increases the cost of additional processing steps, which adds extra cost to the overall starting material cost.

While there are many advantages of using semi-insulating GaAs material for high-precision, high-frequency analog ICs, it has the following limitations:

Poor native oxide
High defect density
Lower thermal conductivity than silicon
Relatively volatile material due to arsenic contents in GaAs

Relatively fragile substrate
Expensive material

1.2 HIGH-FREQUENCY ANALOG CIRCUIT REQUIREMENTS

The choice of input and output gain stages is among the most critical steps in any analog IC design. The differential amplifier, whose basic function is to amplify the difference between the two input signals, represents one of the most widely used class of gain stages in analog circuit design. The bias level and the gain characteristics of a differential stage, by and large, depend on the symmetry between the two branches of the circuit. Since close matching is inherent to the monolithic components, this balanced nature of the differential amplifier makes it ideal as a gain block for ICs. Since the matching and temperature-tracking properties of monolithic components are far better than those of their discrete counterparts, the performance characteristics of an integrated differential-gain stage is, in general, superior to that of a discrete-components-based one. Another important advantage of differential-gain stages is that they can be directly cascaded or coupled to one another, without requiring extensive level shifting or interstage coupling capacitors.

The output stages are the gain blocks or subcircuits designed to provide large signal swings into an output load, with minimum signal distortion. The output stage of an amplifier must be able to deliver a substantial amount of power into a low-impedance load with acceptably low levels of signal distortion. Therefore, in general, output stages in analog circuits are required to have one or more of the following desirable properties:

Large output current swing
Large output voltage swing
Low output impedance
Low standby power
Good frequency response
Built-in short-circuit protection

The monolithic operational amplifier, or op amp, is by far the most widely known and used class of analog ICs. Since its introduction in 1960s, the monolithic op amp has proliferated into numerous designs and has found its way into countless applications. An "ideal" op amp is a differential input, single-ended output voltage amplifier that offers infinite voltage gain with infinite input impedance, infinite bandwidth, and zero output impedance. Although real-life op amps do not exhibit such idealized characteristics, their performance is often sufficient to approximate the properties of the ideal op amp at low frequencies. Operational amplifier specifications are defined below:

1. *Input Offset Voltage*. This is the voltage required at the differential input of an op amp to reduce the output voltage to zero.
2. *Input Offset Current*. This is the mismatch between the bias current at each input terminal of the op amp.
3. *Input Resistance*. This is a function of the input stage configuration and the bias current levels. If the open-loop voltage gain is large, the value of R_{in} has a negligible effect on circuit performance.
4. *Output Resistance*. This depends on the output stage configuration and choice of the current-limiting circuitry used at the output. For general-purpose op amps, R_{out} is in the range of 20–200 Ω.
5. *Input Common-Mode Range*. This is the maximum range of the input voltage that can be simultaneously applied to both inputs without causing the cutoff, saturation, or breakdown of any of the gain stages inside the op amp. Typically, the common-mode range approaches to within a few volts of either supply voltage.
6. *Common-Mode Rejection Ratio*. This is the ratio of the differential open-loop gain to the common-mode open-loop gain.
7. *Power Supply Rejection Ratio*. This is the ratio of the change of input offset voltage for a unit change of any one of the supply voltage.
8. *Open-Loop Voltage Gain*. This is the ratio of the incremental change of the output voltage for a unit incremental change of the differential-input voltage, with no feedback applied. It is measured at low frequencies.
9. *Unity-Gain Bandwidth*. This is small-signal 3 dB (3-decibel) bandwidth for unity-gain closed-loop operation.
10. *Slew Rate*. This is the maximum rate of change of the output voltage for a step input; it is normally specified in volts per microsecond. It is measured at the zero crossing point of the output waveform, when the op amp is compensated for unconditional stability. Slew rates may differ for positive- or negative-going output waveforms.
11. *Full-Power Bandwidth*. This is the maximum frequency of sinusoidal input signal over which the full output swing can be obtained.
12. *Settling Time*. This is the time taken for the output to settle to within $\pm 0.01\%$ of its final value for a step change of the input.

The basic differential-gain stage for an op amp design must fulfill some of the following important requirements:

High input resistance ($>100\ \text{k}\Omega$)
Low input bias current ($<500\ \text{nA}$)

TABLE 1.2 Minimum Requirements for Op Amp and Comparator [4]

Parameter	Unit	Op Amp	Comparator
Gain	dB	>60	>60
Unity-gain bandwidth	GHz	1	-
Slew rate	V/μs	>1000	
Input impedance	GΩ	>100	
Offset voltage over input CMR[a]	mV		<20
Input CMR from V_{dd}, V_{ss}	V	±2	±1
Phase margin, $C_L = 0.2$ pF[b]	degrees	60	
Delay	ns		<1
Output swing from V_{dd}, V_{ss}	V	±1	±2
Power dissipation	mW	<50	<50

[a] Common-mode rejection.
[b] Loading capacitance.

Small input voltage and current offset
High common-mode rejection ratio (>60 dB)
High common-mode range (≥ one-half of total supply voltage)
High differential-input range (≥ one-half of total supply voltage)
High voltage gain (>40 dB)

A general-purpose op amp must fulfill some of the following important requirements (minimum requirements for op amps and comparators are listed in Table 1.2):

High open-loop gain (>1000)
Wide power supply operating range
Common-mode input voltage range that changes with supply voltage
Output voltage swing that changes with supply voltage

Typical requirements for analog, digital, and microwave circuits are listed in Table 1.3.

1.3 GaAs MESFETs AND ANALOG ICs

GaAs exhibits a field-dependent electron velocity that is approximately 4–5 times higher than that in silicon. This corresponds to a much lower drain-to-source voltages for optimum performance of GaAs devices compared to silicon devices of similar geometry. As a result, GaAs circuits can be designed with lower-voltage supplies and to dissipate less power than the

TABLE 1.3 Typical Parameters and Requirements for Digital, Microwave, and Analog ICs [5]

Digital	Microwave	Analog
Noise margin	Concersion gain	Voltage gain
Loading capacitance	Gain compression	Phase margin
Speed–clock rate	Third-order intercept	Resolution
Rise–fall time	VSWR	Linearity
Power dissipation	Noise figure	Offset voltage
	Bandwidth (narrowband to ≤1–2 decades)	Hysteresis
	Operation of device at dc or zero frequency seldom required	Noise (shot, thermal, $1/f$)
		Slew rate
		Bandwidth (typically 5–7 decades)
		Required to operate at or near dc
		Common-mode rejection ratio
		Power supply rejection ratio
		Power dissipation

corresponding silicon devices. Noise performance of GaAs devices is better than for comparable silicon devices, particularly at higher frequencies. Typically, GaAs is two orders of magnitude more radiation-tolerant than silicon devices because of its higher energy bandgap. For the same reason, GaAs devices perform better at both low and high temperature extremes. Gain of GaAs MESFET and silicon-based bipolar transistors varies with changes in device temperature. Although the actual gain variation in an amplifier stage depends on a number of design factors, as a rule of thumb, gain of a GaAs FET amplifier stage varies at a rate of approximately 0.015 dB/°C. Thus a 10-stage amplifier with a nominal gain of 50 dB at 25°C will have a gain of approximately 35 dB at 125°C and similarly will have a gain of about 65 dB at −75°C, if no temperature compensation technique is used. Bipolar transistor stages have nonlinear temperature–gain characteristics, with maximum gain occurring at approximately 0°C as shown in Figure 1.2.

There are other device structures based on GaAs substrate, that are used in designing analog ICs. The performance characteristics of GaAs MESFET, other GaAs device structures, and silicon-based bipolar and CMOS devices are compared in Table 1.4.

Other passive components, which are a critical part of successfully designing high-precision analog circuits, such as monolithic resistor, high-precision NiCr, or other thin-film resistors, MIM capacitors, and spiral inductors are readily available in GaAs-based MESFET process, with minimal increase in the number of processing steps. Also, it is easier to integrate GaAs ICs with the sensing elements on the same substrate as

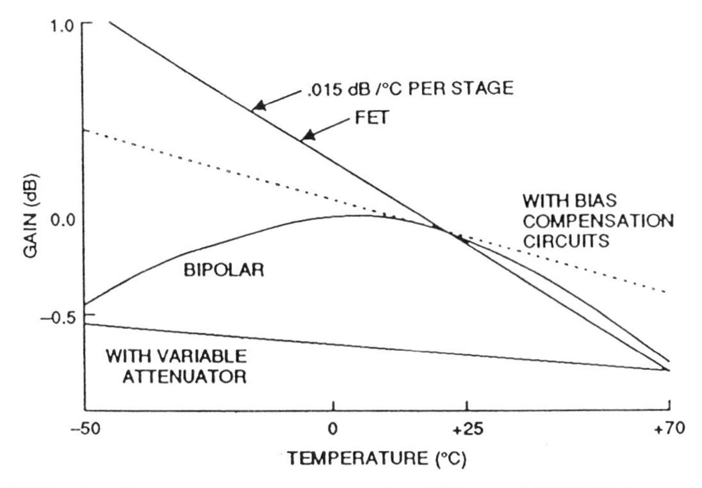

FIGURE 1.2 Gain versus temperature for BJT and MESFET transistors.

TABLE 1.4 Electrical Parameters for Various GaAs Device Structures [6]

			GaAs			Silicon
Parameter	Unit	MESFET	HEMT	HBT	BJT	MOSFET
F_t	GHz	15	20	30–40	5–10	1
F_{max}	GHz	35	50	25	10–15	3
g_m	mS/mm	450	200	2000–4000	3000	10
$1/f$ noise	μV	40	40	3	1	10
$C_{ds} + C_{d,sub}(C_{c,sub})$	pf/mm	0.25	0.25	1	<1	1
$C_{gd}(C_{bc})$	pf/mm	0.2	0.2	2	2	0.3
I_g, I_b ($I_d, I_c = 1$ mA)	μA	≪1	≪1	25–50	12	≪1
V_{gs} (V_{be}) match	mV	5–20	5–20	1	<1	5–20
BV_{ds} (BV_{CE})	V	6–10	6–10	8–10	8	6–12
Hysteresis	mV	20	10	<1	<1	≪1

HEMT, high electron mobility transistor.
HBT, hetero junction bipolar transistor.
BJT, bipolar junction transistor.

GaAs heterostructure devices are widely used for optoelectronics and in microsensors.

Reasons for using GaAs MESFET devices for high-precision and high-performance analog ICs can be summarized as follows:

High speed
Low power consumption
High coupling efficiency with sensing elements
Low noise
High radiation tolerance

Integration with detectors
Extreme-temperature operation safety
Low parasitic capacitance

1.4 DISTORTION IN MESFETs

Application of a device to analog circuit design strongly depends on the linear behavior of the device over a wide range of input voltages. The input signal is distorted at the output, if the amplifying active device has nonlinear behavior. So, it is important to review the linear behavior of MESFETs. The current–voltage I–V characteristics of a MESFET-drain current I_{ds} as a function of the input voltage (gate-to-source voltage) and pinch off voltage of the device, can be represented by a simple relationship as given in Equation (1.1). A complete and detailed MESFET model is described in Chapter 3.

$$I_{ds} = \beta(V_{gs} - V_{po})^n \tag{1.1}$$

where β = transconductance parameter
V_{gs} = gate-to-source voltage
V_{po} = pinchoff voltage
n = device linearity quality factor

Let us consider the simple MESFET amplifier circuit shown in Figure 1.3, with a single MESFET, a load resistor, input dc voltage V_{dc} for biasing the gate with respect to the source terminal, and input alternating-current (ac) signal voltage V_{ac}. Using Equation (1.1), we can express the drain current of

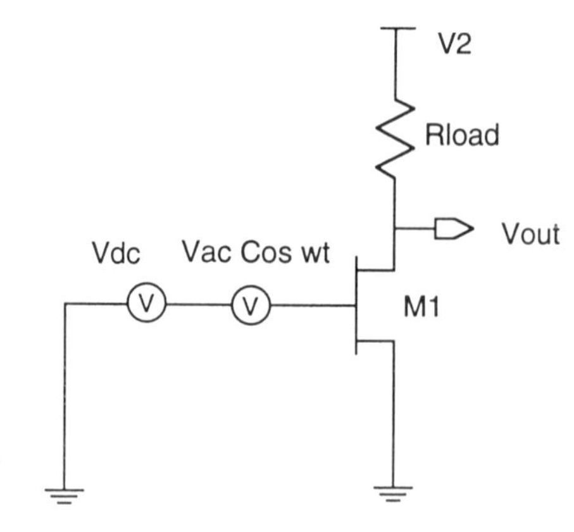

FIGURE 1.3 A Basic small signal amplifier circuit.

MESFET M1 as

$$V_{gs} = (V_{dc} + V_{ac}\cos\omega t)$$

$$I_{ds} = \beta\{(V_{dc} - V_{po}) + V_{ac}\cos\omega t\}^n \tag{1.2}$$

considering

$$a = (V_{dc} - V_{po}), \qquad b = V_{ac}\cos\omega t$$

and using the expansion series

$$(a+b)^n = a^n + na^{n-1}b + \frac{n(n-1)}{2!}a^{n-2}b^2 + \frac{n(n-1)(n-2)}{3!}a^{n-3}b^3 + \cdots$$

Equation (1.2) can be represented as

$$\begin{aligned}\frac{I_{ds}}{\beta} = &(V_{dc} - V_{po})^n + n(V_{dc} - V_{po})^{n-1}V_{ac}\cos\omega t \\ &+ \frac{n(n-1)}{2!}(V_{dc} - V_{po})^{n-2}(V_{ac}\cos\omega t)^2 \\ &+ \frac{n(n-1)(n-2)}{3!}(V_{dc} - V_{po})^{n-3}(V_{ac}\cos\omega t)^3 + \cdots\end{aligned} \tag{1.3}$$

or

$$\begin{aligned}\frac{I_{ds}}{\beta} = &(V_{dc} - V_{po})^n + n(V_{dc} - V_{po})^{n-1}V_{ac}\cos\omega t \\ &+ \frac{n(n-1)}{2!}(V_{dc} - V_{po})^{n-2}V_{ac}^2\frac{(1+\cos 2\omega t)}{2} \\ &+ \frac{n(n-1)(n-2)}{3!}(V_{dc} - V_{po})^{n-3}V_{ac}^3\frac{(3\cos\omega t + \cos 3\omega t)}{4} \\ &+ \frac{n(n-1)(n-2)(n-3)}{4!}(V_{dc} - V_{po})^{n-4}V_{ac}^4\frac{(3 + 4\cos 2\omega t + \cos 4\omega t)}{8} + \cdots\end{aligned} \tag{1.4}$$

Now, assuming $V_{ac} \ll (V_{dc} - V_{po})$, specifically, for small-signal input compared to the bias point of the MESFET, the harmonic distortion contents can be expressed as follows:

Magnitude of second harmonic contents:

$$\frac{(n-1)}{2!}(V_{dc} - V_{po})^{-1}\tfrac{1}{2}V_{ac}$$

Magnitude of third harmonic contents:

$$\frac{(n-1)(n-2)}{3!}(V_{\mathrm{dc}}-V_{\mathrm{po}})^{-2}\ \tfrac{1}{4}V_{\mathrm{ac}}^{2}$$

Magnitude of fourth harmonic contents:

$$\frac{(n-1)(n-2)(n-3)}{4!}(V_{\mathrm{dc}}-V_{\mathrm{po}})^{-3}\ \tfrac{1}{8}V_{\mathrm{ac}}^{3}$$

Harmonic distortion equations indicate that as the magnitude of the input ac signal increases, the distortion contents in the output signal also increase. Also, as the pinchoff voltage of the device increases, the distortion contents in the output signal reduces, which means that larger pinchoff voltage devices are more linear, or have larger dynamic range compared to smaller pinchoff voltage devices. These are well understood and widely used concepts in microwave MESFET design. Further, closely examining the

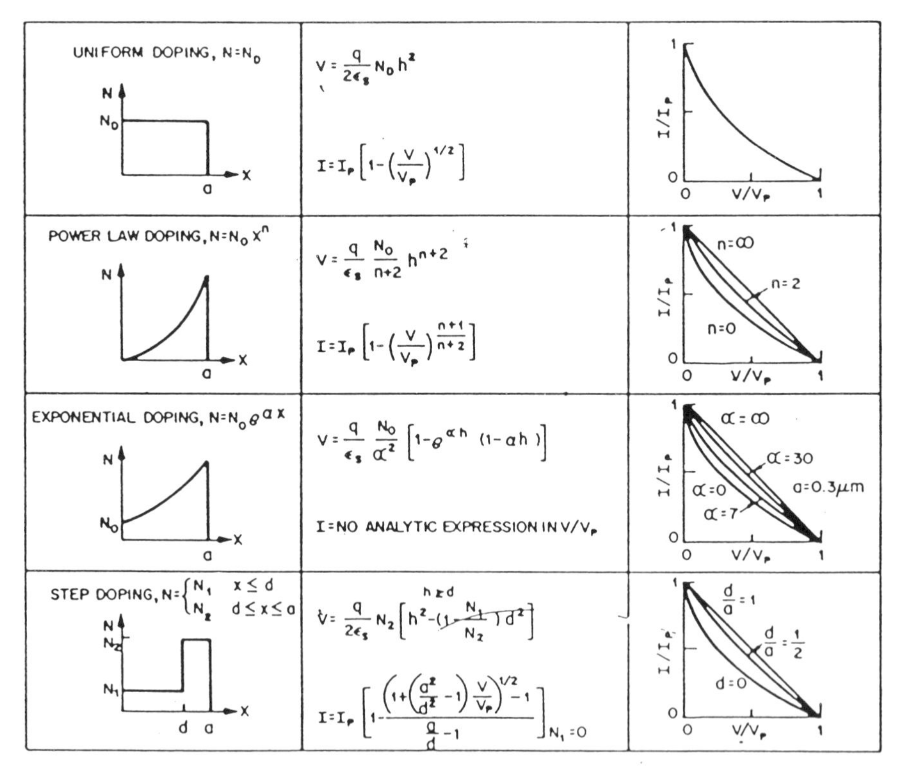

FIGURE 1.4 Typical characteristics of various charge distributions. *Source*: S. M. Sze, *Physics of Semiconductor Devices*, Wiley, New York, 1985.

distortion equations reveal that harmonic distortion contents also depend on n, the device linearity factor. As the value of n increases, the contents of harmonic contents also increase. The optimum value of n is 1, in which case all harmonic distortions are zero. However, it is not practical to fabricate devices with quality factor n of value 1. Table 1.5 indicates the values of harmonic distortion contents for various values of n, which are practically achievable. It can be observed from Table 1.5 that the third harmonic distortion is zero for $n = 2$, and the second harmonic distortion is 6 dB higher than that for $n = 1.5$. Also, second harmonic distortion is largest for $n = 2$ and improves 6 dB each, as the value of n decreases to 1.5 and 1.25 and finally becomes zero for $n = 1$. In short, the closer the value of n to 1, the better the distortion performance of the device, which corresponds to better applicability to analog circuit design.

The quality factor n depends on the doping profile in the channel region of MESFETs. As shown in Figure 1.4 [7], the value of $n = 1$ is achieved when the doping profile in the channel region is linear, which is difficult to achieve, using ion-implantation technique, which is the most commonly used technique for analog IC fabrication. A typical doping profile using ion-implantation technique is Gaussian, which results in a value of n in the range of 1.4–2.0.

TABLE 1.5 Value of *n* and Signal Distortion Contents

n	Second Harmonic Contents	Third Harmonic Contents
1	0	0
1.25	$\frac{1}{16}\frac{V_{ac}}{V_{dc}-V_{po}}$	$\frac{1}{128}\frac{V_{ac}^2}{(V_{dc}-V_{po})^2}$
1.5	$\frac{1}{8}\frac{V_{ac}}{V_{dc}-V_{po}}$	$\frac{1}{96}\frac{V_{ac}^2}{(V_{dc}-V_{po})^2}$
2.0	$\frac{1}{4}\frac{V_{ac}}{V_{dc}-V_{po}}$	0
0.75	$\frac{1}{16}\frac{V_{ac}}{V_{dc}-V_{po}}$	$\frac{5}{384}\frac{V_{ac}^2}{(V_{dc}-V_{po})^2}$
0.5	$\frac{1}{8}\frac{V_{ac}}{V_{dc}-V_{po}}$	$\frac{1}{32}\frac{V_{ac}^2}{(V_{dc}-V_{po})^2}$
0.66	$\frac{1}{12}\frac{V_{ac}}{V_{dc}-V_{po}}$	$\frac{1}{54}\frac{V_{ac}^2}{(V_{dc}-V_{po})^2}$

1.5 GaAs MESFET SHORTCOMINGS

While significant progress has been made for many years in GaAs MESFET-based analog and RF integrated circuit technology and design, problems in GaAs technology have limited the design efforts and advancements of high-precision analog circuits. Progress in analog GaAs circuit design has greatly lagged behind microwave and digital circuit developments, mainly because of the technology immaturity and lack of suitable device models, accurately representing the nonlinear behavior of the MESFET device.

A number of GaAs MESFET-based op amps, which represent a common silicon analog IC building block, have been developed during the last several years. These circuits exhibit wider bandwidth than their silicon counterparts, but some of their low-frequency characteristics are not desirable. The design of these circuits in silicon technology is usually enhanced by the availability of both p- and n-type (positive- and negative-type) devices. Unfortunately, only n-type devices are available in typical GaAs MESFET processes, somewhat inhibiting the flexibility of the designs. Furthermore, since the speed of p-type devices in GaAs-based MESFETs is much worse that their n-type counterparts, their utility for high-speed analog circuit design is questionable, even if they do eventually become available.

A typical GaAs MESFET structure, as described in detail in Chapter 3, is a four-terminal device, which has the back of the active channel floating. Because of the finite substrate resistivity and capacitive coupling between the channels of adjacent devices, through the substrate, the backgate can modulate the depletion region under the channel, similar to the frontgate and hence modulate the drain current; this is known as *backgating* and *sidegating*. Device matching is a major problem because of the backgating and sidegating effects in certain circuit topologies widely used for analog circuit designs. The backgating effect and the device matching requirements place conflicting requirements on the layout. In order to achieve better matching from substrate defect and material variations across the wafer, one would like to place the devices as close to each other as possible (in some cases in bipolar circuits, devices are interleaved for better matching); on the other hand, devices should be placed far from each other to reduce backgating and sidegating effects, let along interleaving the gate fingers of the two devices. Floating backgate also results in another anomaly, known as *looping* or *drain lagging*. When the drain voltage is changed by input ac signal, the drain current is modulated with a long transient decay time of the order of 1 ms–1 s. This is due to the change in device drain conductance, as a function of signal frequency. Drain conductance may vary by a factor of 5–10 in a typical GaAs MESFET device as the frequency of the signal changes from a few kilohertz to several gigahertz. The frequency at which the change in conductance occurs is highly dependent on the nature of the semi-insulating material, the method of growth used in manufacturing the substrate material, and any intentional or unintentional impurities in the substrate.

Electron trapping, surface effects at the semi-insulating-substrate to channel region interface, high-defect-density material, and poor uniformity of the active-layer doping profile contribute to variations of the *I–V* characteristics of MESFETs.

Another problem in using GaAs semi-insulating substrate is related to the nature of electron mobility–electric field curve in GaAs. As shown in Figure 1.1, there is a negative slope of electron velocity around 500 V/cm of the electric field. This results in a negative gradient of electrons and produces low-frequency oscillations in the semi-insulating substrate. As these oscillations happen in the substrate, these can modulate the active channel depletion region and hence modulate the drain current via backgate. In a typical analog and digital circuits operating with typical power supply voltage, achieving 500 V/cm electric fields in the devices are quite typical and hence the low-frequency oscillation phenomenon is quite observable. The frequency of oscillations depends on several factors such as the purity of the semi-insulating substrate, method of growth used in manufacturing the substrate, and temperature of the device. In a typical device the frequency can be as low as few hertz to as high as tens of kilohertz, depending on the temperature. These substrate oscillations contribute to the overall noise of the circuit, which can be as high as a few millivolts. The magnitude of low-frequency device noise has shown dramatic improvement in recent years. However, it is anticipated that this noise will remain relatively high compared to that in silicon bipolar devices for the foreseeable future.

Characteristics of $1/f$ noise is another major consideration in GaAs MESFETs. At lower frequencies, $1/f$ noise in GaAs devices can be several orders of magnitude higher in comparison to $1/f$ noise in silicon bipolar devices. Also, the knee of the $1/f$ noise is at much lower frequency in silicon devices (typically 100 Hz) compared to that in GaAs devices (typically at 100 MHz). The relatively high level of $1/f$ noise and low-frequency oscillations in a typical GaAs MESFET can result in a poor dynamic range in the amplifier circuits.

Finally, the transconductance of GaAs MESFETs is much smaller (in the range of several hundred millisiemens per millimeter) compared to that of silicon bipolar devices (in the range of several millisiemens). This limits the overall gain that can be achieved in analog circuits like op amps, which require extremely large open-loop gains of the order of 100–150 dB which are typically achievable in silicon devices, whereas GaAs-based op amps are limited to 60–80 dB gains only.

The main problems in designing linear or analog circuits using GaAs substrate and MESFET devices can be summarized as follows:

Poor absolute accuracy of electrical parameters of MESFETs across the wafer

Poor pair matching of electrical parameters

Poor device isolation due to backgating and sidegating

Lack of complementary p-channel MESFET device

Frequency-dependent output voltage gain or drain resistance
Low-frequency substrate oscillations
High $1/f$ noise
High gate leakage current at reverse gate bias, due to a Schottky device at the gate terminal
Drain current transients with time constants on the order of milli-seconds to seconds
Piezoelectric effects
Light-sensitive effects
Lack of availability of accurate MESFET device models for time nonlinear circuit analysis.
Lack of availability of suitable high-frequency packages at suitable cost

This is due mainly to the adverse side effects of semi-insulating substrate used in manufacturing GaAs analog ICs. The substrate impurities are mainly responsible for backgating effects that result in hysteresis, poor device matching from one device to the other on a single wafer and from wafer to wafer, low transconductance of the devices to be useful for signal processing circuits, and finally poor yield of the devices, resulting in nonworking circuits, as signal processing circuits tend to use large number of devices that need to be matched closely.

Substrate anomalies also contribute to nonideal behavior, such as hysteresis and long-settling transients, which are especially bothersome in analog ICs. The degree to which individual MESFETs can be matched to each other is relatively poor compared to that for bipolar transistors, which result in high offset voltage in differential amplifiers, and poor uniformity and yield in a variety of matched circuits.

1.6 MESFET-BASED ANALOG ICs

Design of the on-chip biasing circuitry is a starting point of an analog IC design. This first design step is very critical since it determines the internal voltage and current levels over all operating conditions of the IC as well as over all manufacturing process variations. For a linear circuit designer trained in the area of discrete circuits, the basic constraints and limitations of monolithic circuit technology often pose a difficult challenge. This is particularly true with regard to the biasing circuitry. Many of the conventional biasing techniques may not be directly applied to monolithic designs because of the following limitations of IC components:

Poor absolute-value tolerance
Poor temperature coefficients

Limitations on component values
Limited choice of compatible active devices

On the other hand, IC fabrication methods offer a number of unique and powerful advantages to the circuit designer:

Availability of a large number of active devices
Good matching and tracking of component values
Close thermal coupling
Control of device layout and geometry

Low-noise amplifiers in the 1–2 GHz frequency band are now required for IF amplifiers for several applications such as Direct Broadcast Satellite (DBS) converters, cable TV (CATV), and mobile radio systems, optical communications systems, Global Positioning Satellite (GPS) system, and other applications. Cellular phone transmitters require power amplifier at 900 MHz–1 GHz range. For these high-volume applications, traditional hybrid technology is no more competitive, since ICs offer their inherent advantages of smaller size, better reproducibility, higher performance, high-volume manufacturing, and lower cost. Thus a wide variety of monolithic IC RF amplifiers operable in the lower low gigahertz range have been developed by many companies worldwide, and many of these amplifiers are now commercially available and accepted in the system applications. Such amplifiers are typically divided into two broad categories: direct-coupled and capacitively coupled feedback amplifiers. Detailed design considerations for these amplifiers are described in Chapter 5.

GaAs MESFET-based direct-coupled amplifiers have the following general characteristics:

Broadband characteristic from dc (due to direct coupling) to several gigahertz
Smaller in size, so it is suitable for high-volume manufacturing

Topologically these amplifiers are similar to buffered FET logic gates biased into the linear region of operation.

Capacitively coupled amplifiers have the following general characteristics:

Wide bandwidth
Gain flatness over the bandwidth
Excellent input/output voltage standing wave ratio (VSWR)

Some examples of GaAs-based analog ICs are as follows. However, these are just examples and by no means constitute an exhaustive list of such ICs:

An op amp with 63 dB dc gain, 20 GHz gain bandwidth product at 500 MHz, and 550 mW power dissipation, while driving a 50 Ω load. The op amp has 19-mV offset voltage, 600 V/μs slew rate. This circuit was fabricated using 0.2 μm gate length MESFET with 80 GHz cutoff frequency [8].

A 2-Gbit/S comparator with a 1-mV resolution. This uses a self-calibrating scheme. Offsets that are caused by any of the typical reasons in GaAs technology as described earlier such as low-frequency oscillations, higher transconductance at lower frequencies, and dc offset are canceled in this comparator.

A second-order bandpass switched-capacitor filter and a third-order low-pass filter at 100-MHz sampling rates and 60 dB dynamic range [9]. This circuit uses a 1 μm gate length depletion-mode MESFET technology. Using the same technology, a two-pole bandpass filter is also reported at 250 MHz switching frequency and 10 MHz center frequency.

An 8-nS sample-and-hold amplifier uses an operational amplifier with feedback to reduce the effect of low-frequency rolloff in voltage gain. An offset step pedestal and dc offset voltage as low as 20 mV is achieved. The full power bandwidth of 170 MHz, 8 nS acquisition time, and 5 nS sample-to-hold settling time is comparable with high-performance hybrid sample-and-hold amplifiers. High noise of the device limits its application to <10-bit accuracy [10].

An 8-bit 1-Gsample/s (gigasample per second) D/A converter. DC accuracy of several devices may be as high as 12 bits [11]. This circuit uses on-chip current sources using MESFETs with a gate length of 3 μm. The circuit also uses high-precision NiCr resistors for biasing the current source transistors.

A 5-bit A/D converter at 2.2 GSamples/s with effective resolution of 4.5 bits at 400 MHz is reported. This devices dissipates 320 mW.

1.7 APPLICATIONS OF GaAs ANALOG ICs IN WIRELESS COMMUNICATION SYSTEMS

Today, there are several high-volume electronic systems where GaAs-based analog ICs are widely used. Most of these systems are related to telecommunications in ~1 GHz or higher frequency range or in optical communication, high-performance instrumentation, and high-performance signal processing. Applications of GaAs MESFET-based ICs are described in Table 1.6, and some of these system insertions are described below.

Direct Broadcast Satellite (DBS) Receiver. A block diagram of a DBS receiver operating over 11.7–12.2 GHz is shown in Figure 1.5 [12]. The

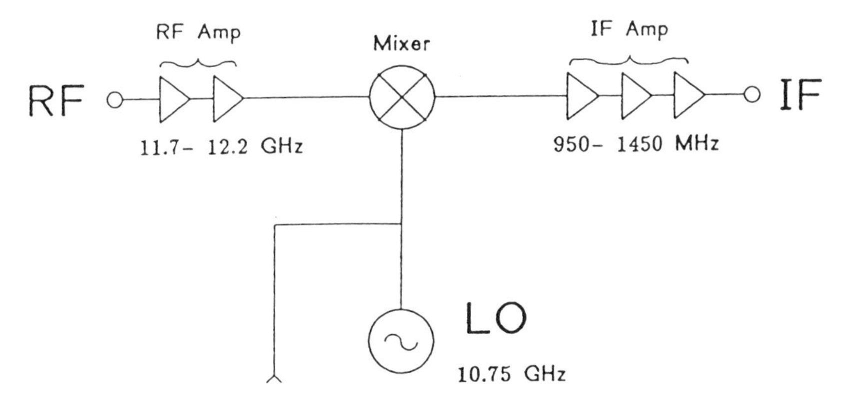

FIGURE 1.5 Block diagram of down-converter.

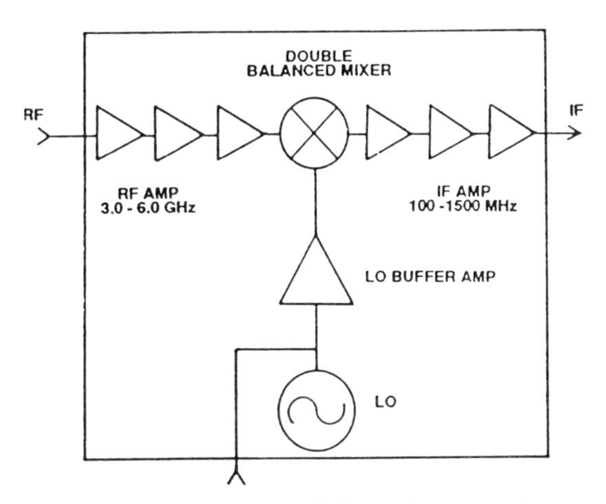

FIGURE 1.6 Block diagram of *C*-band receiver front end.

IF amplifier in the 950–1450 MHz frequency range is ideally suited for GaAs-based circuits. Such circuits are designed using analog design techniques as described in Chapters 5 and 6.

Television Receive Only (TVRO) Receiver. A block diagram of a TVRO receiver operating over 3–6 GHz is shown in Figure 1.6. This architecture is very similar to the DBS receiver. The IF amplifier in the 100–1500 MHz frequency range is also widely designed using GaAs-based circuits.

Optical Telecommunications (Trunk and Local-Area Network). Advances in fiberoptic communications include long-distance networks with repeater spacings in excess of 15 m for trunk applications and also for local-area networks (LANs). GaAs high-frequency analog circuits find application in systems such as for amplifier gain blocks, transimpe-

TABLE 1.6 System Applications of Typical GaAs ICs [5]

	Digital	Microwave	Analog
Types of ICs	AND, OR, XOR Flip-flop, register, divider, Multiplier ALU, mux/demux RAM/ROM	Amplifier FET switch mixer, coupler phase shifter oscillator Transmitter	Op amp Comparator Switched-capacitor filter Analog mux/demux V and I sources, VCO
Types of modules	Digital processor Digital memory DFT, FFT Controller Computer	Receiver Frequency synthesizer Signal generator RF and IF modules	Analog preprocessor Analog memory Sensor convolver PLL correlator
Level of integration	VLSI	SSI	MSI–LSI

[a] *Abbreviations*: ALU—arithmetic–logic unit; mux/demux—multiplexer/demultiplexer; RAM/ROM—random-access/read-only memory; DFT/FFT—delayed/fast Fourier transform; VCO—voltage-controlled oscillator; PLL—phase-locked loop; LSI, MSI, SSI, VLSI— large-, medium-, small-, very-large-scale integrated circuit.

dence amplifiers, and multiplexers and demultiplexers. Optical communication systems operating at high data rates are becoming of increasing interest for high-capacity communication links, particularly in undersea cable systems and high-speed local networks. Most systems that are operational today use 2-Gbit/s data rate, although laboratory experiments have demonstrated transmission at data rates above 8 Gbit/s over 40 miles of fiber [13].

Cellular Radio. Cellular radio is a low-cost and large-volume potential market, with application to land mobile phone services, which is evolving to systems for speech and data communication. The frequency of operation of these systems is typically in the 800–990 MHz band. The need for high-efficiency transmitters in the range of 1-W continuous wave (CW) will require GaAs-based power amplifiers in large volume.

Wireless LAN technology is another high-volume consumer-oriented application for high-frequency analog ICs. FCC has set aside three bandwidths for wireless air-based communications: 902–928 MHz, 2.4 GHz, and 5.72–5.85 GHz. Most commercially available applications are on the 900-MHz bandwidth, although products in the 2.4 GHz bandwidth are also becoming common.

REFERENCES

1. E. W. Mehal and R. W. Wacker, "GaAs Integrated Microwave Circuits," *IEEE Trans. Microwave Theory Tech.*, **MTT-16**, 451–454 (July 1968).
2. W. Jutzi, "A MESFET Distributed Amplifier with 2 GHz Bandwidth," *Proceedings of the IEEE*, **57**, 1195–1196, (1969).
3. R. S. Pengelly and D. Maki, "An X-band Monolithic Amplifier," *Microwaves & RF*, 314–342 (March 1987).
4. P. E. Allen and C. M. Breevoort, "An Analog Circuit Perspective of GaAs Technology," *IEEE International Symposium on Circuits and Systems*, May 1987, pp. 184–187.
5. T. T. Vu and J. N. Vu, "GaAs Analog Integrated Circuits and Their Potential System Insertions," *IEEE International Symposium on Circuits and Systems*, May 1990, pp. 3061–3064.
6. K. de Graff and K. Fawcett, "GaAs Technology for Analog-to-Digital Conversion," *GaAs IC Symp. Tech. Digest*, 205–208 (Oct. 1986).
7. S. M. Sze, *Physics of Semiconductor Devices*, Wiley, New York, 1981.
8. L. E. Larson, K.W. Martin, and Gabor C. Temes, "An Ultrahigh-Speed GaAs MESFET Operational Amplifier," *IEEE J. Solid State Circuits*, **24** (6) (Dec. 1989).
9. L. E. Larson, C. S. Chou, and M. J. Delaney, "GaAs Switched-Capacitor Circuits for High-Speed Signal Processing," *IEEE J. Solid State Circuits*, **Sc-22** (6) (Dec. 1987).
10. R. J. Bayruns et al., "An 8 nS Monolithic GaAs Sample and Hold Amplifier," *ISSCC Digest Tech. Papers* (Feb. 1987).
11. F. Weiss, "A 1 Gs/S 8-Bit GaAs DAC with On-chip Current Sources," *GaAs IC Symp. Tech. Dig.* 217–220 (1986).
12. R. Michels, P. Wallace, R. Goyal, N. Scheinberg, and M. Patel, "A High Performance, Miniaturized X-Band Active Mixer for DBS Receiver Application with On-chip IF Noise Filter," *IEEE Trans MTTs* (Special Issue on Multifunction MMICs and Their System Applications, guest eds. Ravender Goyal and Ed C. Niehenke), **38** (9), 2149 (Nov. 1990).
13. A. H. Gnauck, B. L. Casper, N. K. Dutta, and T. Cella, "8 Gbit/s Transmission over 76 km of Optical Fibre Using a Directly Modulated 1.4 μm DFB Laser," *Electron. Lett.*, **24** (9), 510 (April 28, 1988).

CHAPTER TWO

Integrated Circuit Processing Technology

V. S. RAO GUDIMETLA
Department of Electrical Engineering and Applied Physics
Oregon Graduate Institute of Science and Technology
Beaverton, Oregon

2.1 INTRODUCTION

GaAs is more suitable than silicon as a material for developing high-frequency and high-precision devices because of high peak and saturation velocities of electrons in GaAs [1]. Electron mobility in GaAs can be as high as 6000 cm^2/V-s compared to 1200 cm^2/V-s in silicon. Although other III–V materials may be better suited for low-noise and high-power applications [2,3], GaAs MESFETs are superior in overall performance at RF frequencies; however, a few problems remain. The energy–momentum diagrams of GaAs indicate that the mobilities of electrons and holes are not comparable to each other [4]. This makes GaAs unsuitable for complementary logic devices except in special cases. In addition, thermal conductivity of GaAs is 3 times poorer than that of silicon, and one should therefore pay special attention to the heat-dissipation problem for densely packed integrated circuit (IC) using GaAs. The fabrication technology has some limitations, due to mechanical fragility, stoichiometric problems, and high-vapor pressure of the group V elements [5]. At the device processing level, such limitations lead to poor ohmic contact formation, thermal diffusion of impurities, and development of a stable thermal oxide. Despite these disadvantages, GaAs MESFET technology has matured significantly over

High-Frequency Analog Integrated-Circuit Design, Edited by Ravender Goyal
ISBN 0-471-53043-3

many years as a commercially competitive compared to silicon technology, particularly for high-precision analog applications [5–8] at high frequencies of ~1 GHz and above.

2.2 BULK GROWTH

2.2.1 Horizontal Bridgmann Method

GaAs wafers are grown using the horizontal Bridgman (HB) or liquid-encapsulated Czochralski (LEC) method [5,8,9]. The former leads to high-purity materials. In the HB method, a boat-shaped quartz crucible containing high-purity gallium at 850°C is exposed to arsenic vapor. A specific temperature profile (Fig. 2.1) must be maintained for the molten zone, solidified material, and arsenic vapor to achieve high-quality material of GaAs. Either the temperature gradient is moved electronically across the furnace to allow the crystal to freeze or the boat is moved physically. The purity of GaAs crystals depends on the molar fraction of $AsCl_3$ flowing into the reactor. Grown layers can be doped by adding them directly to a gallium source or introducing gases, containing the dopants such as H_2S, H_2Se and

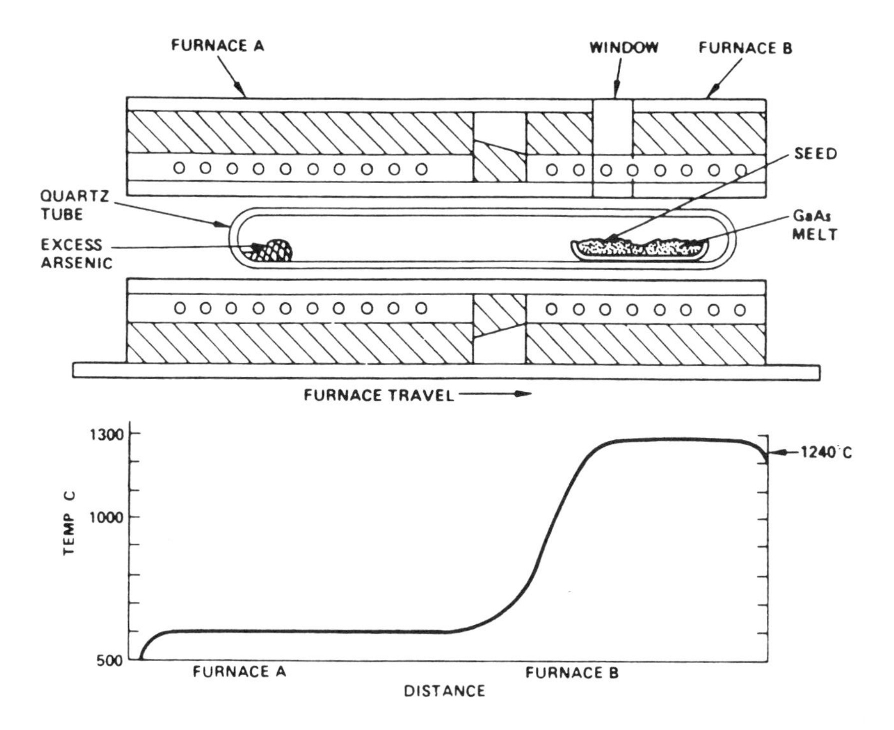

FIGURE 2.1 Horizontal Bridgman growth apparatus [6, p. 26] reproduced by permission.

$Zn + H_2$. The use of quartz crucible makes the GaAs crystal slightly n-doped in the presence of silicon and prevents dislocations. For HB-grown wafers, ingots are along (111) direction and have a dislocation density $<10^3$ due to silicon doping. To achieve high resistivity of 10^8 Ω-cm, Cr doping is needed to compensate for Si.

2.2.2 LEC Method

The LEC method is used to grow wafers for IC processing [8,9]. The technique used here is similar to that for silicon material growth. Figure 2.2 shows an LEC reactor [5]. A pyrolytic boron nitride crucible is filled with elemental gallium and arsenic. Then using a seed GaAs, crystals of GaAs are pulled with the orientation of (100) direction. Boron trioxide is used as the encapsulant material because of its excellent properties as it is not reactive, less dense, and transparent. GaAs crystals form at 800°C and at a pressure of ~60 atm. A seed crystal is introduced into the melt, and the GaAs crystal is pulled from the melt. The crystal is constantly rotated and thermal gradients need to be maintained for growing uniform material. To prevent arsenic from vaporizing, an inert-gas atmosphere on the order of a few atmospheres for low-pressure processing up to 75 atm at high-pressure processing is maintained. For low-pressure growth, gallium resides in the

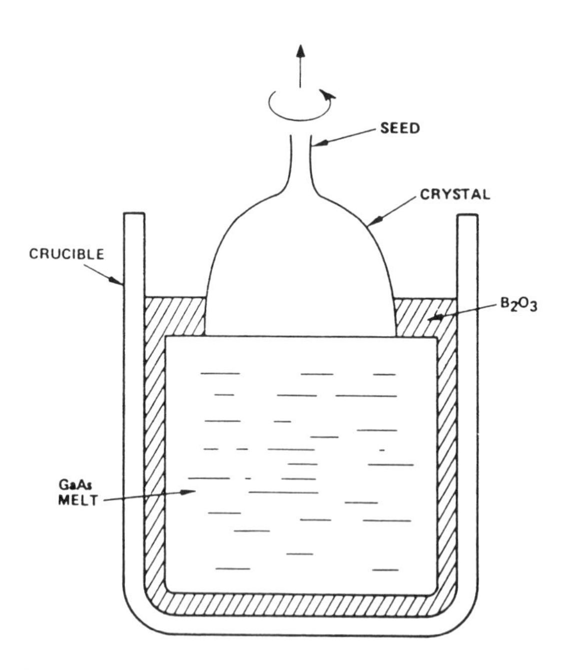

FIGURE 2.2 Liquid-encapsulated Czochralski puller [6. p. 29] reproduced by permission.

crucible while arsenic vapor is injected from a separate source. LEC wafers are generally circular, along (100) direction, have relatively higher dislocation density of 10^4–10^5 cm^{-2}, and have a resistivity higher than 10^7 Ω-cm. Addition of a few percent of indium to the melt hardens the crystal against dislocation formation, but the nonuniform distribution of indium in the wafer leads to nonuniform wafers. Addition of indium also enhances the chances of twinning, limiting the size of the grown crystal. Although the application of the magnetic field reduces twinning, it is not found to improve the dislocation density.

2.3 EPITAXIAL GROWTH

Although the majority of GaAs-based ICs are fabricated by ion-implantation techniques using LEC substrates, for some special applications such as low-noise FETs and high-power FETs, epitaxial techniques are used. Some of the epitaxial techniques are liquid-phase epitaxy, (LPE) vapor-phase epitaxy (VPE), molecular-beam epitaxy (MBE), and organometallic vapor-phase epitaxy.

2.3.1 Liquid-Phase Epitaxy

Liquid-phase epitaxy is the oldest and simplest technique [10,11], still used to grow epitaxial GaAs for optoelectronic devices such as light-emitting diodes (LEDs). Its disadvantages are lack of doping uniformity over large areas, poor surface finish, and poor interface abruptness between layers of different doping levels. Figure 2.3 shows an LPE reactor [8]. GaAs wafers are held face down in a slider and come in contact with one or more wells of molten material, which is supercooled to just below their solidification points. These wells can be p- or n-doped. When the wafer contacts the melt, the material solidifies on the surface.

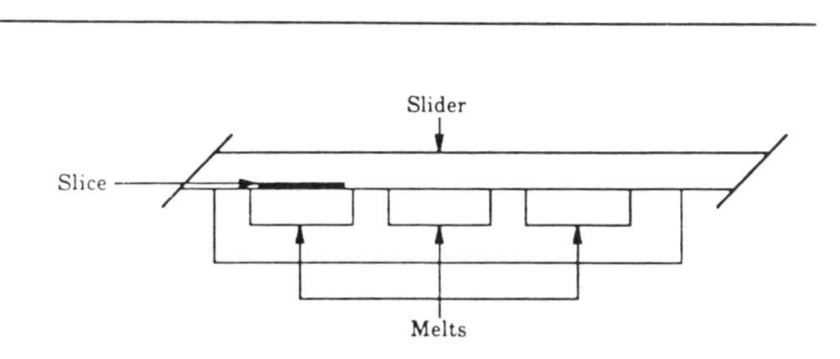

FIGURE 2.3 A liquid-phase epitaxy reactor [8, p. 43] reproduced by permission.

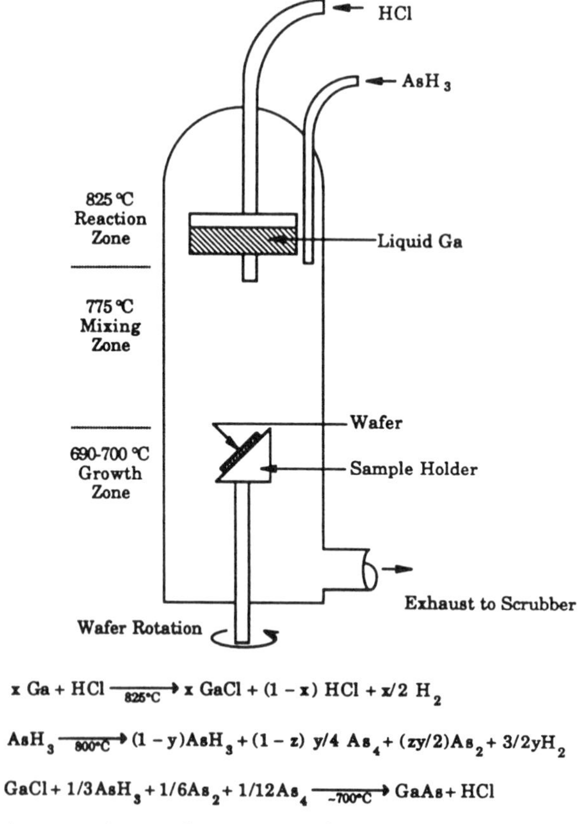

FIGURE 2.4 A vapor-phase epitaxy reactor [5, p. 104] reproduced by permission.

2.3.2 Vapor-Phase Epitaxy

In the VPE process [10,11], gallium and arsenic in the gas state are mixed together and are passed over the GaAs substrate where high-quality GaAs films are deposited as a result of the reaction of the gases. VPE generates crystals with controlled dopant profiles and better surface finish and abrupt junctions. The reactor and the corresponding reactions are shown in Figure 2.4. Although FETs with the lowest possible noise figure have been reported using VPE, some disadvantages exist. Since the film growth is at a high temperature, some silicon is transported into the crystal. The silicon enters the gas through the reaction of the quartz walls of the reactor with the gas at high temperatures. Since gases need to stay in the reactor long enough to interact, abrupt changes in the doping levels cannot be achieved.

2.3.3 MOCVD

Metal–organic chemical vapor deposition is a variant of vapor epitaxy technique [12–14]. A MOCVD reactor is shown in Figure 2.5 [7]. It uses trimethylgallium (TMG) or triethylgallium (TEG) as a source of gallium

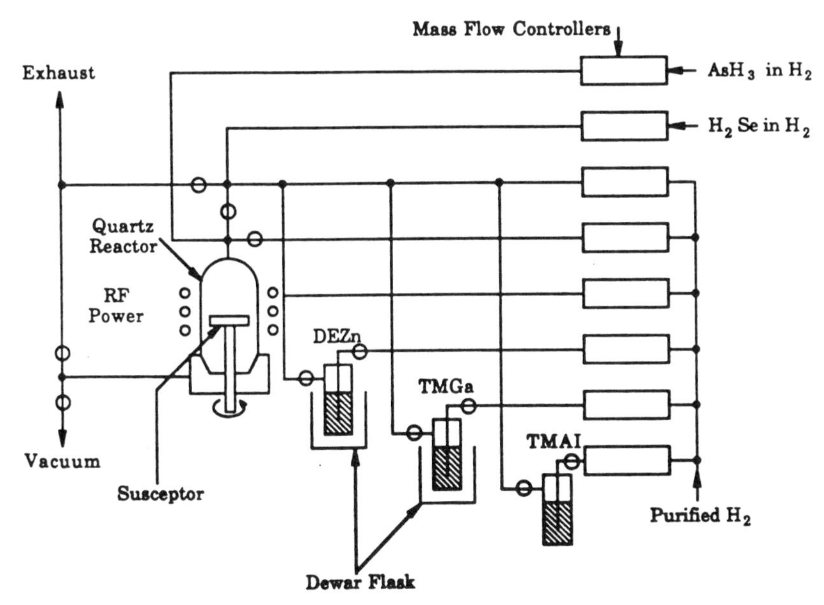

FIGURE 2.5 Cross section of an organometallic vapor-phase epitaxy reactor [5, p. 107] reproduced by permission.

and arsine as a source of arsenic. A GaAs substrate on a graphite susceptor is heated to 600–800°C using a RF coil. The following reaction then proceeds and the resulting GaAs or other material is deposited on the substrate.

$$R_3Ga + AsH_3\text{--------}GaAs + 3RH$$

where R represents CH_3 or C_2H_5.

Growth rates are of order of 5 μm/h, and large areas of crystals (900 cm^2) can be grown. For the films, produced by MOCVD process, background contamination due to metal organics such as zinc, silicon, and magnesium are common. Carbon can be used as an amphoteric dopant, as it can produce n-type or p-type films. A strong negative point for MOCVD is that all gases used in this process are toxic and explosive.

2.3.4 Molecular-Beam Epitaxy

In MBE [15–17], an ultra-high-vacuum steel chamber containing a substrate holder and several effusion cells, from which the element or compounds can be evaporated for deposition or interaction with the substrate when heated (Fig. 2.6). The fusion cells have openings facing the substrate holder. The entire chamber is filled with nitrogen to prevent contamination by other compounds and as well as interactions between the compounds in the effusion cells. The substrate holder and effusion cells are made of alumina, tantalum, or graphite. Monitoring devices such as a mass spectrometer, auger spectrometer, ion gauge, and high-energy diffractometer are located

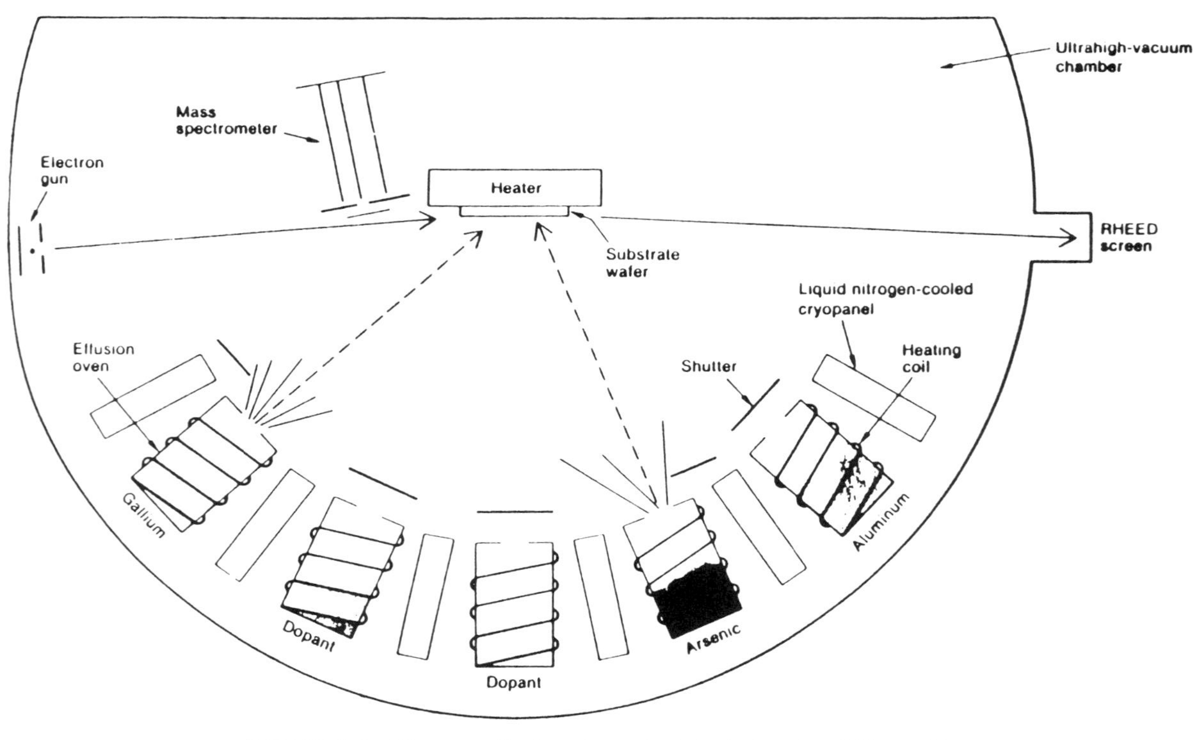

FIGURE 2.6 The growth chamber of a MBE chamber [15, p. 18].

inside the chamber. These are used to characterize molecular content of the gases, layer surface, alloy composition, and impurity composition and profiles. To grow GaAs layers, Ga atoms and about 5 times larger flux of As_2 and As_4 molecules impinge on the substrate. To make it semiinsulating, oxygen should be used. Another method is to grow on AlAs layer and oxidize it with hot water. Silicon is the most often used n-type dopant, and other n-type dopants are Ge and Se. The most often used p-type dopant is beryllium. Since electron concentration in MBE-grown GaAs layers is very high compared with that of other methods, nonalloyed ohmic contacts with a contact resistance of 10^{-7} can be produced. Another advantage of MBE is it leads to smooth surfaces, which are needed for heterojunctions, superlattices, and multilayered structures.

There are two new important masking methods for MBE technology [18]. In one technique, the substate is covered with a thin mask of silicon dioxide. Polycrystalline and simiinsulating material was grown while on the exposed areas a high-quality epitaxial material was grown on the masked parts. Another method is to use a sheet of silicon with precisely dimensioned openings on a wafer and grow the material in the unmasked areas.

One important requirement for MBE is to start with a clean and high-quality surface. This requires extensive chemical treatments and subsequent heating with indium and In–Ga alloys [19]. However, MBE has the potential to make three-dimensional ICs possible, and several new device topologies can be achieved. The major drawbacks for MBE are the slow growth rate and capacity of growing only one wafer at a time and high cost of equipment.

2.3.5 Quality Factors and Defects in GaAs

The DLTS technique shows a deep donor impurity level, called EL2 in GaAs, and this is identified as a structural defect, usually associated with arsenic–gallium antisite. Further investigations are needed to find its definitive cause. Stoichiometry investigations show that the resistivity and hence mobility are influenced by the melt composition. If the melt is rich with gallium, EL2 defects are reduced, but still with the dominance of a new shallow level, mobility decreases. There are many traps in GaAs such as ET1, ET2, ES1, ES2, EL1 through EL16, and so on [20]. These are given in Table 2.1. EL2-level concentration can be measured by detecting the variations in the signal level of 1-μm wavelength radiation after it is focused on the wafer and projected through the wafer, normal to its surface. The typical EL2-level concentrations are $1–2 \times 10^{16}\ cm^{-3}$ with a 2–5% standard deviation across the wafer. The other impurities present are carbon and boron. The dislocations are due to growth-related factors such as temperature gradients and solid–liquid growth interface shape and material-related factors such as stoichiometry and impurity hardening. It was suggested that baking the substrates to 750°C in hydrogen atmosphere for about 20 h

TABLE 2.1 Electron Traps in Gallium Arsenide[a] [20]

	Activation Energy E (eV)	Cross Section σ (cm)	Observation
ET1	0.85	6.5×10^{-13}	BM
ET2	0.3	2.5×10^{-15}	BM
ES1	0.83	1.0×10^{-13}	BM
EF1	0.72	7.7×10^{-15}	Cr-doped BM
EI1	0.43	7.3×10^{-16}	VPEM
EI2	0.19	1.1×10^{-14}	VPEM
EI3	0.18	2.2×10^{-14}	VPEM
EB1	0.86	3.5×10^{-14}	Cr-doped LPEM
EB2	0.83	2.2×10^{-13}	As-grown VPEM
EB3	0.90	3.0×10^{-11}	EIM
EB4	0.71	8.3×10^{-13}	EIM
EB5	0.48	2.6×10^{-13}	As-grown MBEM
EB6	0.41	2.6×10^{-13}	EIM
EB7	0.30	1.7×10^{-14}	As-grown MBEM
EB8	0.19	1.5×10^{-14}	As-grown MBEM
EL1	0.78	1.0×10^{-14}	Cr-doped BM
EL2	0.825	$(0.8\text{–}1.7) \times 10^{-13}$	VPEM
EL3	0.575	$(0.8\text{–}1.7) \times 10^{-13}$	VPEM
EL4	0.51	1.0×10^{-12}	As-grown MBEM
EL5	0.42	$(0.5\text{–}2.0) \times 10^{-13}$	VPEM
EL6	0.35	1.5×10^{-15}	BM
EL7	0.30	7.2×10^{-15}	As-grown MBEM
EL8	0.275	7.7×10^{-15}	VPEM
EL9	0.225	6.8×10^{-15}	VPEM
EL10	0.17	1.8×10^{-15}	As-grown MBEM
EL11	0.17	3.0×10^{-16}	VPEM
EL12	0.78	4.9×10^{-12}	VPEM
EL14	0.215	5.2×10^{-16}	BM
EL15	0.15	5.7×10^{-13}	EIM
EL16	0.37	4.0×10^{-18}	VPEM

[a] *Abbreviations*: BM, bulk material; VPEM, VPE material; LPEM, LPE material; MBEM, MBE material; EIM, electron-irradiated material.

reduces the diffusion of deep acceptor traps into the active layer but makes the substrate p-type conductive to a depth of 0.3–3 μm.

The electrical quality factors include implant activation, mobilities, and the resistivity of the bulk after anneal. The physical quality factors for GaAs wafers are crystal orientation, diameter, thickness, surface–edge finish, flatness, bow, surface topography, subsurface mechanical damage, internal stress, and edge. Besides EL2 defect, wafer stress, which is due to discontinuities in dielectric thickness or other process defects, is another major factor as GaAs is piezoelectric and thus stress induces charge in the lattice. Dislocations are high in GaAs (10^5 cm^{-2}), and this will lead to major problems beyond LSI levels of integration. As remarked earlier, it can be corrected by indium doping [21], although not uniformly from wafer to wafer. In this connection, silicon due to its dislocation density of a few per square centimeter is much superior.

2.4 PROCESSING TECHNOLOGY

The processing of GaAs devices is similar to that of Si devices. However, there are some significant differences as GaAs is a compound semiconductor. Because the vapor pressure of As is much higher than that of Ga, care must be taken during high-temperature steps that As does not vaporize and destroy the material.

2.4.1 Substrate Materials

The often considered materials for substrate for high-frequency IC fabrication are semiinsulating GaAs and GaAs grown on silicon substrate. Semiinsulating GaAs should have low dislocation density and be free from impurities to avoid degradation of the device [22]. The use of Si as a supporting substrate has several advantages [23–25] in that it is a better thermal conductor, lighter in weight, and less expensive for large-area substrates. Having Si as primary substrate provides better strength, and the high-frequency devices fabricated on GaAs grown over silicon substrate, can easily integrate with advanced Si technology. But because of the material interface problems such as large lattice mismatch and antiphase domain formation, semiinsulating GaAs is preferred over GaAs grown over silicon as a primary source of substrate used in high-frequency analog ICs.

2.4.2 Ion Implantation

A schematic illustration of an ion-implantation apparatus is shown in the Figure 2.7. A beam of Si^+ or S^+ ions for n-type dopants is generated from a gas or solid source by heating or passing current through the material [26,27]. The proper ions are selected using a magnetic field, and a beam is accelerated down an evacuated tube at high voltage. The beam is raster-scanned on the wafer that is located at the other end of the tube. The number of implanted ions and their distribution can be determined by the fluency and energy of the ions striking the wafer.

For n-channel active layer and contact implants, Si $(29)^+$ is used [the other choices are Sn and Te; Si $(28)^+$ is not recommended because of possible contamination by nitrogen]; p-implants are needed for complementary devices. Also they are used to enhance the performance of FET devices

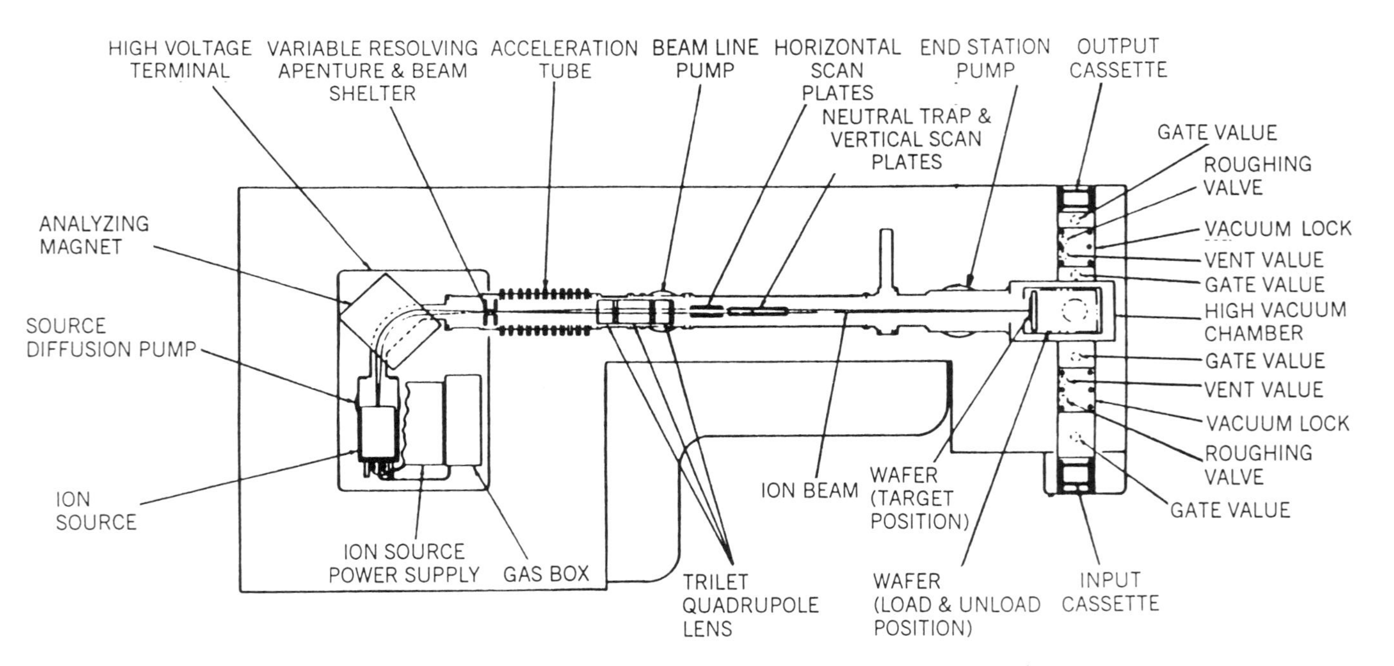

FIGURE 2.7 Ion-implantation apparatus. (Eaton Corporation)

such as backgating characteristics [28] and superior drain characteristics. Isolation implants use oxygen, boron, rare gases, protons, and other materials to improve the insulating nature of the wafer.

A postimplant anneal for about 30 min at a temperature of 750–900°C is required to repair the damage to the lattice by the ion implantation and to move the dopants to active sites. Considering the high vapor pressure of GaAs because of its arsenic contents, to prevent arsenic loss during the high-temperature anneal, either the wafer is encapsulated with dielectric coating of silicon oxide or silicon nitride or the wafer is sealed in an evacuated ampoule that has an auxiliary source of arsenic. The auxiliary source provides enough backpressure to prevent the loss of arsenic from the wafer. The encapsulation technique leads to interface stresses due to different thermal expansion coefficients of silicon dioxide and GaAs, and also GaAs becomes contaminated with silicon atoms due to silicon dioxide transport. Capless techniques eliminate interface stresses but cannot easily provide uniform and reproducible wafers.

A schematic illustration of an ion-implanted doping profile is shown in the Figure 2.8. Silicon dioxide, used primarily as encapsulation material during implant anneal, may be useful later in photolithography, and the energy of implant can be adjusted such that the peak of the profile occurs at the GaAs (surface). The actual profile depends on factors such as the specific dielectric used and its thickness and ion-implant energy and dose and, finally, annealing conditions. The effects of ion implantation of MESFETs are examined in detail elsewhere [29–31].

2.4.3 Dielectric Deposition

The often used dielectrics are silicon dioxide and silicon nitride for GaAs circuits. The two techniques used are sputtering and chemical vapor deposition (CVD). In sputtering, atoms to be deposited are accelerated toward the surface by gas ions such as argon or oxygen. Substrate temperatures are low (<150°C), and sputtered films are of high quality with good adhesion and step coverage. In CVD processing, gases containing the constituent atoms are mixed and the mixture is passed over a heated

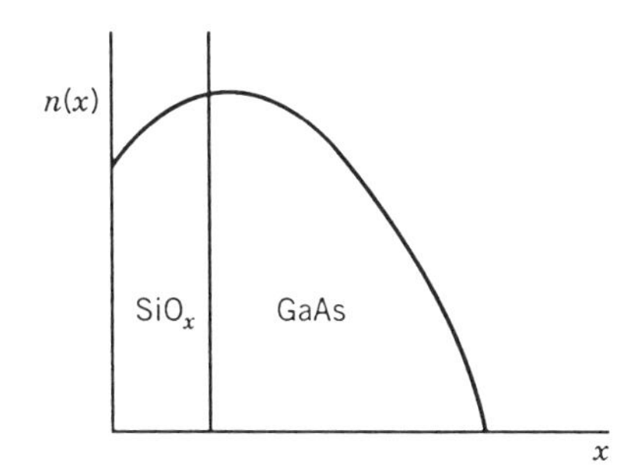

FIGURE 2.8 (*a*) Schematic illustration of an ion-implant profile; (*b*) experimental data showing activation of implanted ions versus implant. [33]

substrate where they deposit the dielectric films. Substrate temperatures for this process step must be kept at ≤350°C, but at those temperatures, deposition rates are slow. Deposition rates can be improved by plasma-enhanced CVD. Silicon dioxide can be deposited using CVD–thermal- or CVD–plasma-assisted techniques. Silicon nitride deposition requires sputtering (reactive or direct) or CVD–plasma-assisted.

2.4.4 Metal Deposition

The generally used metals are gold and nickel for ohmic contacts; Ti, Pd, and Au for Schottky barrier gates; NiCr for thin-film resistors; and Ti–Au for interconnects. The three most often used methods are sputtering, thermal evaporation, and electron-beam deposition. In thermal evaporation, a large current is passed through a refractory-metal boat containing the metal to be deposited until the metal is heated to the point of vaporization. The metal vaporizes and impinges on the GaAs substrate at the other end of the high-vacuum chamber, and the amount and rate of deposition are controlled through a shutter. Although it has a good throughput rate, its main disadvantages are that source temperature is limited by the melting point of the boat and sometimes unwanted metal alloys are formed. Electron-beam evaporation differs from the thermal evaporation in that an electron beam is focused on a solid target to heat the metal. Electron-beam evaporation technique is used for gold, nickel, Ti, Pd and other metals while fabricating ohmic contacts, gates, and interconnects. Its advantages are the small spot size and more efficient and localized heating, and thus the disadvantages of thermal evaporation are eliminated. Sputter deposition technique is the same as that for dielectric deposition and results in a uniform deposition over large areas. It is used for NiCr, TiW, and other compounds and thus for depositing the thin-film resistors and refractory-metal gates.

2.4.5 Etching

Etching is done extensively in GaAs-based IC fabrication [32]. This includes steps such as removing polishing damage from raw wafers, isolation of devices, recessing channels, removing dielectric caps after ion implantation, and cleaning photoresists before metal deposition. A large number of wet chemical etches that oxidize the surface and then dissolve the oxide layer are available. In dry etching (reactive-ion etching), reactive ions, created by striking a plasma in an appropriate gas such as oxygen or chlorine compounds, are accelerated toward the surface and on striking the surface, etch the unmasked regions of the surface. Etching is made very precise by controlling the gas concentration and substrate temperature. Although both wet and dry etching processes are available for removal of dielectric, the dry process is preferred as it allows a more precise transfer of photolithographic

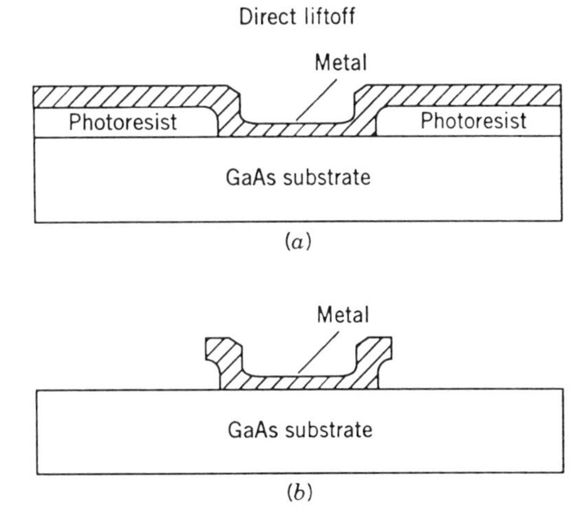

FIGURE 2.9 Schematic illustration of direct liftoff: before liftoff (*a*); after liftoff (*b*). [33]

image. The plasma environment used for drying etching contains chlorinated fluorocarbons. The rate of etching of dielectric by the fluorocarbons is much higher than that for the photoresist material.

Metals can be etched using wet chemical techniques or milled by high-energy ions. Here a high-energy ion beam is directed toward the exposed metal to be removed, and since etch rate of the beam is higher for the metal that it is for the photoresist, the metal is removed much faster with little damage to the photoresist. For submicrometer features, the liftoff technique is preferred to others. Here photoresist is removed in the regions, where the metal is to remain on the wafer as shown in Figure 2.9. Then metal is

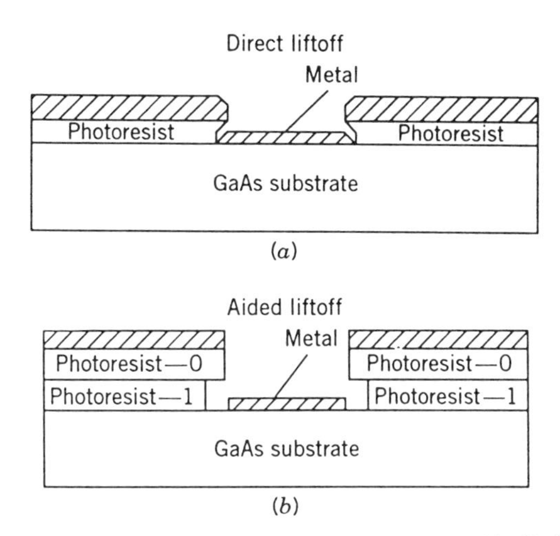

FIGURE 2.10 Swelling of photoresist to aid liftoff. [33]

TABLE 2.2 Properties of Commonly Used Etchants [32]

Etchant	Volume Ratio	Normality Ratio	Etch Type[a]	Etch Rate for {100} GaAs (Å/s)
H_3PO_4–H_2O_2–H_2O	1:9:1	4.1:16	R–D	525
	7:3:3	24.2:4.5	D	330
	1:1:25	1.7:0.7	R	55
	1:9:210	0.21:0.8	R	16
	3:1:50	2.5:0.36	R	13.5
NH_4OH–H_2O_2–H_2O	1:1:16	1:1.1	R	330
	3:1:50	1:0.3	R	100
	3:1:150	0.35:0.13	R	30
H_2SO_4–H_2O_2–H_2O	1:8:1	3.6:15.3	R	1300
	1:1:8	3.6:2.0	R	215
	8:1:1	29:2	D	200
	1:8:40	0.75:3.2	R	200
	1:8:80	0.4:1.75	R	90
	1:8:160	0.22:0.82	R	45
	20:1:1	33:0.9	D	30
	1:8:1000	0.036:0.15	R	7
HCl–H_2O_2–H_2O	1:1:16	1:1.1	R	330
	3:1:50	1:0.3	R	100
	3:1:150	0.35:0.13	R	30

[a] R, reaction-rate-limited; D, diffusion-limited.

deposited on the entire pattern. Then photoresist is dissolved and the hanging out excess metal is removed. One problem with this method is the presence of the wings, which may become a major problem in processing later. This can be avoided by swelling of the photoresist as shown in Figure 2.10.

Table 2.2 gives the properties of some of the commonly used wet chemical etchants for GaAs [32]. Table 2.3 gives major characteristics differences between diffusion-rate-limited and reaction-rate-limited wet etching processes [32].

2.5 MESFET DEVICE PROCESS

A MESFET consists of a moderately doped channel of short length (<1 μm) between a source and a drain [18] as shown in Figure 2.11*a*, and thus will conduct current when a proper bias is applied between them. Source and drain areas are heavily doped before depositing source and drain metal contacts to provide good ohmic contacts. A gate metal is deposited on the channel between the source and drain, thus establishing a Schottky barrier with the moderately n-doped channel, by depleting a part of the channel. A voltage of proper polarity applied to the gate controls the width of the depletion region and thus controls the current flow between source and

TABLE 2.3 Characteristic Differences between Diffusion-Rate-Limited and Reaction-Rate-Limited Wet Etching Processes [32]

Parameter	Diffusion-limited	Reaction-rate-limited
Agitation dependence	Agitation increases supply of etchants and increases etch rate	No increases observed
Mask edge trenching	Supply of fresh etchants from the masked side causes excessive etching at mask edges to form trenches	No such effect
Preferential (anisotropic) etching	Not pronounced; can be completely eliminated by increasing solution viscosity	Very pronounced effect; $E\{111\}$ As > $E\{100\}$ > $E\{111\}$ Ga (E = etch rate)
Temperature dependence	Lower activation energies, hence relatively insensitive to temperature	Higher activation energies, hence more sensitive to temperature
Time dependence	Etch rate is roughly proportional to the square root of time and, if not, is sublinear	Etch rate constant with time; it is sublinear if the etching reagents, such as H_2O_2, NH_4OH, dissociate during etching.
Galvanic effect	Not very pronounced, except at metallization edges	Galvanic cell formation enhances etch rates everywhere on wafer
Doping type (n or p) and concentration dependence	Very small	If each rate is surface-oxidation-limited, then etch rate could increase with increased electron concentration
Chemical polishing effects	Surface protrusions experience higher diffusion current, hence higher etch rates; surface becomes smoother	Surface topography is well maintained, unless affected by crystal anisotropic effects (iii)
Etch rate magnitudes	Highly viscous solutions have low etch rates, but very high rates in the transition region	Usually rather low as result of diluted solutions used in practice
Residual oxide-layer thickness and composition	High etch rates leave thinner oxides; highly viscous, low-etch-rate solution would leave thicker, partially complexed and/or oxidized residues	Low etch rates cause thicker oxide buildup that resembles native oxides, rarely <40 Å thick

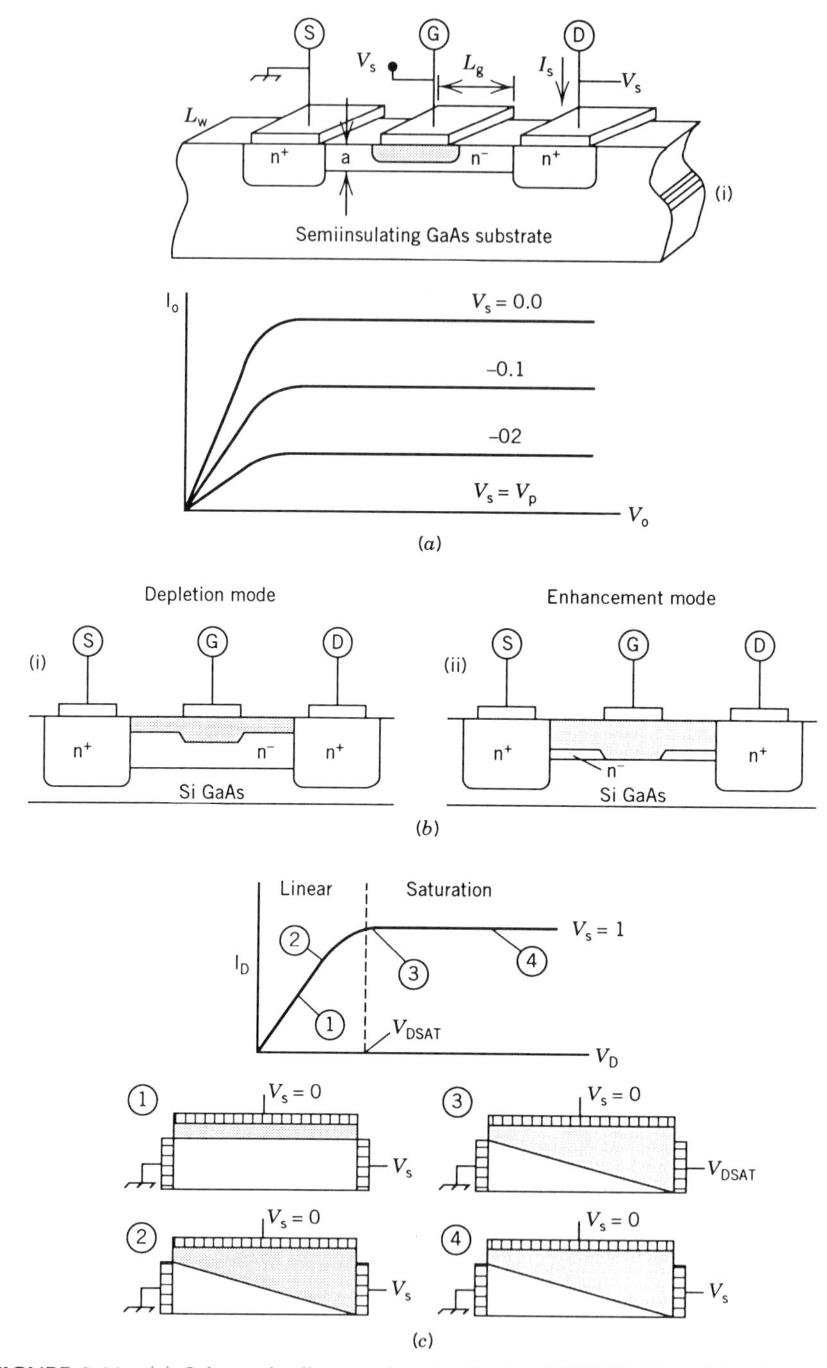

FIGURE 2.11 (*a*) Schematic diagram for the GaAs MESFET (i) and its drain I–V characteristics (ii); (*b*) fundamental differences between depletion-mode (i) and enhancement-mode (ii) device structures; (*c*) I–V characteristics and mode-depletion region for MESFET operation. [33]

drain. In depletion mode (D-mode) devices, the channel is moderately doped and thus the depletion region is not large enough to completely pinch off the channel. When the gate bias is zero of floating and there is nonzero drain-to-source current (Fig. 2.11*b*,*c*), the magnitude of the current depends on the voltage between drain and source terminals. As the bias of property polarity on the gate terminal increases, the width of the depletion region increases and for sufficient bias, the width of the depletion region is large enough to cut off the current flow. In enhancement mode (E-mode) devices, as a result of low doping of the channel, the depletion region is quite wide and the channel is pinched off when the gate is at zero potential with respect to source. Application of positive bias on the gate opens the channel (Fig. 2.11*b*,*c*), and this facilitates the flow of current. The D-mode device has a finite amount of current, flowing between source and drain even when the device is pinched off, but this is not so in E-mode devices.

A typical equivalent circuit for MESFET is shown in the Figure 2.12 [33]. In the figure, C_{gs} is the gate to source capacitance and C_{gd} is the gate to drain capacitance, and both are due to depletion region under the gate. Similarly R_{in} is the resistance between the gate and source and R_{ds} is the channel resistance; C_{ds} is the channel capacitance, R_s is the source resistance, and finally g_m is the transconductance of the channel. An important figure of merit of the device is unity current gain frequency f_t, which is given as

$$F_t = \frac{g_m}{C_{gs}} = \frac{\mu}{L_g^2}$$

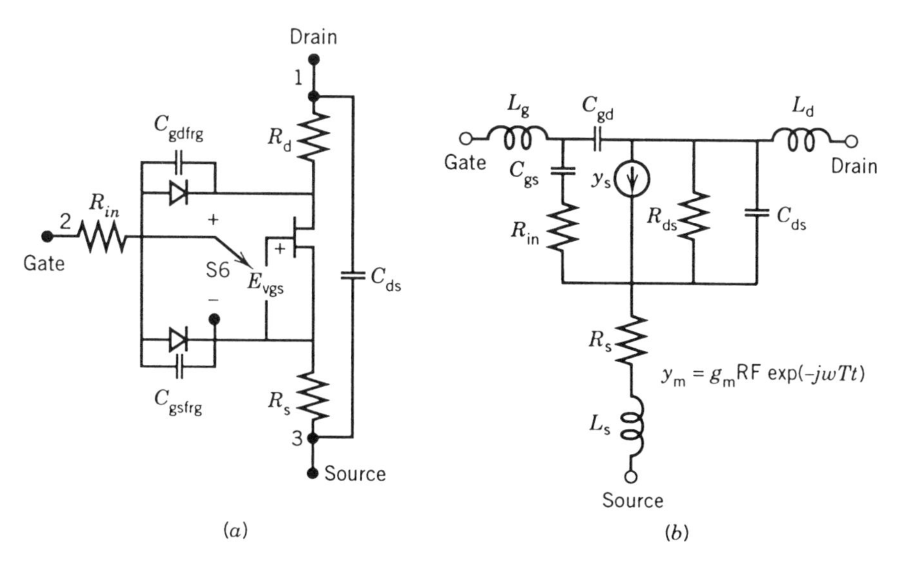

FIGURE 2.12 GaAs MESFET equivalent circuit models: SPICE (Simulation Program for IC Emphasis) equivalent circuit (*a*), lumped-element equivalent circuit (*b*) [33].

where L_g is the gate length and μ is the electron mobility in the channel. One way to improve f_t is to reduce the gate length or increase the mobility. Reduction of gate length has short-channel effects such as increased drain conductance, pinchoff reduction, subthreshold leakage, and other problems. Some of these problems can be alleviated by using highly doped short channels, but such high dopings reduce the mobility of electrons.

The deep level and other defects in GaAs substrate result in adverse electrical characteristics of the devices [33]. Growing the active layer directly on the substrate leads to several dc and low frequency as well as RF problems such as low-frequency oscillations, looping in drain characteristics, high channel current noise and backgating in GaAs ICs [34,35]. These are believed to be due to the space charge between the substrate and the active layer. Anisotropy of GaAs crystals will give different pinchoff voltages for devices along different directions.

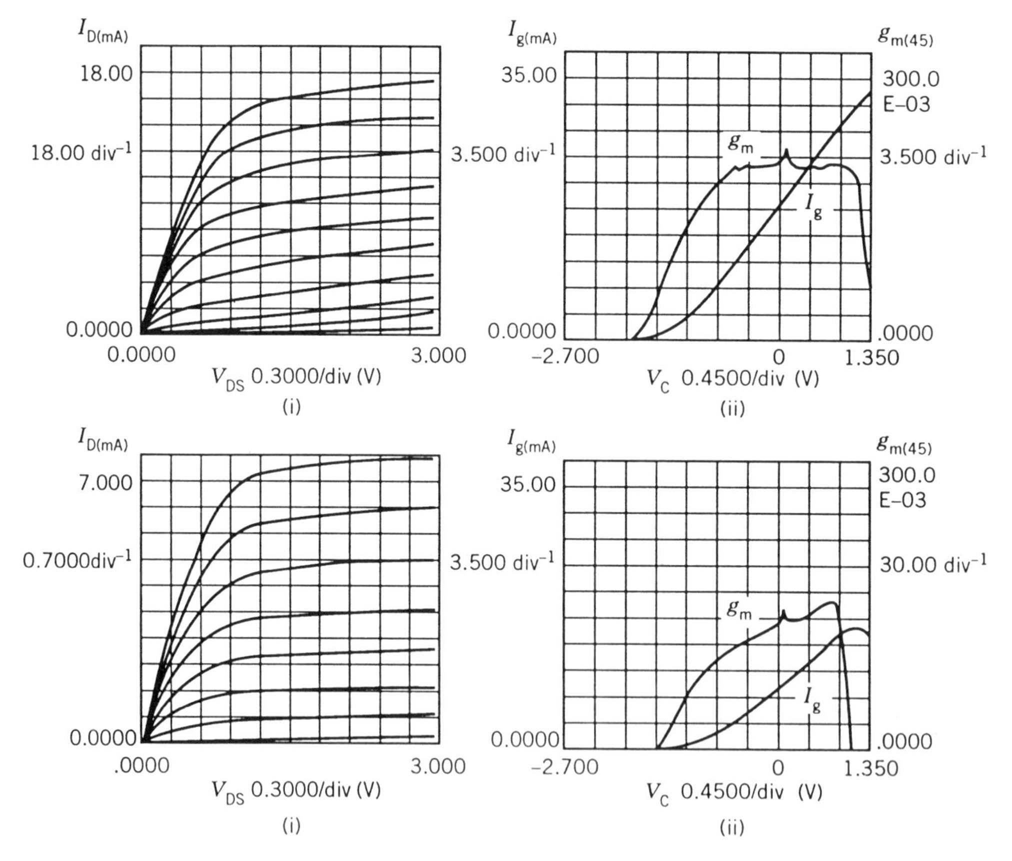

FIGURE 2.13 (*a*) Experimental drain I–V (i) and transconductance and gate (ii) characteristics for 0.25×60-μm gate MESFET [33]; (*b*) experimental drain I–V (i) and transconductance and gate (ii) characteristics for 1.00×60-μm gate MESFET [33].

Figure 2.13*a*,*b* show I–V curves for 1.0- and 0.25-μm gate length devices, respectively. The 0.25-μm device clearly shows the short-channel effects such as high output conductance and poor pinchoff characteristics for submicrometer gate lengths due to high-energy channel electron injection into the substrate. These problems can be avoided by having a thick buffer layer of undoped GaAs or p-type GaAs [36,37] using low-temperature MBE technology. The buffer layer serves as barriers for electron flow due to higher bandgap or discontinuity at the conduction band edge. In addition, 0.25-μm structure has less gate-to-source capacitance and hence larger f_t.

High-performance GaAs MESFETs have been fabricated using ion implantation or epitaxy for the channel layer. In terms of mass production, where uniformity of the electrical characteristics of the devices over a wide area is important, direct ion implantation is the bet method for doping the channel. Ion-implanted GaAs MESFETs fabricated on (211) GaAs substate have shown low noise figures [38] but are very process-sensitive. High-performance GaAs MESFETs have also been fabricated using MBE and atomic layer epitaxy techniques because of excellent control on layer thickness and dopant distribution profile.

The doping level and profile of a MESFET channel depend strongly on the specific application requirements. For example, to achieve a higher output power, a constant thickness of the epitaxial layer of a nonuniform doping is used. Since current is proportional to the product of doping and thickness, more heavily doped layers carry larger currents, and this leads to higher output power. When a good linearity between the drain current and gate bias is needed, step-doped MESFETs [39,40], which contain a highly doped channel sandwiched between undoped GaAs layers is used. Such a design results in higher transconductance at low-drain currents, suppression of short-channel effects, and small gate capacitance and leakage current.

The source and drain contact areas in MESFET must be doped as high as possible to achieve low ohmic resistance through the mechanism of tunneling. Although most contacts are made by alloying contact metals, other choices are nonalloyed ohmic contacts, low-bandgap materials such as Ge, or delta doping structures [41–43].

Schottky barrier at the gate should be as high as possible to reduce gate leakage current and allow for a larger ac voltage swing. In addition, the contact should have good thermal expansion, compatible with that of GaAs, low chemical reactions with the substrate, and no diffusivity into GaAs even at high temperatures and high-temperature stability. To fabricate a good Schottky contact, the sample is dipped into an alkaline solution such as NH_4OH for a few seconds to remove any impurity deposited on the surface of gate channel area. This dissolves native arsenic and gallium oxides, and a better surface is presented in the evaporation chamber for further processing [44]. Commonly used gate metals are Ti–Pt–Au or Al. Also refractory metals such as W and Mo with their nitrode and silicides are used. Because of it high-temperature stability, WSi_x is used in self-aligned GaAs MESFET technology.

Figure 2.14 shows several GaAs MESFET structures that were proposed and fabricated over years for high-frequency application and better process yield. The two main categories of GaAs MESFET fabrication are gate-recess technology and self-aligned gate technology. Gate-recess technology is useful for high-frequency operation because it maintains thin channels for high transconductance while allowing for the adjustment of the threshold voltage. To achieve high-speed performance, device design should ensure that it provides a high transconductance, low gate-to-source capacitance, and low parasitic resistances. Parasitic resistance is controlled by the distance between the gate and source pads and the conductivity of the channels. Hence the gate is aligned as close as possible to the source. While this also increases transconductance, it produces significant changes in the electric field distribution in the channel that makes gate-to-drain capacitance

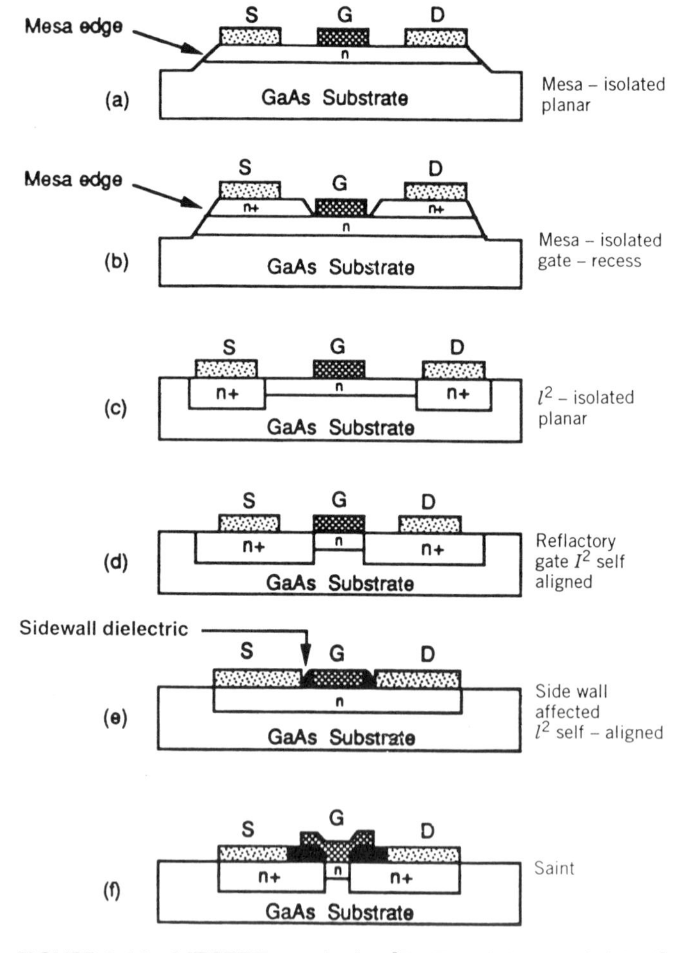

FIGURE 2.14 MESFET topologies [D. Estreich, unpublished]

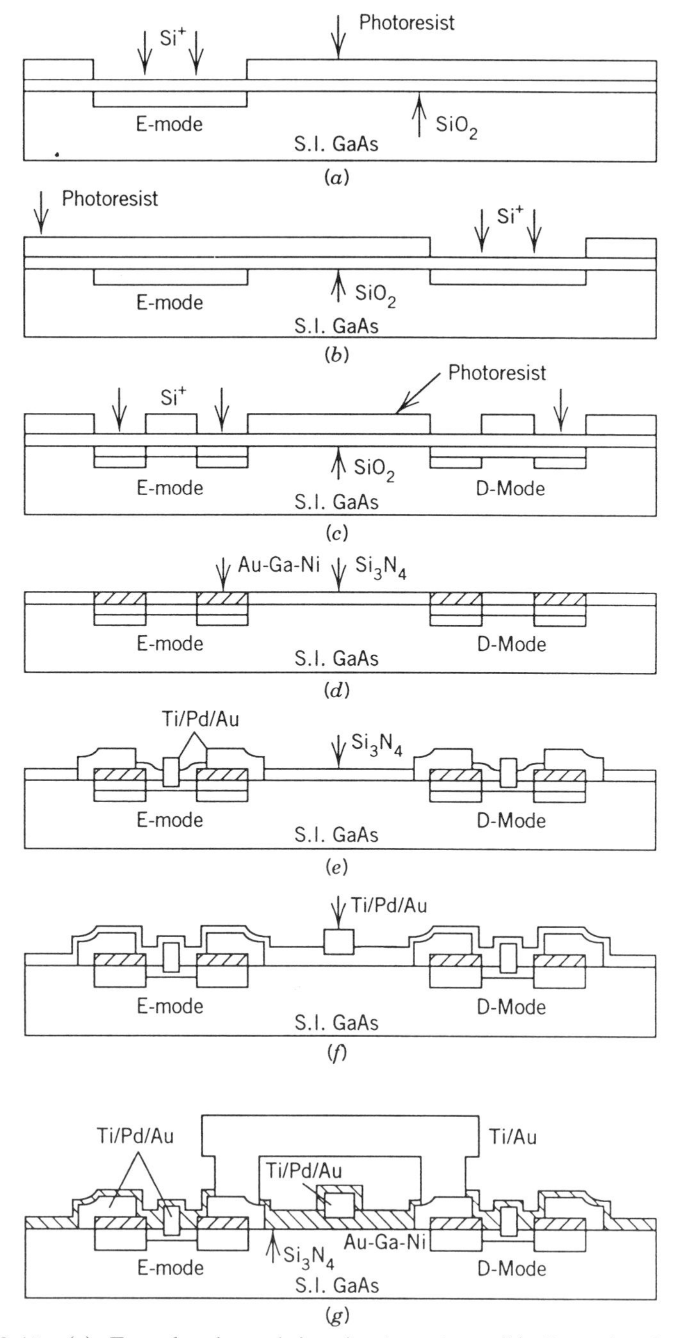

FIGURE 2.15 (*a*) E-mode channel ion implantation; (*b*) D-mode channel ion implantation; (*c*) source–drain N^+ ion implantation; (*d*) ohmic metallization; (*e*) gate metallization; (*f*) first-level interconnections; (*g*) air-bridge interconnections [49].

C_{gd} large, thereby making the feedback large and increases the output conductance [45–47]. Both these features are undesirable. The shape and depth of the gate recess influences the dc I–V characteristics, gate–drain breakdown voltage and noise performance. However, gate-recess technology can not provide the uniformity of threshold voltage across a wafer and from wafer to wafer, which is required for GaAs ICs. This is due to the fact that it is difficult to control the amount of etching of a small channel area of 0.5 μm or less.

To overcome some of the undesirable characteristics of gate-recess technology and to provide uniform threshold voltage from wafer to wafer, the complex self-aligned gate FET process technology, which is a planar process, was developed [48]. In this process, the gate is formed prior to source–drain metalization on a lightly doped channel, and this prevents gate leakage and low breakdown voltages. However, source and drain contacts, which need a high level of doping, is obtained through selective ion implantation.

Figure 2.15 shows various key steps in GaAs MESFET processing [49]. First the wafer surface is cleaned and a dielectric layer is deposited. A photoresist pattern defines the E-mode, D-mode, and contact regions. Then implantations takes place with photoresist as the mask. Then it is annealed for repair of the damage. The wafer is patterned with photoresist, and openings at source and drain are etched into the dielectric at the contacts and ohmic metallization is deposited and lifted off after the pattern. The metal is alloyed further at high temperature to achieve a good contact. Then the gate opening is defined in the photoresist and dielectric is etched until a predetermined saturation current is achieved (monitored while etching). Then the gate metal is deposited and lifted off. The first level of metallization for ohmic contacts is defined in the photoresist and metals are overlayed to improve the contact resistance. Dielectric passivation is applied to protect metallizations. Enhancement MESFET technology also requires a second channel implant and have increased complexity of processing.

Besides MESFET structure, there are other GaAs-material-based devices, occasionally used for high-frequency analog IC designs. Figure 2.16 shows a heterostructure FET. It consists of an AlGaAs layer on top of a

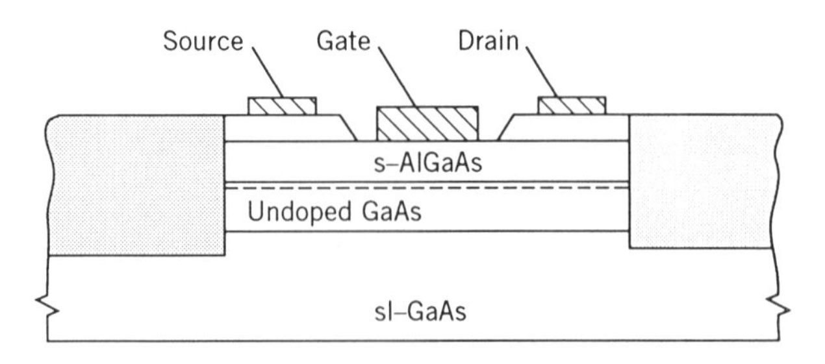

FIGURE 2.16 Schematic diagram of the generic heterostructure FET device [33].

layer of undoped GaAs. The substrate is semiinsulating GaAs. Electrons from the doped AlGaAs move to the undoped GaAs as a result of band-edge discontinuity until charge equilibrium is established and band bending takes place. The electrons can now move freely through undoped GaAs, which results in high mobilities and hence larger transconductance due to lack of impurity scattering. It also allows high positive swing for the gate, and improvements in noise figure and pinchoff voltage. But the technology is not yet mature. There are several other structures in use such as resonant tunneling device, hot-electron transistor, heterojunction bipolar transistor (HBT), and modulation-doped field-effect transistor (MODFET).

2.6 ANALOG IC PROCESS

In this section, we describe step-by-step fabrication of various individual circuit elements used in analog ICs. This process is based on depletion-mode MESFET fabrication process [5]. Figure 2.17*a* shows the devices to be fabricated and semi-insulating GaAs substrate. A low-level n-doping (N^-) ion implantation is done to form MESFET channel (Fig. 2.17*b*). The

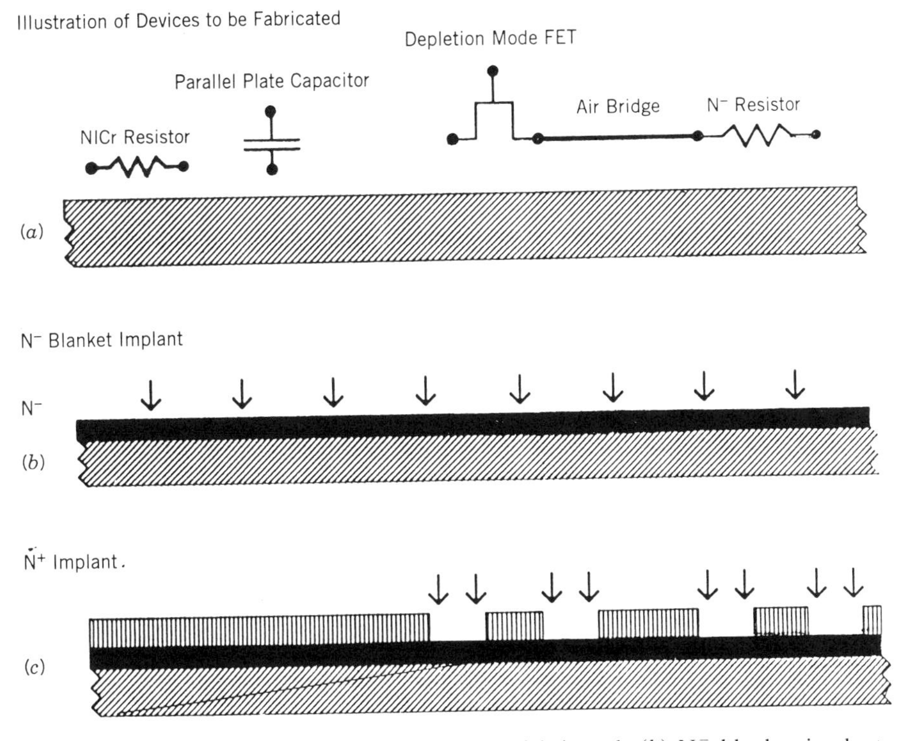

FIGURE 2.17 (*a*) Illustration of devices to be fabricated; (*b*) N^- blanket implant; (*c*) N^+ implant.

implant can also be used to fabricate Schottky switching diodes where capacitance needs to be minimized. A photoresist-defined pattern is used for N^+ ion implantation to establish ohmic contacts and for level anode of shifting Schottky diodes, if needed, and monolithic resistor contacts (Fig. 2.17*c*).

Figure 2.18*a* shows the substrate after cleaning and high-temperature annealing, which is required to repair the damage and to activate the implanted ions. The substrate is capped with a protective dielectric layer to prevent dislocation of GaAs at the surface, and after annealing the cap is stripped and a layer of SiN is deposited. Figure 2.18*b* shows a photoresist pattern to define mesa etch, and Figure 2.18*c* shows the device after mesa etching and dissolving photoresist. The dark regions are N^+-doped regions, and the remaining area is semiinsulating GaAs.

Figure 2.19*a* is again a photoresist definition to establish the metallic contacts, and the metal is deposited everywhere on the wafer. After that, photoresist is dissolved and metallization is lifted off, leaving metallization at desired locations only. The first two locations of ohmic contacts will be used as source and drain contacts of MESFET (Fig. 2.19*b*). In Figure. 2.19*c*, gate metallization using a liftoff process is shown.

Figure 2.20*a* shows the photoresist pattern for resistor etching for developing monolithic resistors, and Figure 2.20*b* shows the photoresist pattern after etching is peformed and the wafer is cleaned. Another pattern

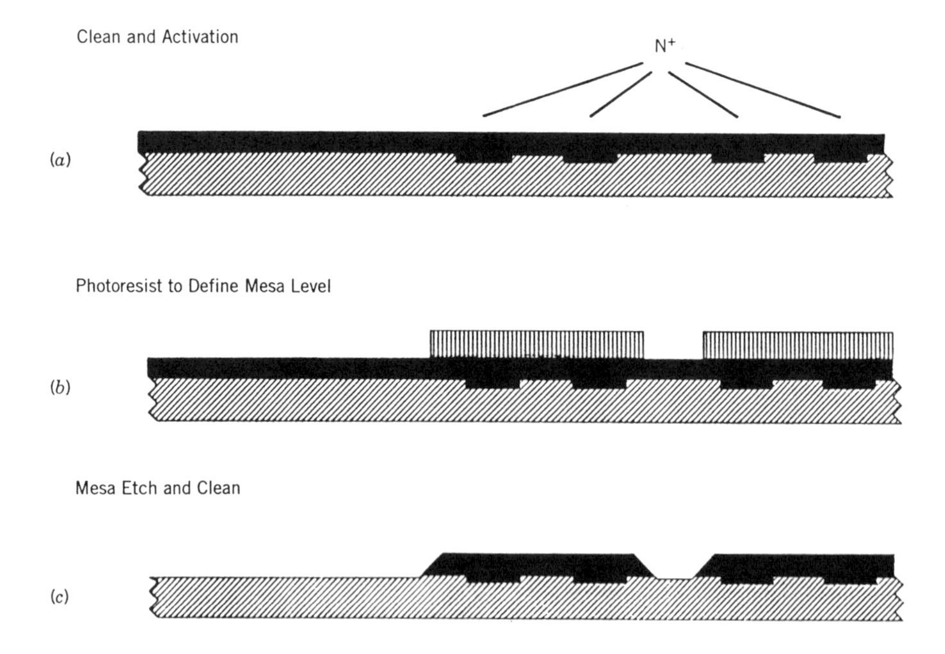

FIGURE 2.18 (*a*) Clean and activation; (*b*) photoresist to define mesa level; (*c*) mesa etch and clean.

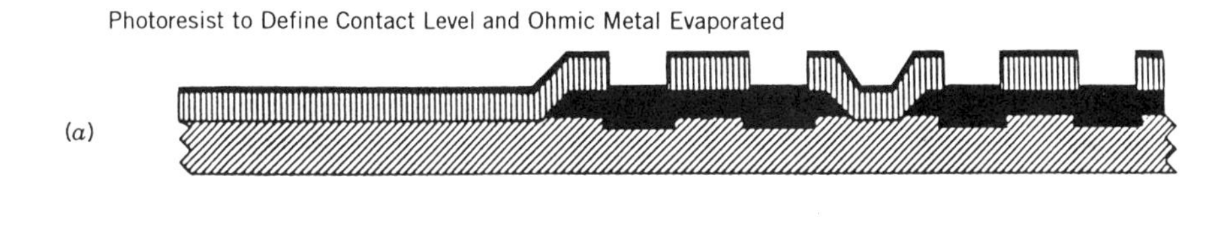

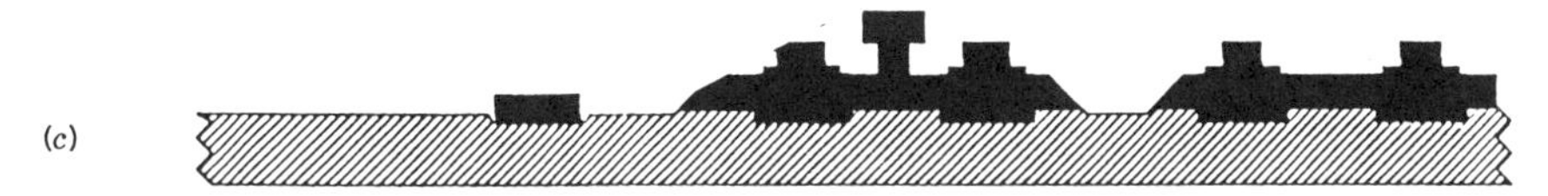

FIGURE 2.19 (*a*) Photoresist to define contact level and ohmic metal evaporated; (*b*) liftoff, wafer clean, and ohmic metal alloy; (*c*) gate metallization.

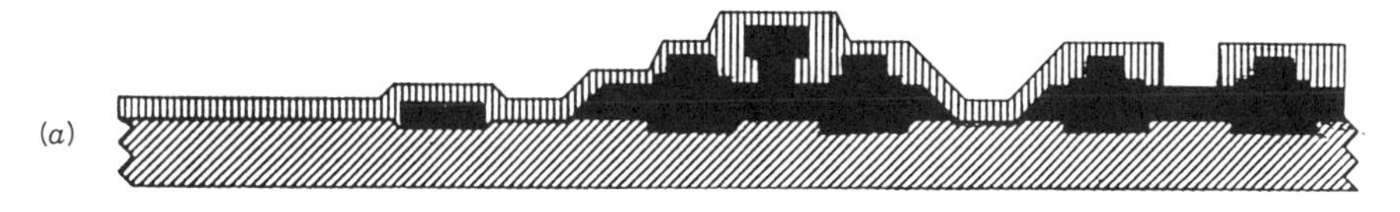

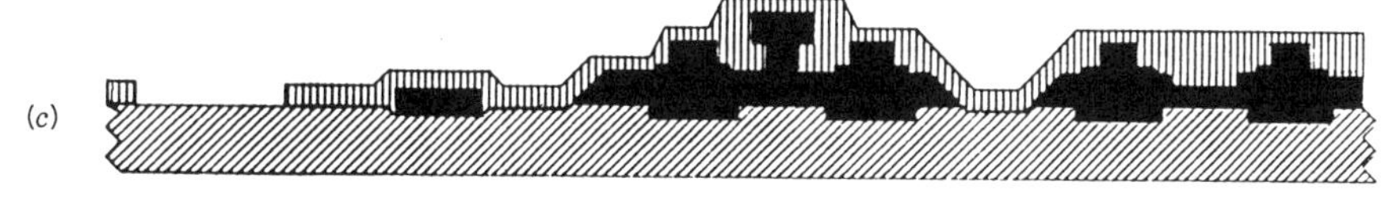

FIGURE 2.20 (*a*) Photoresist to define N^- resistor etch; (*b*) N^- resistor etch and wafer clean; (*c*) Photoresist to define NiCr resistors.

to define NiCr resistors is shown in Figure 2.20*c*. Figure 2.21 shows steps for NiCr and dielectric deposition. After NiCr is deposited, the wafer is once more cleaned to show the pattern. The dielectric is deposited everywhere after cleaning the wafer. Figure 2.21*c* shows a photoresist layer to define vias through the dielectric. The dielectric is now etched, and again a photoresist pattern is defined to define post–metal deposition Figure 2.22*a*,*b*. Figure 2.22*c* shows metal deposition on the photoresist. Without dissolving the photoresist, another photoresist pattern is defined on the top of the wafer to help define the location of air-bridge metal. Metal is again deposited. Now the photoresist is removed and wafer is cleaned (Fig. 2.23). The air-bridge contact is clearly shown. Wafer is now thinned and a photoresist is now defined on the backside of the wafer (Fig. 2.24*a*). Via holes are etched, and metallization is deposited everywhere on the backside of the wafer. To separate ICs from each other, a grid is defined using the photoresist (Fig. 2.25*a*) and backside metal is plated. Photoresist is now removed and the process is completed.

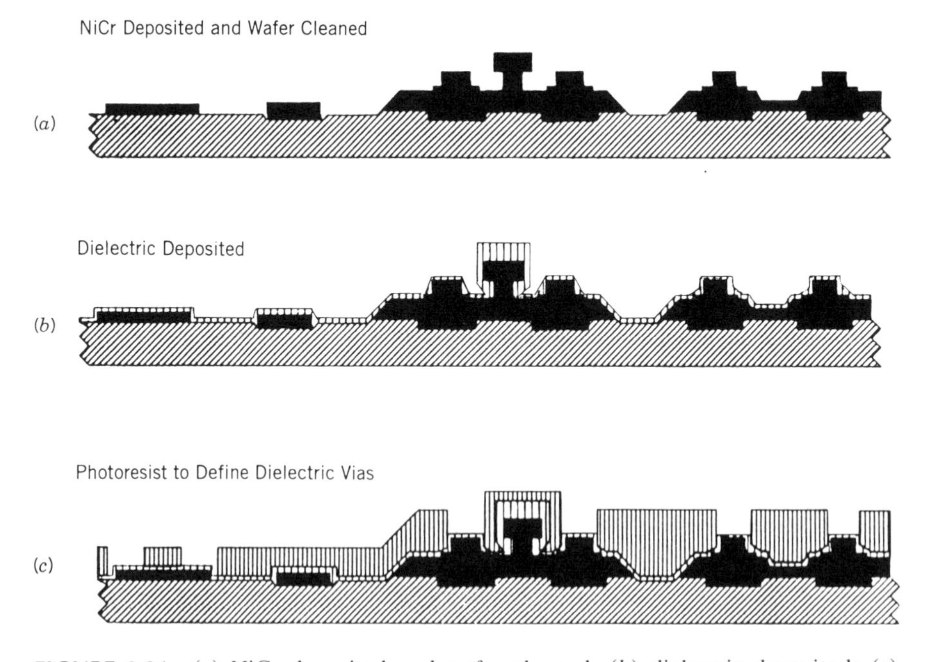

FIGURE 2.21 (*a*) NiCr deposited and wafer cleaned; (*b*) dielectric deposited; (*c*) photoresist to define dielectric vias.

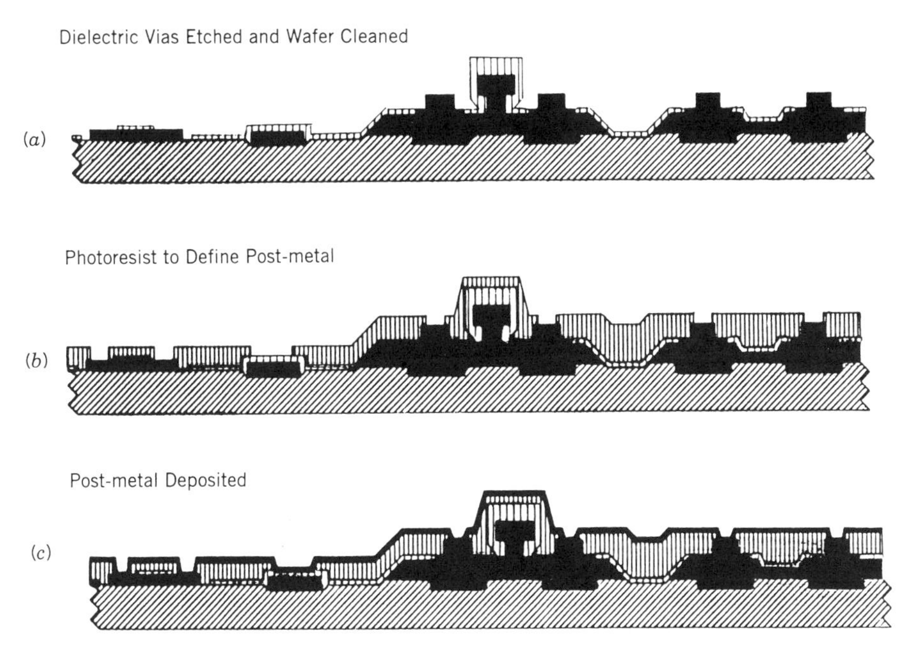

FIGURE 2.22 (*a*) Dielectric vias etched and wafer cleaned; (*b*) photoresist to define post-metal; (*c*) post-metal deposited.

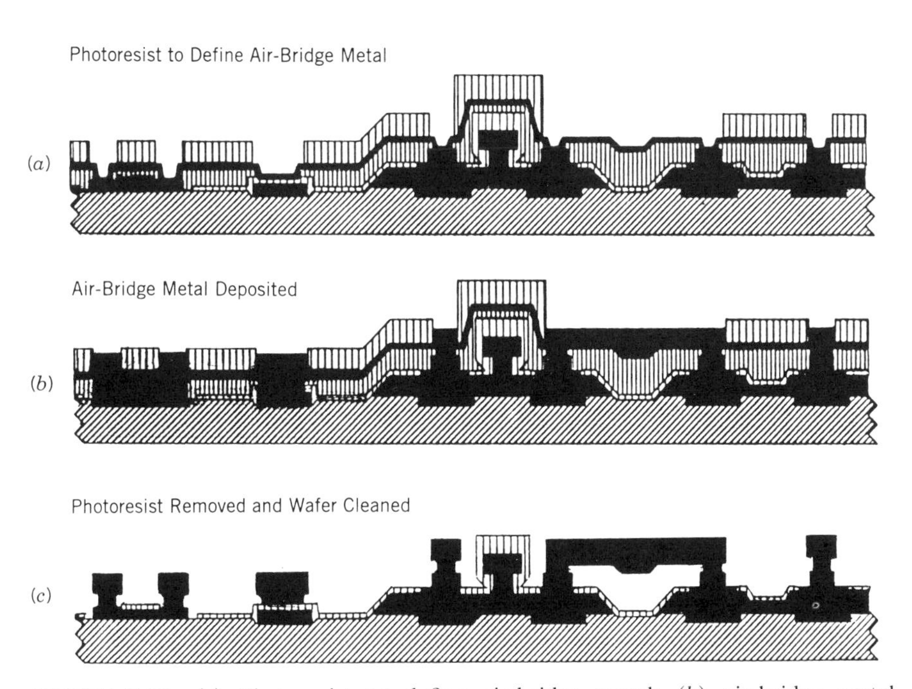

FIGURE 2.23 (*a*) Photoresist to define air-bridge metal; (*b*) air-bridge metal deposited; (*c*) photoresist removed and wafer cleaned.

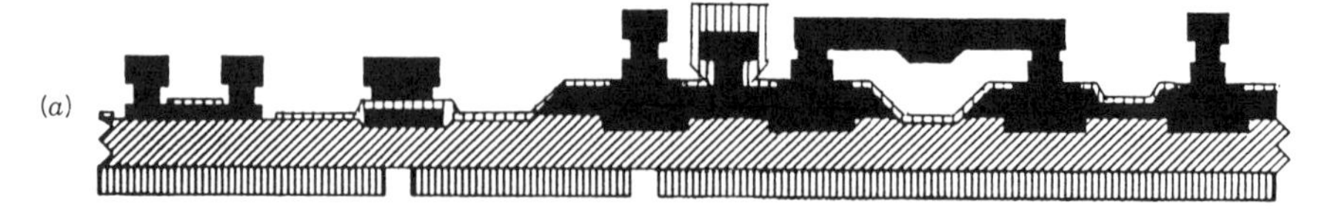

FIGURE 2.24 (*a*) Wafer thinning and photoresist to defined via holes; (*b*) via holes etched and photoresist removed; (*c*) Backside metallization.

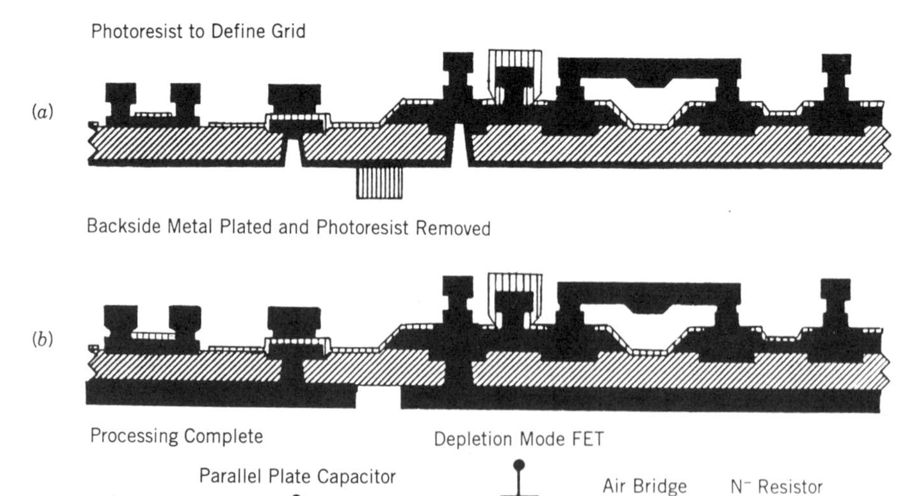

Processing Complete

Depletion Mode FET

Parallel Plate Capacitor

Air Bridge

N⁻ Resistor

NiCr Resistor

(c)

FIGURE 2.25 (*a*) Photoresist to define grid; (*b*) backside metal plated and photoresist removed; (*c*) processing complete.

REFERENCES

1. S. M. Sze, ed., *High Speed Semiconductor Devices*, Wiley, New York, 1990.
2. J. M. Golio and R. J. Trew, *IEEE Trans.*, **ED-27**, 1256 (1980).
3. A. Cappy, B. Carnez, R. Fauquemlorques, G. Salmer, and E. Constant, *IEEE Trans.*, **ED-27**, 2158 (1980).
4. S. M. Sze, *Physics of Semiconductor Devices*, Wiley, 1969, p. 22.
5. R. Goyal, ed., *Monolithic Microwave Integrated Circuits*, Artech House, Norwood, MA, 1989, Chapter 3.
6. J. V. DiLorenzo and D. D. Khandelwal, *GaAs FET Principles and Technology*, Artech House, Norwood, MA, 1984.
7. H. Thomas, D. V. Morgan, B. Thomas, J. E. Aubrey, and G. B. Morgan, eds., *Gallium Arsenide for Devices and Integrated Circuits*, Peter Peregrinus Ltd., London, 1986.
8. R. E. Williams, *Gallium Arsenide Processing Techniques*, Artech House, Norwood, MA, 1984.
9. I. R. Grant, "Bulk Growth of GaAs," in *Gallium Arsenide For Devices and Integrated Circuits*, H. Thomas, D V. Morgan, B. Thomas, J. E. Aubrey, and G. B. Morgan, eds., Peter Peregrinus Ltd., London, 1986.
10. P. A. Houston, "Liquid Phase and Vapor Phase Epitaxy of GaAs and Related Compounds," in *Gallium Arsenide For Devices and Integrated Circuits*, H. Thomas, D V. Morgan, B. Thomas, J. E. Aubrey, and G. B. Morgan, eds., Peter Peregrinus Ltd., London, 1986.
11. L. Hollan and J. Hallais, "Vapor Phase Epitaxy for GaAs FETs," in *GaAs FET Principles and Technology*, J. V. DiLorenzo and D. D. Khandelwal, eds., Artech House, Horwood, MA, 1984.
12. W. G. Herrenden-Harker and R. H. Williams, "Epitaxial Growth of GaAs: MBE and MOCVD," in *Gallium Arsenide For Devices and Integrated Circuits*, H. Thomas, D V. Morgan, B. Thomas, J. E. Aubrey, and G. B. Morgan, eds., Peter Peregrinus Ltd., London, 1986.
13. C. E. C. Wood, "Molecular Beam Epitaxy for Microwave Field Effect Transistors," in *GaAs FET Principles and Technology*, J. V. DiLorenzo and D. D. Khandelwal, eds., Artech House, Horwood, MA, 1984.
14. K. Hiruma, E. Yanokura, M. Mori, H. Mizuta, and S. Takohasi, in *Proceedings of 15th International Symposium on GaAs and Related Copmounds*, Georgia, 1988, p. 489.
15. M. G. Panish and A. Y. Cho, "Molecular Beam Epitaxy," *IEEE Spectrum* (April 1980).
16. H. Morkoc, W. F. Kopp, T. J. Drummand, S. L. Su, R. E. Thorne, and R. Fischer, *IEEE Trans.*, **ED-29**, 1013 (1982).
17. H. P. Shih, B. Kim, K. Bradshaw, H. Q. Tseng, J. M. Anthony, D. L. Farrington, T. S. Kim, and T. M. Moore, in *Proceedings of 15th Int. Symposium on GaAs and Related Compounds*, Georgia, 1988, p. 503.
18. M. Shur, *GaAs Devices and Circuits*, Plenum Press, New York, 1987.
19. C. E. C. Wood, "Progress, Problems and Applications of MBE," in *Physics of Thin Films*, G. Hass and M. Francone, eds., Academic Press, New York, 1981.

20. G. M. Martin, *Electron. Lett.* **13**, 192 (1977).
21. C. Jacob et al., "Dislocation-Free GaAs and InP Crystals by Isoelectronic Doping," *J. Crystal Growth*, **61**, 417–421 (1983).
22. R. Anholt and T. W. Sigmon, "Substrate Impurities Effects on GaAs MESFETs," *J. Electron. Materials*, 5 (1988).
23. H. K. Choi, B. Y. Tsaur, G. M. Metze, G. W. Turner, and J. C. C. Fan, *IEEE Trans.*, **EDL-5**, 205 (1984).
24. R. Fischer et al., *Electron. Lett.*, **20**, 945 (1984).
25. H. Shichijo, R. J. Matyi, and A. H. Taddiken, in *Proc. 15th International Symposium on GaAs and Related Compounds*, Georgia, 1988, p. 171.
26. R. Anholt, P. Balasingam, S. Chou, T. W. Sigmon, and M. D. Deal, "Ion Implantation into GaAs," *J. Appl. Phy.* (1989).
27. H. Ryssel and I. Ruge, *Ion Implantation*, Wiley, New York, 1986.
28. E. P. Finhem, W. A. Vetanen, B. Odekirk, and P. C. Canfield, *1988 IEEE GaAs IC Symposium Technical Digest*, IEEE Press, Piscataway, NJ, 1988.
29. R. Anholt and T. W. Sigmon, "A Process and Device Model for GaAs MESFETs," *IEEE Trans. Computer Aided Design* (April 1989).
30. R. Anholt and T. W. Sigmon, "Model of Threshold Voltage Fluctuations in GaAs MESFETs," *IEEE Electron Devices Lett.*, 16 (1987).
31. R. Anholt and T. W. Sigmon, "Ion-Implantation Effects on GaAs MESFETs," *IEEE Trans. Electron Devices* (1988).
32. S. D. Mukharjee and D. W. Woodward, in *GaAs Materials, Devices, and Circuits*, M. J. Howes and D. V. Morgan, eds., John Wiley and Sons, Chapter 4.
33. R. Koyama, "The GaAs MESFET (and Other III–V) Integrated Circuit Technology," *Proceedings of Microwave Engineering Colloquium on High Frequency Transistors and Circuit Design*, Oregon Center for Advanced Technology Education, 1988.
34. S. Makram-Ebied and P. Minondo, *IEEE Trans.*, **ED-32**, 632 (1985); D. Miller, M. Bujatti, and D. Estreich, *IEEE GaAs IC Symposium*, 1985, p. 31.
35. M. Feng, T. R. Lepkowski, G. W. Wang, C. L. Lau, and C. Ito, in *Proceedings of 15th International Symposium on GaAs and Related Copounds*, Georgia, 1988, p. 513.
36. C. L. Ghosh and R. L. Layman, *IEEE Electron. Devices Lett.*, **EDL-5**, 3 (1984).
37. F. W. Smith, A. R. Calawa, C. L. Chen, M. J. Manfra, and L. J. Mahoney, *IEEE Electron. Devices Lett.*, **EDL-9**, 77 (1988).
38. I. Benerjee, P. W. Chye, and P. E. Gregory, *IEEE Electron. Devices Lett.*, **EDL-9,** 10 (1984).
39. J. J. M. Dekkers, F. Ponse, and H. Beneking, *IEEE Trans.*, **ED-28**, 1065 (1981).
40. U. K. Mishra et al., *IEEE Electron. Device Materials*, 829 (1986).
41. J. R. Stall, C. E. C. Wood, and L. F. Eastman, *Electron. Lett.*, **15** 800 (1979).
42. K. Ploog, M. Hauser, and A. Fischer, *Appl. Phys.*, **A45**, 233 (1988).
43. M. M. Hashemi, J. Ramdani, A. C. Gossard, and W. Weigman, *Appl. Phys. Lett.*, **56**, 964 (1990).
44. A. C. Adams and B. R. Pruniaux, *J. Electrochem. Soc.*, **120** 408 (1973).

45. T. Furutsuka, T. Tsuiji, and F. Hasegawa, *IEEE Trans.*, **ED-25**, 563 (1978).
46. K. Ohata, H. Itoh, F. Hasegawa, and Y. Fujiki, *IEEE Trans.*, **ED-27**, 1029 (1980).
47. P. H. Ladbrooke, *MMIC Design GaAs FETs and HEMTs*, Artech House, Norwood, MA, 1989.
48. K. Uetake, F. Katano, M. Kamiya, T. Misaki, and A. Higashisaka, *International Symposium on GaAs Related Compounds*, Karuizawa, Japan, 1985.
49. A. G. Rohde and J. G. Roper, "GaAs Digital IC Processing—Manufacturing Perspective," *Solid State Technol.* (Feb. 1985).

CHAPTER THREE

MESFET Design and Modeling

TZU-HUNG CHEN
Hexawave Inc.
1F, 2 Prosperity Road II
Science-Based Industrial Park
Hsinchu, Taiwan, ROC

3.1 INTRODUCTION

The acronym *MESFET* has been adopted because of its similarity to MOSFET (metal–oxide semiconductor field-effect transistor). If the field-effect transistor is constructed by a metal–semiconductor Schottky barrier junction the device is called a *MEtal–Semiconductor Field-Effect Transistor* (MESFET). The substrate material can be either silicon (Si) or gallium arsenide (GaAs), and the channel may be either n-type or p-type. However, p-channel devices are less popular because hole mobility is lower than electron mobility for both Si and GaAs. GaAs MESFETs are more commonly used in high-frequency analog IC designs as compared to Si counterparts because of higher gain and cutoff frequency as well as lower noise figure. The higher gain is due to the higher mobility and higher saturation drift velocity of electrons. The higher cutoff frequency is because of higher gain and smaller parasitic capacitance resulting from the semiinsulating substrate. The lower noise figure is due partially to the higher mobility of the electron carriers. Moreover, few noise sources are present in the FET (no shot noise) as compared with the bipolar transistor.

This chapter describes the basic operation of MESFETs, the channel

High-Frequency Analog Integrated-Circuit Design, Edited by Ravender Goyal
ISBN 0-471-53043-3

characteristic of ion-implanted devices, analytical models for the calculation of the current–voltage characteristics of ion-implanted and epitaxial devices, the effects related to parasitic resistance, the empirical formula for the I–V characteristics and characterization technique to extract parameters from measured dc I–V curves, and equivalent circuits of the MESFET in Section 3.6.7.

3.2 PRINCIPLE OF OPERATION

The planar structure of GaAs MESFET is shown schematically in Figure 3.1. We will use this figure to explain the basic principle of operation of GaAs MESFET. Beginning with Ohm's law, the current density between the ohmic contacts (source and drain) of a linear resistor is

$$J = qN_d v = q\mu_0 N_d F \tag{3.1}$$

for n-type epitaxial material. The low field mobility μ_0 is related to the drift velocity by $v = \mu_0 F$. For an epitaxial layer of thickness A and width W, the total current through the resistor becomes

$$I_{ds} = JWA = qN_d WAv \tag{3.2}$$

As the electric field is increased in the resistor, the velocity of the carrier (and therefore I_{ds}) saturates. Next, a Schottky barrier gate electrode is fabricated between the ohmic source and drain contacts. With the source-to-gate potential at zero volts, the drain-to-source current again has a linear and saturation region with a reduced channel thickness. The applying of the

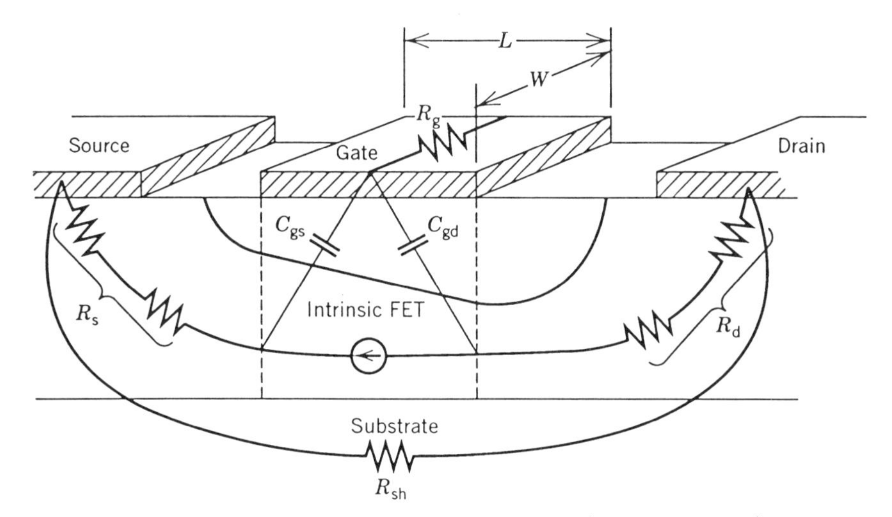

FIGURE 3.1 MESFET schematic diagram (planar structure).

Schottky barrier gate bias allows the thickness of the conduction channel t and, hence, the drain-to-source current to be modulated according to

$$I_{ds} = qN_d Wtv \tag{3.3}$$

An increasing negative voltage between gate and source will eventually pinch off the channel and reduce the channel current to essential zero. The three-terminal MESFET is a drain–source resistor that has a variable current controlled by the gate-to-source bias voltage.

Basically, there are two metallurgical technologies for the fabrication of GaAs MESFETs: epitaxy and ion implementation as described in Chapter 2. The devices fabricated by the epitaxy have a nearly uniform doping concentration within the active layer. The current–voltage characteristics of GaAs MESFETs with uniform profile have been widely investigated [1–5] since the basic principle of operation of the field-effect transistor was first described by Shockley in 1952 [6]. As to the devices fabricated by implantation, their doping profile is very close to Gaussian distribution. The current–voltage characteristics of GaAs FETs with nonuniform doping profiles were analyzed elsewhere [7–10]. However, the important effects of backgating and the change in the ion implantation due to the capping during the ion-implantation process have not been taken into account. Also, only the saturation current was calculated, which makes it difficult to use these models in the circuit simulation. Further investigations have been carried out to provide much more practical models [11–13].

3.3 ION-IMPLANTED MESFET MODEL

Ion implantation has been applied to the fabrication of GaAs FETs and ICs [14–17] and is widely used in the fabrication of GaAs analog ICs [14–20]. The motivation for these applications of ion implantation to GaAs lies in the ability to control the thickness and doping level of implanted layers by varying the energy and dose of the implanted ions and in the very good uniformity of doped layers that can be reproducibly fabricated. The reproducibility and uniformity of doping that can be achieved using ion implantation are essential to the attainment of acceptable yields of ICs.

The ion-implanted doping profile for low dose ($<5 \times 10^{12}\ \text{cm}^{-2}$) silicon or selenium implants in GaAs can be approximated by Gaussian function. For multiple ion implantation with different doses and ion energies, the net profile can be expressed as the sum of the profiles of the individual ion implantations.

MESFET channels require dopant profiles with $1–3 \times 10^{17}\ \text{cm}^{-3}$ peak concentration and 500–2000 Å depth for digital application and larger for analog application because of requirements for large currents. Channel implantation is usually performed directly into semiindulating GaAs substrate, at $1\text{–}5 \times 10^{12}\ \text{cm}^{-2}$ doses with silicon (100–200 kV) or selenium (300–400 kV). The postimplantation annealing temperature, required to obtain good activation of implanted n-type dopants in GaAs, must be sufficiently high (~800–850°C) to ensure that maximum activation of implanted impurities occurs. The most common technique employed in device fabrication to prevent this dissociation involves encapsulation of the implanted material with a dielectric layer such as silicon nitride or silicon dioxide as described in Chapter 2. This technique is also used to locate the peak of the implanted profile at a desired depth in order to optimize the device performance. Figure 3.2 shows a typical ion-implanted doping profile with a capping depth t_{cap}

The doping profile and the energy diagram of an ion-implanted GaAs MESFET with capping depth t_{cap} are shown in Figure 3.2. The channel concentration is described by the difference between the Gaussian function and the effective background doping concentration N_a:

$$N(x) = N_p \exp\left\{\frac{-[(x - R_p)]^2}{[(\sigma\sqrt{2})]^2}\right\} - N_a \tag{3.4}$$

where $N_p = Q/[\sigma\sqrt{(2\pi)}]$ is the peak concentration, $R_p = R_{po} - t_{cap}$ is the effective projected range of the Gaussian profile and Q, R_{po}, and σ are the dose, the projected range, and the standard deviation of the implant, respectively.

The decrease in the drain-to-source current when a negative voltage is applied to the substrate is termed *backgating* [21–26]. The actual pinchoff voltage V_{po}, taking into account backgating, is defined by

$$V_{po} = \frac{q}{\varepsilon}\int_0^t N(x)x\, dx \tag{3.5}$$

where t is the depletion width of the Schottky gate junction when the

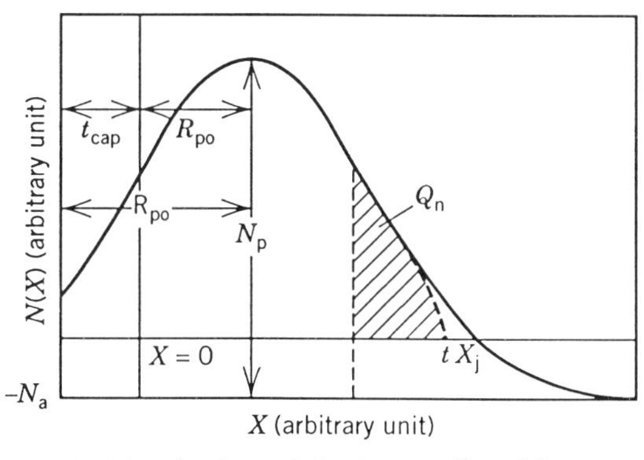

FIGURE 3.2 A typical ion-implanted doping profile with a capping depth t_{cap}.

conducting channel is pinched off (see Fig. 3.2). If the total implantation dose Q is large enough or the effective background doping density N_a is small enough so that they satisfy the inequality

$$\frac{N_a t}{Q} < \frac{N_a}{Q}\left[R_p + \sigma\sqrt{2}\ln\left(\frac{N_p}{N_a}\right)\right] \ll 1 \tag{3.6}$$

(which is a typical case for GaAs MESFETs), then the pinchoff voltage for the Gaussian profile [see Eq. (3.4)] can be shown to be

$$V_{po} \sim 4V_z \frac{Q_a}{Q}\left[\frac{Q_a}{Q} + \alpha - \mathrm{erf}(z_p)\right] \tag{3.7}$$

where

$$V_z = \frac{qQ\sigma}{\varepsilon\sqrt{(2\pi)}} \tag{3.8}$$

$$\alpha = \frac{R_p}{2\sigma}\left(\frac{\pi}{2}\right)^{1/2} \tag{3.9}$$

and

$$Q_a = \int_0^t N(X)\,dx \sim \frac{Q}{2}[\mathrm{erf}(z_t) + \mathrm{erf}(z_p)] \tag{3.10}$$

Here $z_p = R_p/(\sigma\sqrt{2})$, $z_t = (t - R_p)/(\sigma\sqrt{2})$, and erf($z$) is the error function [the error function is defined by the integral, $\mathrm{erf}(z) = (2/\sqrt{\pi}) \int_0^z \exp(-x^2)\,dx$].

3.4 INTRINSIC *I–V* MODEL

The MESFET can be divided into two parts: the intrinsic device and the extrinsic parasitics. The intrinsic device represents the active region under the gate, and its characteristics vary with biasing conditions whereas the extrinsic parasitics are not necessary for device operation and their values do not vary with biasing conditions. In this section, the physical model for the calculation of the current–voltage characteristics of the intrinsic MESFET will be discussed in detail. Since the intrinsic MESFET includes only the active portion of the channel (see Fig. 3.3), the drain-to-source current is fully controlled through the modulation of the channel depth by applying gate voltage. The effect of the parasitic drain and source resistances is to reduce the effective drain-to-source and gate-to-source voltages and will be incorporated in the model later. The *I–V* model presented here will be generalized so that it can be applied for both epitaxial and ion-implanted MESFETs.

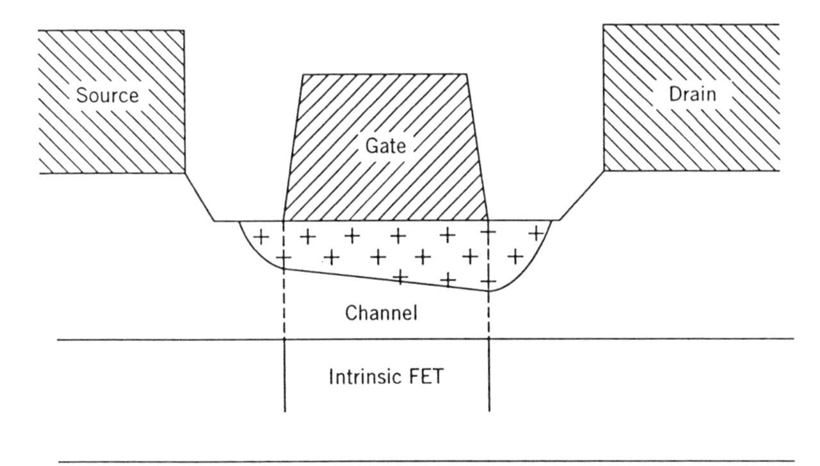

FIGURE 3.3 Cross-sectional view of a planar MESFET.

In GaAs, the equilibrium electron velocity versus electric field reaches a peak value at about 3 kV/cm, then decreases and levels off at a saturated velocity that is about equal to the limiting velocity in silicon (see Fig. 3.4). The existence of a bulk negative differential mobility will cause the formation of a stationary dipole-layer in the channel at the drain end of the gate, which we shall discuss later. However, neglecting transient effects in electron transport and basing their conclusions on a three-piece linear approximation of the velocity-field characteristic, Dawson and Frey have found that fields at which a negative differential mobility occurs exist over only a very small portion of the channel [27]. Therefore, we shall neglect the existence of the negative differential mobility and make use of a two-piece linear approximation of the velocity-field characteristic as shown in Figure 3.4.

This velocity-field characteristic divides the intrinsic FET model into two

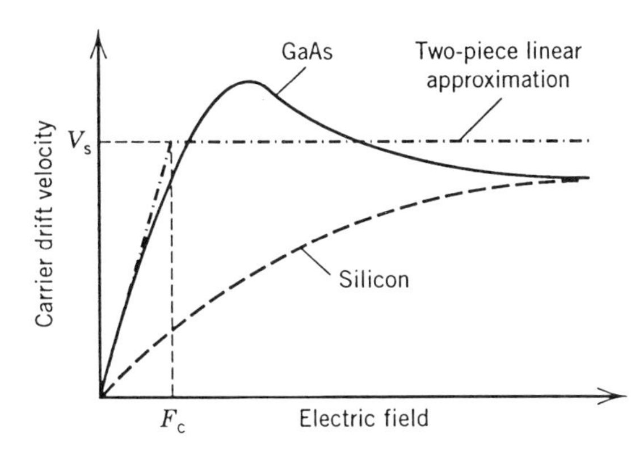

FIGURE 3.4 Carrier drift velocity versus electric field characteristics for GaAs and silicon.

regions: region I of length L_1, the constant mobility region where the velocity is field-dependent; and region II of length L_2, the constant velocity region where the electric field has exceeded the critical field F_c and the carriers travel at the saturation velocity v_s, independent of field strength. The piecewise linear approximation of the velocity-field characteristic necessarily involves compromise in the choice of values used for F_c, μ_0, and v_s.

The charge in the conducting channel at a point with channel potential V is found by the use of the depletion approximation as

$$Q_n = Q_a\left[1 - \left(\frac{V_{bi} - V_{gs} + V}{V_{po}}\right)^2\right] \tag{3.11}$$

for a uniform doping profile device. Here $Q_a = N_d t$ is the total accessible charge per unit gate area, V_{bi} is the built-in potential of the gate Schottky junction, and V_{gs} is the gate-to-source voltage. For a Gaussian doping profile, the conducting charge can be shown to be [28]

$$Q_n = Q'_a - (Q/2)\left\{[\alpha - \mathrm{erf}(z_p)]^2 + \frac{V_{bi} - V_{gs} + V}{V_z}\right\}^{1/2} \tag{3.12}$$

where $Q'_a = Q_a + (Q/2)[\alpha - \mathrm{erf}(z_p)]$ is the effective accessible dose. If we define the effective built-in voltage and the effective pinchoff voltage as

$$V'_{bi} = V_z[\alpha - \mathrm{erf}(z_p)]^2 + V_{bi} \tag{3.13}$$

$$V'_{po} = V_z[\alpha - \mathrm{erf}(z_p)]^2 + V_{po} \tag{3.14}$$

then Equation (3.11) becomes

$$Q_n = Q'_a\left[1 - \left(\frac{V'_{bi} - V_{gs} + V}{V'_{po}}\right)^2\right] \tag{3.15}$$

which has the same form as for the uniform doping profile [see Eq. (3.11)]

The current in the conducting channel of region I is given by Ohm's law, using a gradual channel approximation, so that

$$I_{ds} = q\mu_0 W Q_n(x)|F(x)| \tag{3.16}$$

where $|F(x)| = dV/dx$ is the longitudinal electric field in the channel. After an integration over the length L and the use of Equation (3.11) or (3.15), we find

$$I_{ds} = g_0\left\{V_1 - \frac{2}{3\sqrt{V_p}}[(V_b - V_{gs} + V_1)^{3/2} - (V_b - V_{gs})^{3/2}]\right\} \tag{3.17}$$

where $V_1 = V(L_1)$ is the channel potential at the boundary between region I and region II and $g_0 = q\mu_0 WQ_a/L_1$, $V_b = V_{bi}$ and $V_p = V_{po}$ for uniform doping profile (or $g_0 = q\mu_0 WQ'_a/L_1$, $V_b = V'_{bi}$ and $V_p = V'_{po}$ for Gaussian doping profile).

In region II, immediately at the boundary with region I, the current can be calculated as

$$\begin{aligned} I_{ds} &= qv_s WQ_n(V_1) \\ &= I_{fc}\left[1 - \left(\frac{V_b - V_{gs} + V_1}{V_p}\right)^{1/2}\right] \end{aligned} \tag{3.18}$$

where $I_{fc} = qv_s WQ_a$ for uniform doping profile (or $I_{fc} = qv_s WQ'_a$ for a Gaussian doping profile) is the full channel saturation current.

The value of L_1 can be found by requiring that the electric field along the channel at the boundary of region I with region II reach the velocity saturation field F_c. This leads to the following equation:

$$\gamma = \frac{u_1 - 2[(u_g + u_1)^{3/2} - u_g^{3/2}]/3}{1 - (u_g + u_1)^{1/2}} \tag{3.19}$$

where $u_1 = V_1/V_p$, $u_g = (V_b - V_{gs})/V_p$, and $\gamma = F_c L_1/V_p$. Equation (3.19) is equivalent to the condition of the current continuity in the conduction channel, which requires the equality of Equations (3.17) and (3.18). As shown in Shur [29], Equation (3.19) can be approximated by

$$u \sim \frac{\gamma(1 - u_g)}{\gamma + 1 - u_g} \tag{3.20}$$

Equation (3.19) gives the normalized drain-to-source saturation voltage (V_{dss}/V_p) if we substitute L for L_1 in γ.

In dimensionless units ($i_{ds} = I_{ds}/[g_0 V_p]$), Equation (3.18) becomes

$$i_{ds} = \gamma[(1 - (u_1 + u_g)^{1/2}] \tag{3.21}$$

For short-channel and high pinchoff voltage devices ($\gamma = F_c L/V_p \ll 1$), we can expand u_1, given by Equation (3.19), in Taylor's series and neglect second-order and higher terms; then we have $u_1 \sim \gamma$. This is consistent with the result of Equation (3.20) at $\gamma \to 0$. Therefore, Equation (3.21) reduces to

$$i_{ds} \sim \gamma[(1 - (\gamma + u_g)^{1/2}] \tag{3.22}$$

For most MESFETs Equation (3.22) is well satisfied. Figure 3.5 summarizes the key features of a MESFET operated far in the saturated current region

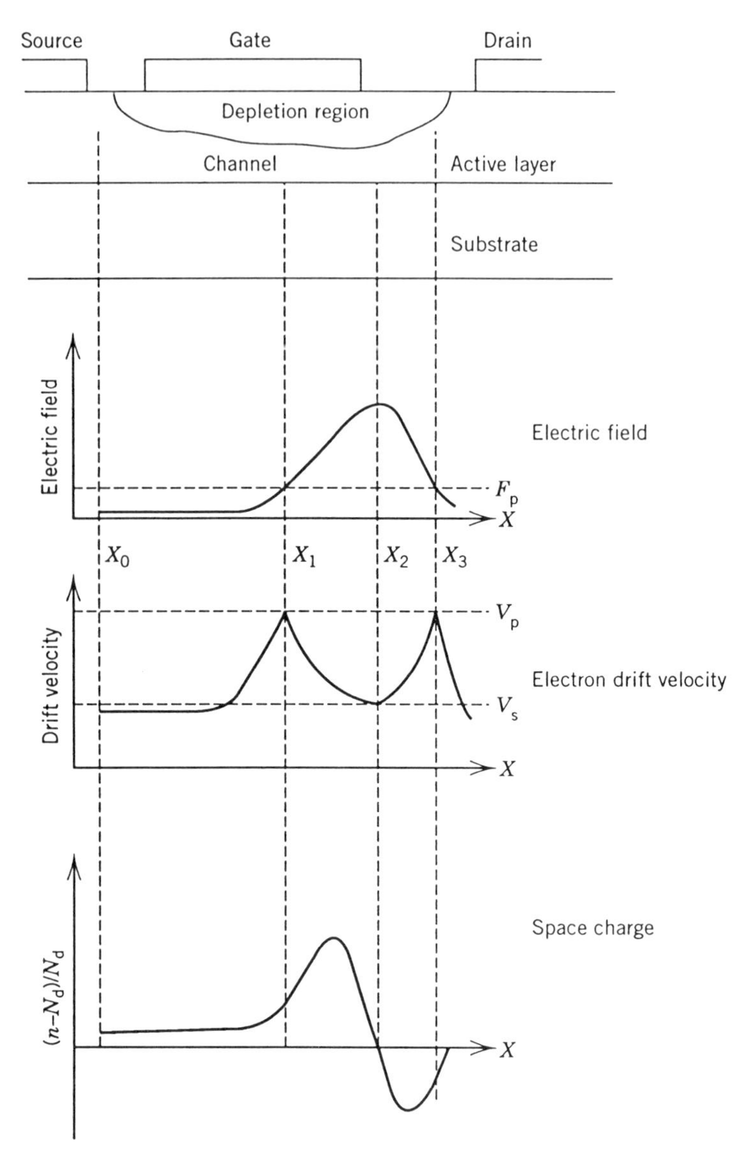

FIGURE 3.5 The channel cross section, electric field, electron drift velocity, and space-charge distribution in the channel are illustrated for a GaAs MESFET operated in the current-saturated region.

[30]. The narrowest channel cross section is located under the drain end of the gate. The drift velocity rises to a peak at x_1, close to the center of the channel, and falls to the low saturated value under the gate edge. To preserve current continuity according to Equation (3.3), heavy electron accumulation has to form in this region because the channel cross section is

narrowing and, in addition, the electrons are moving progressively slower with increasing x. Exactly the opposite occurs between x_2 and x_3. The channel widens and the electrons move faster, causing a strong depletion layer. The charges in the accumulation and the depletion layers are nearly equal and most of the drain voltage drops in this stationary dipole layer [31–33]. Because of the formation of this high field domain in the channel on the drain side of the gate, the boundary between region I and region II moves further toward the source side of the gate. Therefore, the length of region I is smaller than the metallurgical length of the gate. This effect is termed the *channel length modulation*. For MESFET operated far in the saturated current region as shown in Figure 3.5, the length L_1 of region I becomes much smaller than the actual gate length L. In this case γ can even be neglected from Equation (3.22) because γ ($\gamma = F_c L/V_p$) becomes much smaller than u_g except near threshold region, where u_g is also much less than 1.

For low-pinchoff-voltage (low-V_{po}) devices, the dipole layer cannot form [31–33] and the free-carrier accumulation in the channel is negligibly small [34]. Hence, the longitudinal electric field within region II is determined entirely by free charges on the drain electrode. In this case, the field strength and the potential distribution in region II can be found from the approximate solution of the Laplace equation as follows [34]:

$$F(x) = F_c \cosh\left[\frac{\pi}{(2t)(x - L_1)}\right] \tag{3.23}$$

$$V(x) = V_{1s} + \frac{2t}{\pi} F_c \sinh\left[\frac{\pi}{(2t)(x - L_1)}\right] \tag{3.24}$$

Then the voltage drop across region II is found by substituting L and x in Equation (3.23) as

$$V_{2s} = V(L) - V_{1s} = \frac{2t}{\pi} F_c \sinh\left[\frac{\pi}{(2t)(L - L_1)}\right] \tag{3.25}$$

In general, Equation (3.21) can be approximated by

$$i_{ds} \sim \frac{\gamma(1 - u_g)^2}{1 + 3\gamma} \tag{3.26}$$

as proposed in Shur [11]. Equation (3.26) can be rewritten in the square-law form [35]

$$I_{ds} = \beta(V_{gs} - V_T)^2 \tag{3.27}$$

where $V_T = V_b - V_p$ and

$$\beta = \frac{qWQ_a v_s}{V_p(V_p + 3F_c L_1)} \tag{3.28}$$

The term Q_a in Equation (3.28) should be replaced by Q'_a for a Gaussian doping profile. Finally, Equations (3.25)–(3.29) must be solved simultaneously to give the saturation drain-to-source current as a function of drain-to-source voltage $V(L)$.

Under low drain bias conditions ($V_{ds} < V_{dss}$), the saturation region, region II, does not exist and I_{ds} has to be calculated from Equation (3.17), where L and V_{ds} are substituted for L_1 and V_{1s}, respectively. The channel conductance in this linear region is found by differentiating I_{ds} in Equation (3.17) with respect to V_{1s} and by keeping V_{gs} constant:

$$G_{ch}(V_{ds}) = \frac{qW\mu_0}{L} Q_n(V_{ds}) \tag{3.29}$$

where we have made the substitution of V_{ds} and L for V_1 and L_1, respectively.

To obtain the current–voltage characteristics of MESFETs, the drain-to-source saturation current I_{dss} and the channel conductance at $V_{ds} = 0\ V$, G_{ch}, need to be calculated first. Then, an interpolation formula, proposed elsewhere [35,36] can be used for I_{ds} in the entire range of the drain-to-source voltage V_{ds}:

$$I_{ds} = I_{dss}(1 + \lambda V_{ds}) \tanh(\eta V_{ds}) \tag{3.30}$$

where η is chosen in such a way that at $V_{ds} \rightarrow 0$ V. Equation (3.30) converts into the corresponding equation of the Shockley model [6]:

$$\eta = \frac{G_{ch}}{I_{dss}} \tag{3.31}$$

Constant λ in Equation (3.30) is an empirical parameter that account for the output conductance in addition to the channel length modulation. The saturation current I_{dss} and the channel conductance G_{ch} are related to the conducting channel charge $Q_n(0)$:

$$I_{dss} = qWv_s Q_n(0) \tag{3.32}$$

$$G_{ch} = \frac{qW\mu_0 Q_n(0)}{L} \tag{3.33}$$

Here W is the gate width, L is the metallurgical gate length, μ_0 is the low field mobility, and v_s is the electron saturation velocity. Both Equations (3.32) and (3.33) are valid for devices with arbitrary doping profiles;

however, Equation (3.32) is applicable only for high-pinchoff-voltage devices. For low-pinchoff-voltage devices, the saturation drain-to-source current I_{dss} is calculated by the square law [see Equations (3.27) and (3.28)].

For a Gaussian doping profile, the conducting channel charge at the source end of the gate is given by Equation (3.12) with $V = 0$:

$$Q_n(0) = Q'_a - \frac{Q}{2}\left\{[\alpha - \mathrm{erf}(z_p)]^2 + \frac{V_{bi} - V_{gs}}{V_z}\right\}^{1/2} \tag{3.34}$$

For a uniform profile, the conducting channel charge at the source end of the gate is given by Equation (3.11) with $V = 0$:

$$Q_n(0) = Q_a\left[1 - \left(\frac{V_{bi} - V_{gs}}{V_{po}}\right)^2\right] \tag{3.35}$$

where $Q_a = N_d t$. Here t is the effective thickness of the active layer.

3.5 PARASITIC EFFECTS

For a realistic description of GaAs MESFETs, the effects related to the source and drain series resistances R_s and R_d as well as the drain-to-source shunt resistance R_{sh} should be incorporated into the model. The source series resistance represents the total resistance of the source ohmic contact and the bulk resistance between the source contact and the active channel, while the drain series resistance represents the similar resistance on the drain side. As to the shunt resistance R_{sh}, it accounts for the additional output conductance beyond that which is related to the gate length modulation and is due to the leakage current through the substrate underneath the conducting channel. The effect of the source and drain series resistances can be taken into account as follows. The external measurable gate-to-source voltage V_{GS} and drain-to-source voltage V_{DS} are given by

$$V_{GS} = V_{gs} + I_{ds} R_s \tag{3.36}$$

$$V_{DS} = V_{ds} + I_{ds}(R_s + R_d) \tag{3.37}$$

Combining Equations (3.36) and (3.34) with Equation (3.32) in the Gaussian profile case and solving the resulting equation for $I_{ds} = I_{dss}$, we find

$$I_{dss} = \frac{I_0}{2}\left[h_1 - \left\{h_1^2 - \frac{4[V_{po} - (V_{bi} - V_{GS})]}{V_z}\right\}^{1/2}\right] \tag{3.38}$$

where

$$I_0 = \frac{qWv_sQ}{2} \tag{3.39}$$

$$h_1 = 2\left\{[\alpha - \text{erf}(z_p)]^2 + \frac{V_{po}}{V_z}\right\}^{1/2} + \frac{R_S I_0}{V_z} \tag{3.40}$$

Similarly, combining Equations (3.36) and (3.35) with (3.32) and solving the resulting equation for $I_{ds} = I_{dss}$ in the uniform profile case, we obtain

$$I_{dss} = \frac{I_{fc}}{2}\left[h_2 - \left\{h_2^2 - \frac{4[V_{po} - (V_{bi} - V_{GS})]}{V_{po}}\right\}^{1/2}\right] \tag{3.41}$$

where

$$h_2 = 2 + \frac{R_s I_{fc}}{V_{po}} \tag{3.42}$$

and I_{fc} has the same definition as in Equation (3.18).

Both Equations (3.38) and (3.41) are used to compute the saturation current of the high pinchoff voltages devices based on the external bias voltages. A similar expression for low-pinchoff-voltage devices has been derived in Shur [11]. The result is

$$I_{dss} = \frac{1 + 2\beta R_s(V_{GS} - V_T) - [1 + 4\beta R_s(V_{GS} - V_T)]^{1/2}}{2\beta R_s^2} \tag{3.43}$$

To describe the drain-to-source current as a function of the external drain-to-source voltage, a similar interpolation formula as used in the intrinsic model [see Eqn. (3.30)] is used:

$$I_{DS} = I_{dss}(1 + \lambda V_{DS}) \tanh(\eta V_{DS}) + \frac{V_{DS}}{R_{sh}} \tag{3.44}$$

where I_{dss} is given by Equations (3.38)–(3.42) for a high-pinchoff-voltage device and by Equation (3.43) for a low-pinchoff-voltage device, constant λ is the same empirical constant that is used in Equation (3.30), and R_{sh} is the effective parasitic drain-to-source resistance shunting the intrinsic device. Parameter η in Equation (3.44) is chosen in th same way as in Equation (3.30); that is, at $V_{DS} \rightarrow 0$ V Equation (3.44) converts into the corresponding equation of the Shockley model:

$$\gamma = \frac{G_{CH}}{I_{dss}} \tag{3.45}$$

where G_{CH} is the output conductance excluding the contribution due to the shunt resistance $1/R_{sh}$ at $V_{DS} \to 0$:

$$G_{CH} = \lim_{(V_{DS}\to 0)} \left\{ \left. \frac{dI_{ds}}{dV_{DS}} \right|_{V_{GS}} \right\} \tag{3.46}$$

Here I_{ds} is the channel current when the device is biased in the linear region and is given by

$$I_{ds} = \int_{I_{ds}R_s}^{V_{DS}-I_{ds}R_s} Q_n(V)\, dV \tag{3.47}$$

where $Q_n(V)$ is the conducting channel charge for an arbitrary doping profile. From Equations (3.46) and (3.47) we find

$$G_{CH} = \frac{G_{ch}}{1 + (R_s + R_d)G_{ch}} \tag{3.48}$$

where G_{ch} is given by Equations (3.33) and (3.35) with the substitution of V_{gs} by V_{GS}.

Figure 3.6 compares the measured and calculated I–V curves of a 300-μm MBE-grown MESFET with a 1-μm gate length. A similar comparison is illustrated in Figure 3.7 for a 300-μm ion-implanted MESFET with a 1-μm

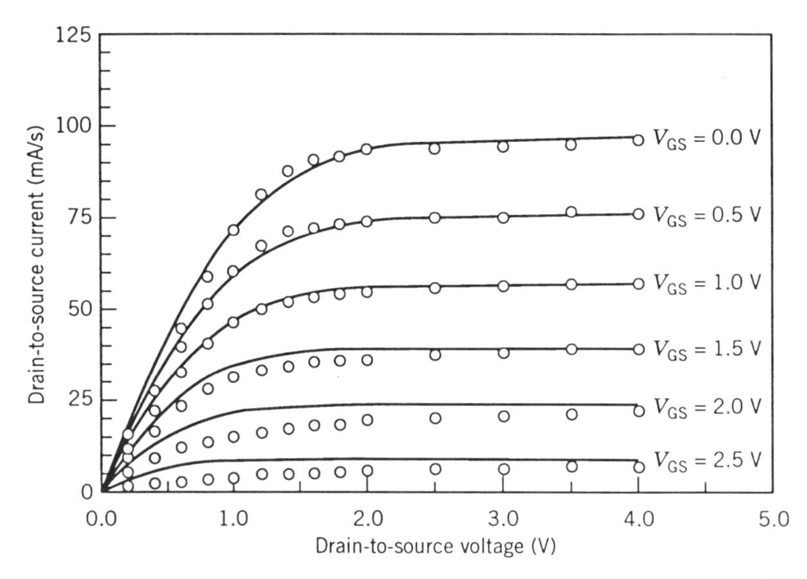

FIGURE 3.6 Comparison between measured and calculated I–V curves of a 300-μm MBE-grown MESFET with a 1-μm gate length.

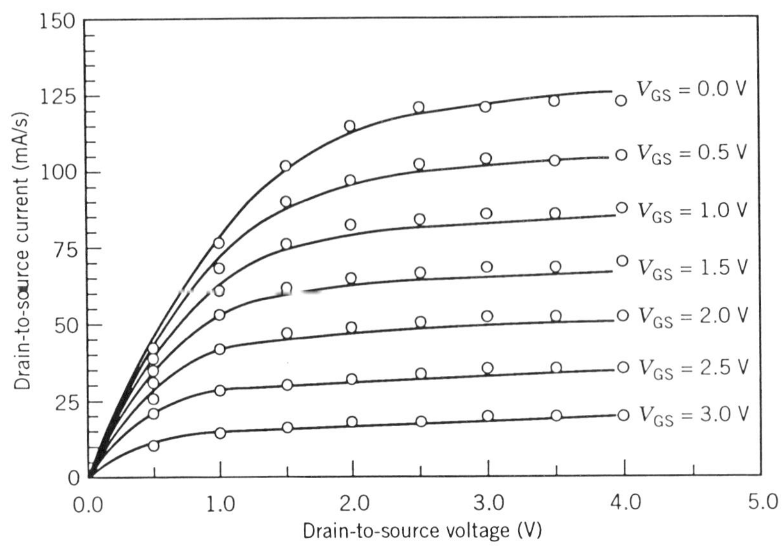

FIGURE 3.7 Comparison between measured and calculated *I–V* curves of a 300-μm ion-implanted MESFET with a 1-μm gate length.

gate length. Both figures show that the generalized model can be equally well applied to both types of MESFET.

3.6 EMPIRICAL MODELS AND PARAMETER EXTRACTION

The *I–V* model discussed in the previous sections is derived from physical equations, and its model parameters are obtained from process considerations. The physical model is useful for device design since it directly relates the physical and process parameters to the device *I–V* characteristics. However, in most situations, the physical and process parameters of the devices are not known; therefore, it is not practical to apply the physical model for the circuit simulation. To resolve this difficulty, empirical equations of measurement-based models are often used to describe the device *I–V* characteristics. The empirical models are based on experimental data, and their model parameters can be easily and accurately extracted. There are several empirical models, such as the Materka model [37], the Curtice cubic model [38], and the Statz model [39], which are commonly used and have been implemented in commercially available circuit simulation tools. An important advantage of empirical models is flexibility, simplicity and computational performance. The concise mathematical representation of empirical models can fit the drain–source current characteristics of MESFETs with a variety of doping profiles; only the numeric values of the parameters will differ.

The Materka model assumes a square-law relationship between drain–source current I_{DS} and gate–source voltage V_{gs} at saturation. The formulas that describe the drain–source current are

$$I_{DS} = I_{dss}\left(1 - \frac{V_{gs}}{V_p}\right)^2 \tanh\left(\frac{\alpha V_{ds}}{V_{gs} - V_p}\right) \tag{3.49}$$

$$V_p = V_{po} + \gamma V_{ds} \tag{3.50}$$

where I_{dss}, V_{po}, α, are γ are the model parameters.

In the Curtice cubic model, the drain–source current is given by

$$I_{DS} = (a_0 + a_1 V_1 + a_2 V_1^2 + a_3 V_1^3)\tanh(\gamma V_{ds}) \tag{3.51}$$

$$V_1 = V_{gs}[1 + \beta(V_{ds0} - V_{ds})] \tag{3.52}$$

where V_1 is an effective gate–source voltage that takes into account the drain voltage dependence of the pinchoff voltage. The cubic polynomial in Equation (3.51) determines the gate voltage dependence of the saturated current, while the hyperbolic-tangent term determines the current slope below saturation. The Curtice cubic model can give good fits to experimental data for a large variety of MESFETs and is most frequently used. However, the main weakness of the Curtice cubic model is its computational efficiency, due to the hyperbolic-tangent term, and it does not adhere to charge conservation.

The Statz model improves the computational efficiency by approximating the hyperbolic tangent with a cubic polynomial below saturation and unity above saturation. The following equations describe the drain–source current I_{DS}:

$$I_{DS} = I_{dss}(1 + \lambda V_{ds})\left[1 - \left(1 - \frac{\alpha V_{ds}}{3}\right)^3\right] \quad \text{for} \quad 0 < V_{ds} < \frac{3}{\alpha} \tag{3.53}$$

$$I_{DS} = I_{dss}(1 + \lambda V_{ds}) \quad \text{for} \quad V_{ds} > = \frac{3}{\alpha} \tag{3.54}$$

$$I_{dss} = \frac{\beta(V_{gs} - V_t)^2}{1 + b(V_{gs} - V_t)} \tag{3.55}$$

where V_t (the threshold voltage), β, b, λ, and α are model parameters. For small values of $V_{gs} - V_t$, the expression of I_{dss} is quadratic, while for large values, I_{dss} becomes almost linear in $V_{gs} - V_t$. This feature gives better fits at high $V_{gs} - V_t$ are compared with the square-law approximation.

The gate current due to either forward-biased conduction or reverse-biased avalanche breakdown can significantly change the device characteristics. Its effect is commonly modeled by two diodes, the gate–source and gate–drain junction diodes. The gate–source and gate–drain currents are normally predicted using the standard ideal-diode equation

$$I_{GS} = I_s\left[\exp\left(\frac{qV_{gs}}{nkT}\right) - 1\right] - I_{bk}\exp\left(\frac{q(V_{gs} - V_b)}{nkT}\right) \tag{3.56}$$

$$I_{GD} = I_s\left\{\exp\left[\frac{qV_{gd}}{nkT}\right] - 1\right\} - I_{bk}\exp\left[\frac{q(V_{gd} - V_b)}{nkT}\right] \tag{3.57}$$

where I_s is the reverse saturation current, q is the electron charge, n is the ideality factor, k is Boltzmann's constant, I_{bk} is the leakage current associated with the breakdown, V_b is the breakdown voltage, and T is the absolute temperature. In the Curtice cubic model, the gate–source and gate–drain currents are given by the two-piece linear approximation for better computational efficiency. Namely, the gate–source current is due to the forward-biased conduction and taken to be

$$I_{GS} = \frac{V_{gs} - V_{bi}}{R_f} \quad \text{for} \quad V_{gs} >= V_{bi} \tag{3.58a}$$

$$I_{GS} = 0 \quad \text{for} \quad V_{gs} < V_{bi} \tag{3.58b}$$

where V_{bi} is the built-in voltage of the gate Schottky contact and R_f is the effective forward-bias resistance. The gate–drain current is due to the avalanche breakdown and is taken to be

$$I_{GD} = -\frac{(V_{dg} - V_b)}{R_{bk}} \quad \text{for} \quad V_{dg} >= V_b \tag{3.59a}$$

$$I_{GD} = 0 \quad \text{for} \quad V_{dg} < V_b \tag{3.59b}$$

$$V_b = V_{b0} + R_x I_{DS} \tag{3.60}$$

where V_{b0} is the drain–gate breakdown voltage, R_{bk} is the breakdown resistance, and R_x is the resistance relating breakdown voltage to channel currents.

So far, the Curtice cubic model is the most frequently used model in the RF and microwave community because of its availability and simplicity. Here, we will discuss the technique and procedure to extract model parameter values from the measured dc I–V curves. As an example, the Curtice cubic model will be considered in the following illustration. This parameter extraction procedure consists of eight steps.

3.6.1 Determination of Source and Drain Series Resistances (R_S, R_d)

The resistances R_s and R_d are usually determined from Fukui measurement [40] or curve fit of S parameters generated from equivalent circuit to measured. The Fukui measurement can also provide the values of R_g, while the curve-fit method can also provide the values of R_g, C_{gs}, C_{gd}, and other parasitics, such as series gate, drain and source inductances. The detail of the curve-fit technique will be addressed in the next section.

3.6.2 Calculation of *I–V* Curves of the Intrinsic Device

The internal node voltages can be calculated from the measured I–V data and the values of R_s and R_d by using Equations (3.36) and (3.37).

3.6.3 Selection of V_{ds0} and Determination of *a* Coefficients

Since the Curtice cubic model gives the best fit to the measured I—V curves at $V_{ds} = V_{ds0}$, it is naturally to select the dc drain–source bias V_{dsdc} for the values of V_{ds0}. However, in some cases, it may need to choose a value other than V_{dsdc} for V_{ds0} to give a better overall fit of the I–V curves. For a given value of V_{ds0}, the a coefficients can be determined from a curve fit using Equations (3.51) and (3.52) and the approximation of $\tanh(\gamma V_{ds0}) \sim 1$ to the measured I_{DS}/V_{gs} relationship at $V_{ds} = V_{ds0}$.

3.6.4 Estimation of β

For a given value of V_{ds}, say, V_{ds1}, the value of β can be estimated by substituting the calculated values from Section 2.6.2 for I_{DS} and V_{gs} and V_{ds} in Equations (3.51) and (3.52). The value of V_{ds1} is chosen to be close to the maximum knee voltage in the current saturation region of the I–V curves so that the assumption of $\tanh(\gamma V_{ds1}) \sim 1$ is valid.

3.6.5 Estimation of γ

Knowing the values of a coefficients β and V_{ds0}, we can estimate the value of γ by substituting the calculated output conductance at $V_{ds} = 0$ from the measured I–V data for G_{ds0} in the following equations:

$$\gamma = \frac{G_{ds0}}{a_0 + a_1 V_1 + a_2 V_1^2 + a_3 V_1^3} \tag{3.61}$$

$$V_1 = V_{gs}(1 + \beta V_{ds0}) \tag{3.62}$$

Generally, the calculated value of γ varies with V_{gs}; therefore, it is necessary to take an average for γ over a given range of V_{gs}.

3.6.6 Curve Fit of *I–V* Curves between Calculated and Measured Data

Using the parameter values obtained from the previous steps as the initial values of Equations (3.51) and (3.52), we can use an optimization routine to get the best fit between the calculated and measured *I–V* curves. The four *a* coefficients are usually kept constant during the optimization; however, in certain situations, it may be beneficial to let them vary as well.

3.6.7 Determination of V_{bi} and R_f

Figure 3.8 illustrates the determination of V_{bi} and R_f. The exponential curve in Figure 3.8 represents the gate current–voltage characteristic of a MESFET whose drain and source are electrically connected; I_{Gm} is the estimated maximum gate current and V_{GSm} is its corresponding gate voltage. A tangential straight line at point (I_{Gm}, V_{GSm}) is drawn to intercept the V_{GS} axis at a point V_{bi}. The reciprocal of the slope of the tangential line gives half of

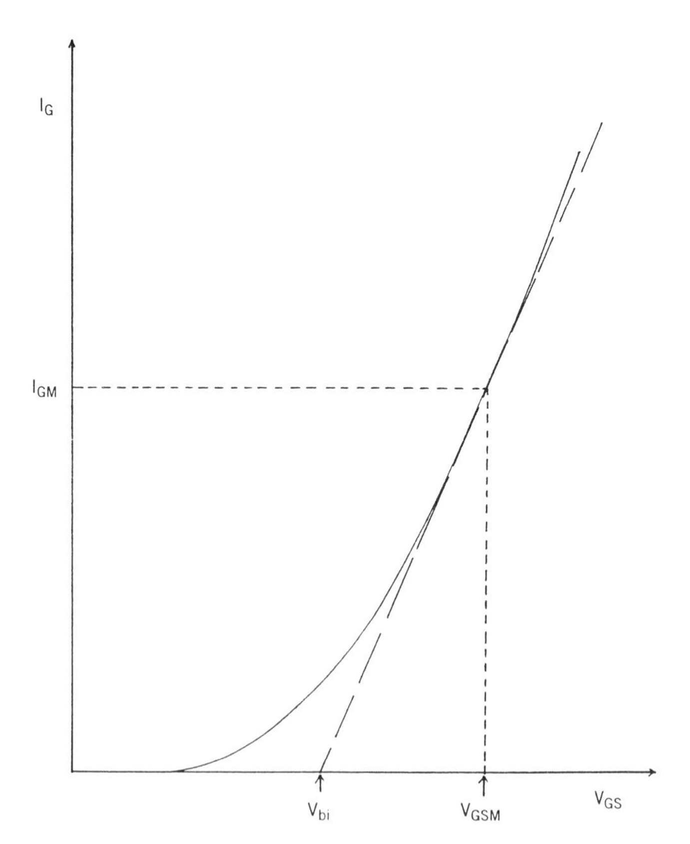

FIGURE 3.8 Graphic illustration for the determination of V_{bi} and R_f.

the R_f value. This is because R_f represents the gate–source junction whose area is half of the total gate junction area.

3.6.8 Determination of V_{b0}, R_{bk}, and R_x

To determine the values of V_{b0}, R_{bk}, and R_x, the I–V curves of the device need to be measured beyond its breakdown voltage. Therefore, the current limit of the power supply need to be set properly during the measurement to protect the device from permanent damage. The values of V_{b0}, R_{bk}, and R_x can be calculated from the I–V curves as illustrated in Figure 3.9. First two different gate–source voltages, V_{GS} and V'_{GS}, are chosen and their corresponding current–voltage curves are used for the calculation of V_{b0}, R_{bk}, and R_x. To cover the gate–source voltage range as wide as possible, V_{GS} should be chosen at 0 V or higher and V'_{GS} at pinchoff or lower. Second, two asymptotes are drawn and intercept each other at the drain–source breakdown, (BI_{DS}, BV_{DS}) for the V_{GS} curve and (BI'_{DS}, BV'_{DS}) for the V'_{GS} curve. Third, the value of R_x can be calculated from the following equation:

$$R_x = \frac{(BV_{DS} - BV'_{DS}) - (V_{GS} - V'_{GS})}{BI_{DS} - BI'_{DS}} \tag{3.63}$$

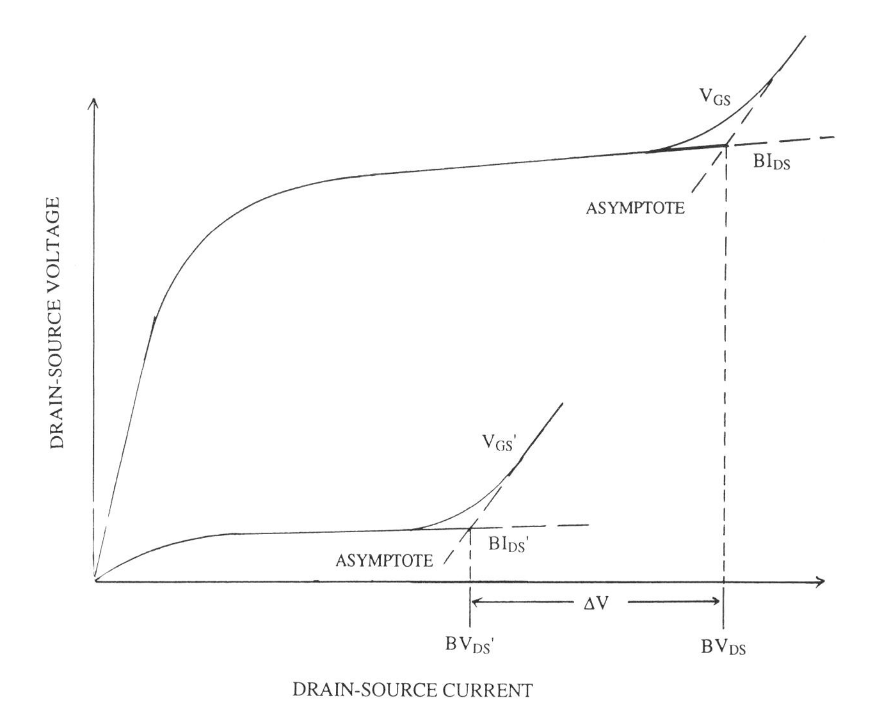

FIGURE 3.9 Graphic illustration for the determination of V_{b0}, R_{bk}, and R_x.

Finally, the value of R_{bk} are given by the reciprocal of the slope of the asymptote associated with the breakdown current, and the value of V_{b0} can then be calculated from either of the following two equations:

$$V_{b0} = BV_{DS} - R_x I_{DS} - V_{GS} \tag{3.64}$$

$$V_{b0} = BV'_{DS} - R_x I'_{DS} - V'_{GS} \tag{3.65}$$

At the end of the last step, the parameter extraction procedure is complete and the model parameters are ready for circuit simulations. However, it is wise to view the I–V curves generated from the model before using the model in any circuit simulation because it is much easier to identify a mistake in a model alone than in a circuit. Furthermore, it is worth mentioning that the breakdown voltage calculated from the dc I–V curves in Section 3.8 (this section) is always lower than the RF breakdown. To obtain a more realistic RF breakdown voltage, it is advised that pulsed I–V curves be used.

3.7 MESFET EQUIVALENT CIRCUITS

Although either a set of s parameters or an equivalent circuit can be used to represent the small-signal behavior of the MESFET when designing a circuit, the equivalent circuit is preferable. This is because the equivalent circuit makes the representation of the FET more manageable in terms of scaling device size, analyzing circuit yield and predicting the device's characteristics beyond the measured frequency range, and so on. Moreover, the equivalent circuit is no longer just an abstraction and simplification for the representation of the MESFET but a necessary tool when dealing with large-signal or nonlinear characteristics of the FET. Since the validity of the predicted result from a circuit simulation depends mostly on the accuracy of the device model, the selection of the equivalent circuit model and the determination of model parameters are extremely important. In this section three different equivalent circuit models of the FET, small-signal, large-signal, and noise models will be discussed.

3.7.1 Small-Signal Equivalent Circuit

For high-frequency and high-speed applications, the effect due to the reactive parts of a MESFET can significantly degrade its performance. Therefore, these reactive elements need to be taken into account in the circuit design and simulation in order to faithfully represent device high-frequency characteristics. Figure 3.10 shows the small signal equivalent circuit of a MESFET. This equivalent circuit is constructed on the basis of the physical structure of the device and is applicable up to microwave

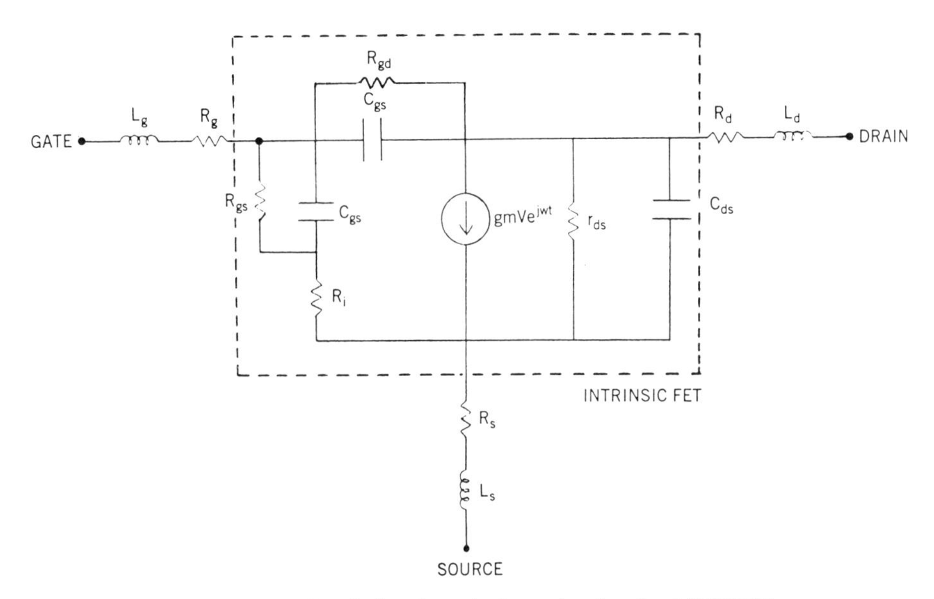

FIGURE 3.10 Small-signal equivalent circuit of a MESFET.

frequency (several tens of gigahertz). Each circuit element has its own physical origin and represents the electric character of a particular region of the device. For instance, C_{gs} and C_{gd} represent charge storages in the gate–source and gate–drain depletion regions; R_s and R_d represent the bulk resistance together with the ohmic contact resistance of the gate–source and gate–drain regions, respectively; L_g, L_s, and L_d respectively represent the self-inductances of the gate, source, and drain electrodes; R_g represents the resistance of the gate metallization; and R_i represents the channel resistance between gate and source. The circuit elements within the square frame in Figure 3.10 represent the intrinsic device while the series resistances (R_g, R_s, and R_d) and series inductances (L_g, L_s, and L_d) are extrinsic parasitics.

The values of the intrinsic and extrinsic elements depend on the channel structure, doping concentration, device size, device layout, and process techniques. The main intrinsic parameters are the transconductance g_m, the input capacitance C_{gs}, the output resistance R_{ds}, and the feedback capacitance C_{gd}. These parameters are related to the current and charge in the active region of the channel and vary with the gate and drain voltages. The time delay τ associated with the drain current or the transconductance with respect to the input signal corresponds to the transit time needed for the electrons to travel in the channel across the active region. Strictly speaking, τ is the time required for the charge exchange with the depletion layer in the velocity saturation region of the channel. Therefore, τ is zero until velocity saturation in the channel is attained, and it can be expected to increase with drain voltage and decrease for increases in gate bias.

The extrinsic parameters are the undesired parts of the device; to a first-order approximation, their values can be assumed to be independent of biasing conditions. The most important extrinsic parameters are the source series resistance and inductance. In a common source FET configuration, the most frequently used, these two elements are in the common path of the input and output circuits and provide a series feedback. This negative feedback reduces the maximum available gain and cutoff frequency of the FET. The other important extrinsic parameter is the gate resistance. It has been shown that the total gate resistance of a device consisting of N fingers and having a total width W is given by [41]

$$R_g = \frac{WR_{g0}}{3N^2} \tag{3.66}$$

where R_{g0} is the dc resistance per unit length of the gate metallization (in ohms per millimeter). Other parasitic elements also degrade the performance of the device, but to a less severe degree.

The equivalent circuit model of the device is generally based on parameter extraction from experimental data. Although theoretical prediction of the parameter values is possible, its accuracy is generally not good enough for the application of high-frequency circuit simulation. So far, the most commonly used parameter extraction method is the curve-fit technique based on the measured small-signal s parameters. For better accuracy of the derived device model, the s parameters of the device should be measured using on wafer probing techniques for frequency range as wide as possible. Moreover, because of the inherent ambiguity among model elements, s parameters taken at different biasing conditions are required to improve the consistency of the model parameter values obtained from the curve-fit technique [42]. The s parameters for cold FET biasing conditions (zero drain–source voltage) are especially useful in determining the extrinsic parameter values [43]. This is because the equivalent circuit of the cold FET has few element and it is easier to distinguish among elements when curve-fitting the s parameters. On the other hand, it also requires model parameter values at various biasing conditions to construct voltage dependences of the intrinsic parameters, such as C_{gs} and C_{gd}. As a consequence, it is necessary to take s-parameter measurements for biasing conditions covering the gate and drain voltage ranges as wide as possible. In general, at least 25 data points are required to give a faithful representation of the bias dependence of model parameters

Mathematically, there are many local minima in searching for the optimal fit of the calculated s parameters from a set of model parameters to the measured s parameters. During the optimization process, the optimized set of parameters is most often trapped in one of the local minima. Generally, this results in physically unmeaningful parameter values that can cause inaccurate results when the model scales for different device size or extrapolates the model for higher frequencies. To improve the consistency of the optimized set of parameters, the number of local minima can be reduced by adding more constraints to the optimization. To do so, one can perform the curve-fit process simultaneously for several sets of *s* parameters with properly selected biasing conditions. Finally, an optimization process started with a set of close-in initial values and guided with boundaries set by physical limitation is essential to obtaining consistent solution.

Figure 3.11 illustrates the agreement between measured and modeled *s* parameters from 0.5 to 25.5 GHz for a 300 μm device biased at 50% I_{DSS} and $V_{ds} = 5$ V (I_{DSS} is the drain–source current when $V_{GS} = 0$ V). The values S_{11}, S_{22}, and S_{21} are shown together in a mixed polar–Smith chart while S_{12} is shown in a polar chart. The measured *s* parameters that were taken directly on the wafer using an HP8510C vector network analyzer and Cascade Microtech RF probes are very smooth up to 26 GHz. In addition, the excellent close fit of the modeled *s* parameters with measured data implies both reliable measured data and an accurate model. The bias dependences of the four most important intrinsic equivalent circuit elements, g_m, g_{ds}, C_{gs}, and C_{gd}, are presented in Figures 3.12–3.15 for a 0.8×400-μm MESFET [44]. These bias-dependent curves are typical for MESFET devices; however, the actual shape of each curve may vary from device to device slightly.

When designing monolithic ICs, it is often desirable to have the freedom of choosing different device sizes. Unfortunately, device model parameter values are usually available for only one or two fixed-size devices. Therefore, it is necessary to scale the model parameter values of a fixed-size device for a desired-size device. The device size can be scaled by changing the width of individual gate finger *W*, the number of gate fingers *N*, or both. The scaling of each equivalent circuit element is governed by the following rules:

$$X = X_0 \frac{N_0 W}{N W_0} \tag{3.67}$$

for $X = R_g$, L_g, L_s, and L_d, and

$$Y = Y_0 \frac{N_0 W_0}{N W} \tag{3.68}$$

for $Y = R_s$, R_d, R_i, R_{ds}, R_{gs}, and R_{gd}, and

$$Z = Z_0 \frac{N W}{N_0 W_0} \tag{3.69}$$

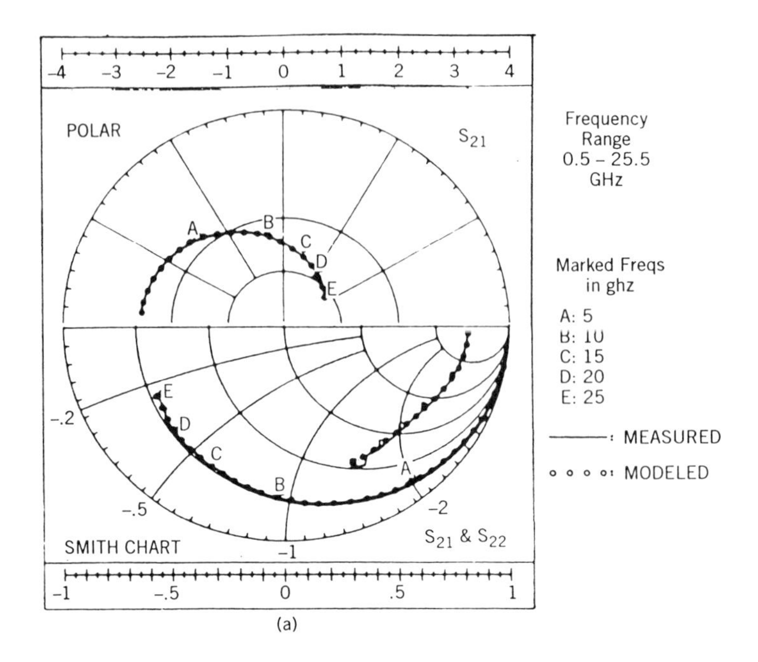

(a)

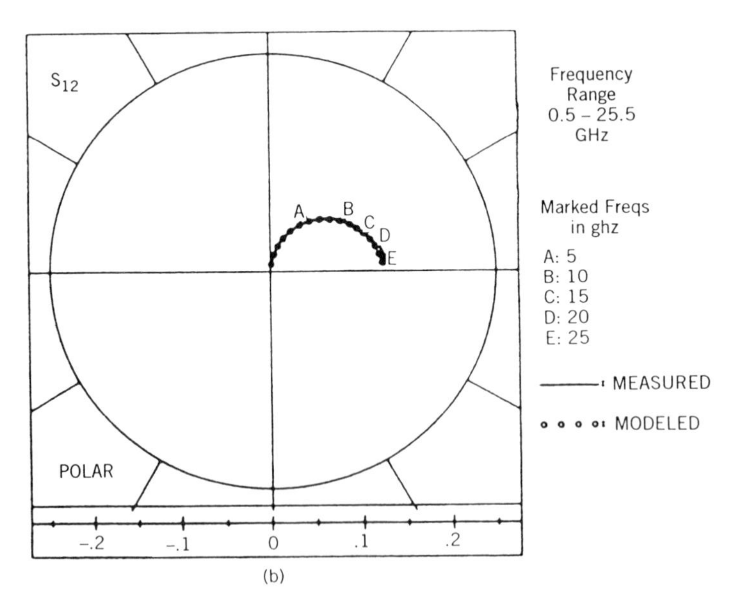

(b)

FIGURE 3.11 Comparison between measured and modeled *s* parameters for a 300-μm FET biased at 50% I_{DSS} and $V_{DS} = 5$ V.

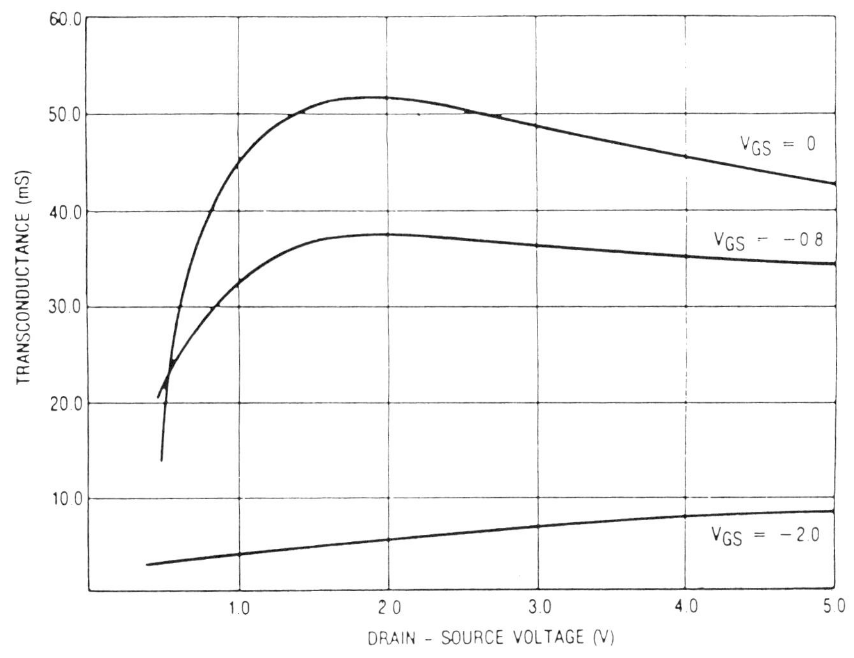

FIGURE 3.12 Transconductance as a function of bias for a 0.8×400-μm MESFET.

for $Z = C_{gs}$, C_{gs}, C_{gd}, and G_m, and

$$\tau = \tau_0 \tag{3.70}$$

Here X_0, Y_0, Z_0, and τ_0 are the element values of a reference device having finger width W_0 and number of fingers N_0, while X, Y, Z, and τ are the element values of a device having finger width W and number of fingers N. The time delay $\tau(\tau_0)$ depends only on the gate length and hence is not a function of the device size. In most cases, L_s and L_d cannot be scaled linearly according to Equation (3.67) because of the use of air bridges and/or two-dimensional structure when interconnecting source or drain strips. Furthermore, the above-mentioned linear scaling rules can result in significant discrepancy if the change of the scaled finger width is more than 100% of the original.

A more accurate method for predicting the equivalent circuit element model is using interpolation based on at least three measured devices with identical properties and structure except for gate width. Similar to the previous simple scaling rules, this sort of scaling can also be done by changing either the individual gate finger width or the number of gate fingers. A family of devices that have the same number of gate fingers but different finger widths can be measured and modeled. Another family of

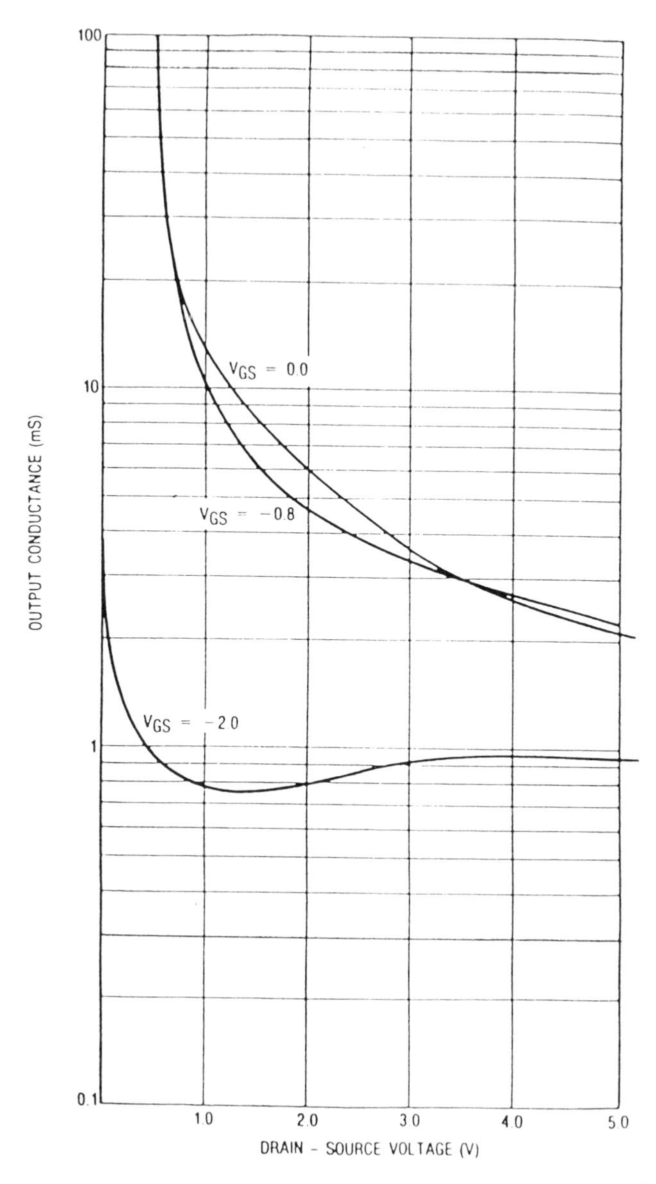

FIGURE 3.13 Output conductance as a function of bias for a 0.8×400-μm MESFET.

devices that differ from one another by only the number of gate fingers can also be characterized. Once equivalent circuit element values have been determined for all measured devices, an interpolation or extrapolation technique can be applied for each circuit element on the basis of its

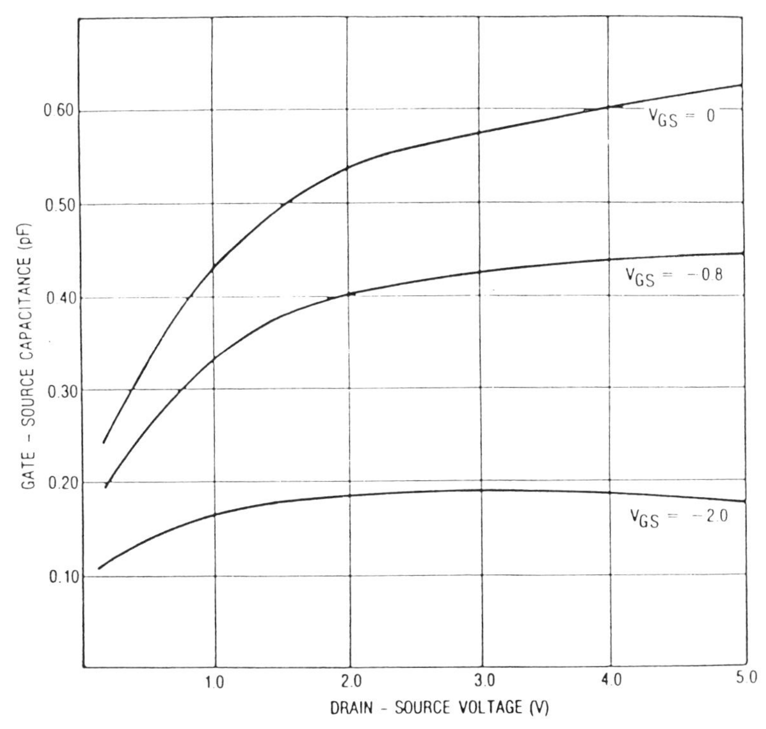

FIGURE 3.14 Gate-source capacitance as a function of bias for a 0.8×400-μm MESFET.

functional relationships with respect to finger width and number of gate fingers to determine the element value of any desirable device size.

3.7.2 Large-Signal Equivalent Circuit

The GaAs MESFET has been used overwhelmingly in power amplifiers from low L-band and up because of its superior performances in linearity and efficiency as compared with the silicon bipolar transistor. In addition, it also shows potential in other nonlinear applications, such as microwave mixers and oscillators. Large-signal device models are required for circuit simulations that are involved in predicting either large signal or nonlinear performances. For example, predicting output power, two-tone intermodulation distortion of an amplifier, frequency conversion and spurious responses of a mixer, or oscillation frequency and output power of an oscillator are some typical applications of the large-signal model. To serve for all these applications, a large-signal model for MESFETs should be accurate for all operating conditions in terms of I–V characteristics and s parameters as a function of bias and frequency. Particularly, it should be

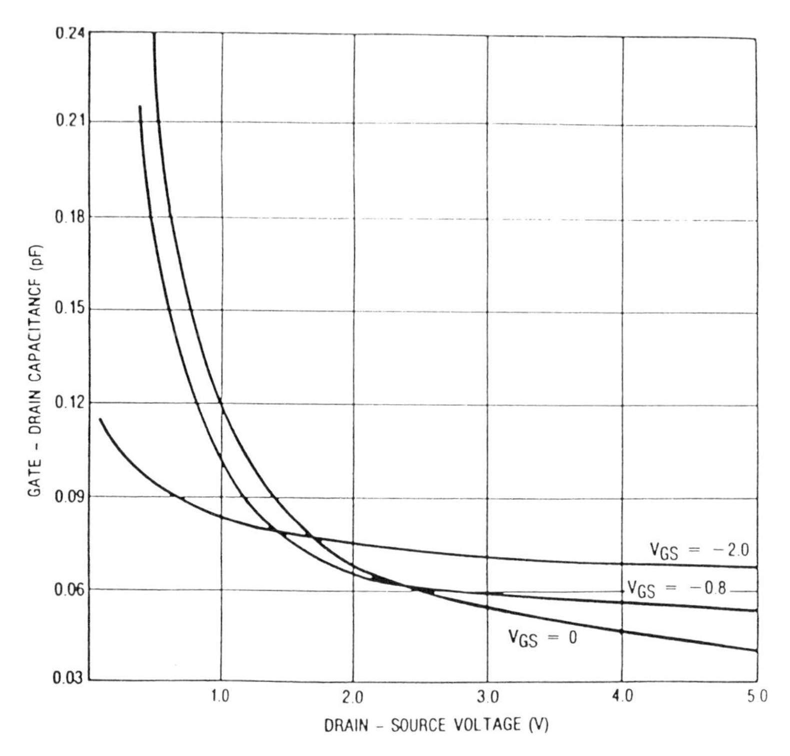

FIGURE 3.15 Gate–drain capacitance as a function of bias for a 0.8×400-μm MESFET.

consistent in the static case, the small-signal case, and the large-signal case. To meet these requirements, the model must contain an accurate description of all important device nonlinearities. For MESFETs, there are six major device nonlinear sources: g_m, g_{ds}, C_{gs}, C_{gd}, g_{gs}, and g_{gd}. In a large-signal model, g_m and g_{ds} are modeled by the current source I_{ds} g_{gs} and g_{gd} are described by the current sources I_{gs} and I_{gd}, respectively; and C_{gs} and C_{gd} are represented by the nonlinear gate–source and gate–drain diode capacitances.

Figure 3.16 shows the equivalent circuit of the large-signal model. It is similar to the small-signal model shown in Figure 3.10 except that g_m, g_{gs}, and g_{gd} are replaced by three current sources I_{ds}, I_{gs}, and I_{gd}; C_{gs} and C_{gd} become nonlinear capacitances; and a series RC network, R_{ds} and C_x, is substituted for R_{ds}. Here I_{ds} represents the drain–source current and is given by Equations (3.38)–(3.48) for a physical model or by Equations (3.49)–(3.55) for various empirical models. The terms I_{gs} and I_{gd} represent the gate–source and gate–drain junction diode currents and are given by (3.55) and (3.56) for the ideal-diode model or (3.58) through (3.60) for the simplified model. As to C_{gs} and C_{gd}, they represent the gate–source and

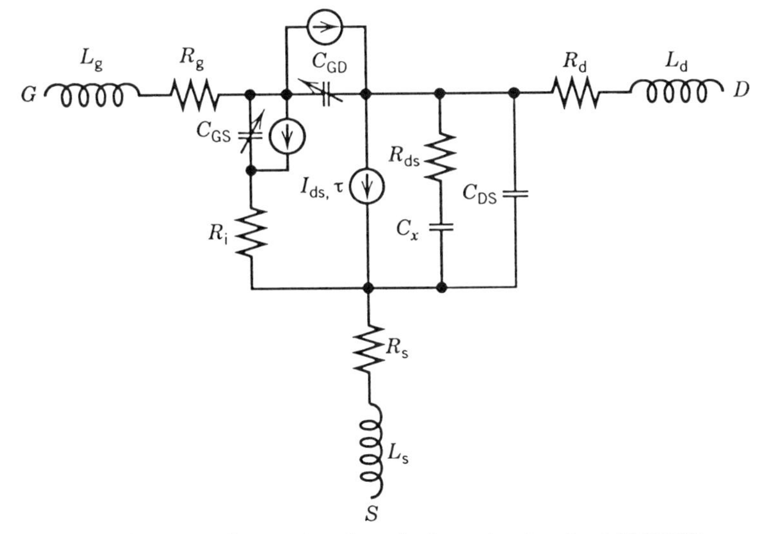

FIGURE 3.16 Large-signal equivalent circuit of a MESFET.

gate–drain junction diode capacitances and are given by

$$C_{gs} = C_{gs0}\left(1 - \frac{V_{gs}}{V_{bi}}\right)^{-1/2} \quad \text{for} \quad V_{gs} < 0.8V_{bi} \tag{3.71a}$$

$$C_{gs} = C_{gs0}\left(5.59\left[\frac{V_{gs}}{V_{bi}} - 0.8\right] + 2.236\right) \quad \text{for} \quad V_{gs} >= 0.8V_{bi} \tag{3.71b}$$

$$C_{gd} = C_{gd0}\left(1 - \frac{V_{gd}}{V_{bi}}\right)^{-1/2} \quad \text{for} \quad V_{gd} < 0.8V_{bi} \tag{3.72a}$$

$$C_{gd} = C_{gd0}\left(5.59\left[\frac{V_{gd}}{V_{bi}} - 0.8\right] + 2.236\right) \quad \text{for} \quad V_{gd} >= 0.8V_{bi} \tag{3.72b}$$

for the Materka model

$$C_{gs} = C_{gs0}\left(1 - \frac{V_{gs}}{V_{bi}}\right)^{-1/2} \quad \text{for} \quad V_{gs} < F_C V_{bi} \tag{3.73a}$$

$$C_{gs} = C_{gs0}(1 - F_C)^{-1/2}\left[1 + \frac{V_{gs} - F_C V_{bi}}{2V_{bi}(1 - F_C)}\right] \quad \text{for} \quad V_{gs} >= F_C V_{bi} \tag{3.73b}$$

and

$$C_{gd} = C_{gd0}\left(1 - \frac{V_{gd}}{V_{bi}}\right)^{-1/2} \qquad \text{for} \quad V_{gd} < F_C V_{bi} \tag{3.74a}$$

$$C_{gd} = C_{gd0}(1 - F_C)^{-1/2}\left[1 + \frac{V_{gd} - F_C V_{bi}}{2V_{bi}(1 - F_C)}\right] \qquad \text{for} \quad V_{gd} >= F_C V_{bi} \tag{3.74b}$$

for the Curtice symmetric model

$$C_{gs} = C_{gs0}\left(1 - \frac{V_{gs}}{V_{bi}}\right)^{-m} \qquad \text{for} \quad V_{gs} < F_C V_{bi} \tag{3.75a}$$

$$C_{gs} = C_{gs0}(1 - F_C)^{-m}\left[1 + \frac{m(V_{gs} - F_C V_{bi})}{V_{bi}(1 - F_C)}\right] \qquad \text{for} \quad V_{gs} >= F_C V_{bi} \tag{3.75b}$$

and

$$C_{gd} = C_{gd0}\left(1 - \frac{V_{gd}}{V_{bi}}\right)^{-m} \qquad \text{for} \quad V_{gd} < F_C V_{bi} \tag{3.76a}$$

$$C_{gd} = C_{gd0}(1 - F_C)^{-m}\left[1 + \frac{m(V_{gd} - F_C V_{bi})}{V_{bi}(1 - F_C)}\right] \qquad \text{for} \quad V_{gd} >= F_C V_{bi} \tag{3.76b}$$

for models based on the generalized ideal–diode model and

$$C_{gs} = C_{gs0} f_1 f_2 \left(1 - \frac{V_C}{V_{bi}}\right)^{-1/2} + C_{gd0} f_3 \tag{3.77}$$

and

$$C_{gd} = C_{gs0} f_1 f_3 \left(1 - \frac{V_C}{V_{bi}}\right)^{-1/2} + C_{gd0} f_2 \tag{3.78}$$

where

$$f_1 = \frac{1 + (V_{eff} - V_T)/[(V_{eff} - V_T)^2 + 0.2^2]^{1/2}}{2} \tag{3.79}$$

$$f_2 = \frac{1 + (V_{gs} - V_{gd})/[(V_{gs} - V_{gd})^2 + \alpha^{-2}]^{1/2}}{2} \tag{3.80}$$

$$f_3 = \frac{1 - (V_{gs} - V_{gd})/[(V_{gs} - V_{gd})^2 + \alpha^{-2}]^{1/2}}{2} \tag{3.81}$$

$$V_C = V_A \qquad \text{for} \quad V_A < 0.5 \tag{3.82a}$$

$$V_C = 0.5 \qquad \text{for} \quad V_A >= 0.5 \tag{3.82b}$$

$$V_A = \frac{V_{eff} + V_T + [(V_{eff} - V_T)^2 + 0.2^2]^{1/2}}{2} \tag{3.83}$$

$$V_{eff} = \frac{V_{gs} + V_{gd} + [(V_{gs} - V_{gd})^2 + \alpha^{-2}]^{1/2}}{2} \tag{3.84}$$

for the Statz model. Here V_{bi} is the barrier potential, C_{gs0} and C_{gd0} are the junction capacitances at zero bias of C_{gs} and C_{gd}, respectively, and F_c, V_T and α are empirical parameters.

The series *RC* network, R_{ds} and C_x, placed in parallel with the drain–source current source accounts for output resistance variation with frequency. In GaAs MESFETs, output resistance drops as frequency increase above a certain frequency and then levels off at a higher frequency. The characteristic frequency where the output resistance starts dropping varies from device to device in the range between 100 Hz and several hundred kilohertz. Figure 3.17 illustrates a typical variation in output resistance with frequency. The output resistance change at low frequency is attributable to deep-level traps that can respond only to low-frequency signals. In the large-signal equivalent circuit the dc output resistance is embodied in the drain–source current source, and the series *RC* network, R_{ds} and C_x, is added to lower the output resistance at higher frequency. The *RC* time constant of R_{ds} and C_x corresponds to the characteristic frequency and typically yields the capacitance value of C_x in the order between nanofarads and microfarads.

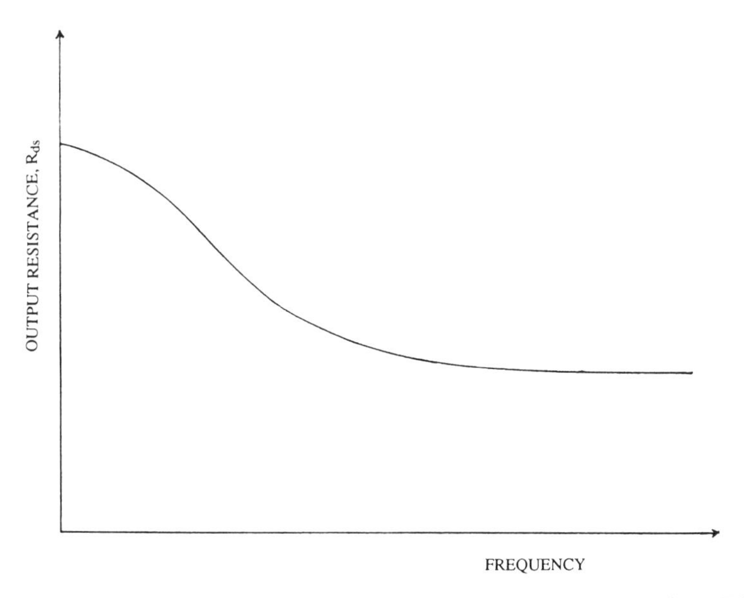

FIGURE 3.17 A typical variation in output resistance with frequency of a MESFET.

3.7.3 Noise Equivalent Circuit

The GaAs MESFET is not being used to replace the silicon bipolar transistor in low-noise amplifiers because it outperforms its counterpart in gain and noise figure. For low-noise amplifier design, the small-signal equivalent circuit is insufficient and must be supplemented by noise model. The intrinsic noise in the MESFET can be characterized by the $1/f$ noise at low frequency and the thermal noise and diffusion noise at higher frequency. The low-frequency $1/f$ noise is caused by the generation–combination cycle owing to the traps in the depletion layer below the gate electrode and the surface states. At higher frequency the noise is due mainly to fluctuations in the electron velocity. For low-drain–source biasing conditions, the electric field in the channel is low and the fluctuations in electron drift velocity give rise to "thermal noise." On the other hand, the high-drain–source biasing conditions, the electric field is high so that the electron velocity saturates and the fluctuations in saturation velocity results in "diffusion noise."

The intrinsic noise can be described by two correlated white-noise sources: one in the drain circuit and one in the gate circuit. The drain noise source $\langle i_{\mathrm{d}}^2 \rangle$ represents the noise generated along the channel due to current fluctuations, and the gate noise source $\langle i_{\mathrm{g}}^2 \rangle$ represents the noise induced on the gate electrode by charge fluctuations associated with the fluctuations in the drain current. Since they are generated from the same origin, the gate and drain noise sources are partially correlated. These two noise sources can be written [45]

$$\langle i_{\mathrm{d}}^2 \rangle = 4k_{\mathrm{B}} T g_{\mathrm{m}} P \, \Delta f \tag{3.85}$$

$$\langle i_g^2 \rangle = 4k_{\mathrm{B}} T \omega^2 C_{\mathrm{gs}}^2 g_{\mathrm{m}}^{-1} R \, \Delta f \tag{3.86}$$

where k_{B} is the Boltzmann constant, T is the absolute temperature, Δf is the frequency bandwidth, and P and R are two dimensionless coefficients. Both P and R depend on the biasing conditions as well as the material and structural features of the FET. The coupling between these two noise sources, represented by the correlation coefficient C

$$jC = \frac{\langle i_{\mathrm{g}}^* i_{\mathrm{d}} \rangle}{(\langle i_{\mathrm{g}}^2 \rangle \langle i_{\mathrm{d}}^2 \rangle)^{1/2}} \tag{3.87}$$

In addition to the intrinsic noise sources, the parasitic resistances introduce thermal noise. The thermal noise associated with a resistance R can be expressed according to the Nyquist formula

$$\langle e^2 \rangle = 4k_{\mathrm{B}} TR \, \Delta f \tag{3.88}$$

To account for the noise effect of the resistance, a voltage source representing the thermal noise of the resistance is inserted in series with the resistance in the equivalent circuit. Some equivalent circuit elements of the FET have only a small effect on the noise figure. These elements include the drain resistance R_d, the feedback gate–drain capacitance C_{gd}, and the drain–source capacitance C_{ds}, as well as all parasitic series inductances. As a first approximation, the effects of these elements can be neglected.

The noise equivalent circuit of the FET can be deduced from the small-signal equivalent circuit shown in Figure 3.10 by the incorporation of the intrinsic and extrinsic noise sources. It has been shown that by a simple circuit analysis [46] this noise equivalent circuit can be equally described by two correlated equivalent noise sources preceding the noiseless FET, which is represented by small-signal equivalent circuit. For simplicity, these two correlated noise sources can be further decorrelated by introducing two noiseless "correlation impedance." The resulting noise equivalent circuit of the noisy FET is shown in Figure 3.18, where all noise properties of the FET are embodied in a simple noisy network and characterized by r_n, g_m, and Z_c. The noise resistance r_n represents a thermal noise voltage source at the reference temperature, the noise conductance g_n represents a shunt thermal noise current source at the same temperature, and the correlation impedance Z_c is an impedance at absolute zero (noiseless).

In terms of r_n, g_m, and Z_c the noise figure of the noisy FET (at the temperature T) can be derived from the noise equivalent circuit shown in Figure 3.18:

$$F = 1 + \frac{r_n + g_n|Z_{sg} + Z_c|^2}{R_{sg}} \tag{3.89}$$

where Z_{sg} is the source impedance (assumed to be at the reference temperature T_0), R_{sg} is the real part of Z_{sg} and r_n, g_m, and Z_c are given by [47]

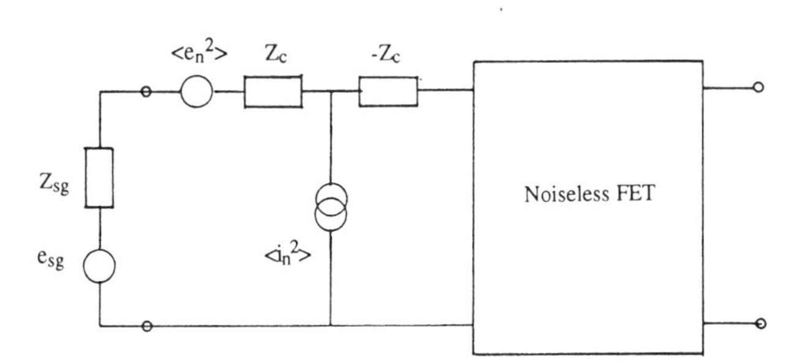

FIGURE 3.18 Noise equivalent circuit of a noisy FET.

$$r_n = \frac{(R_s + R_g)T}{T_0} + \frac{K_r(1 + \omega^2 C_{gs}^2 R_i^2)}{g_m} \tag{3.90}$$

$$g_n = \frac{K_g \omega^2 C_{gs}^2}{g_m} \tag{3.91}$$

$$Z_c = R_s + R_g + K_c \left(R_i + \frac{1}{j\omega C_{gs}} \right) \tag{3.92}$$

The parameters K_r, K_g, and K_c are numeric noise coefficients that represent the properties of the intrinsic noise sources i_g, i_d and their correlation. By simple algebraic manipulations, Equation (3.89) can be written

$$F = F_{min} + g_n \frac{(R_{sg} - R_{sg0})^2 + (X_{sg} - X_{sg0})^2}{R_{sg}} \tag{3.93}$$

where

$$F_{min} = 1 + 2g_n(R_c + R_{sg0}) \tag{3.94}$$

$$R_{sg0} = \left(R_c^2 + \frac{r_n}{g_n} \right)^{1/2} \tag{3.95}$$

$$X_{sg0} = -X_c \tag{3.96}$$

where R_c and X_c are respectively the real and imaginary parts of Z_c and X_{sg} is the imaginary part of Z_{sg}. By inspecting Equation (3.93), we see that the noise figure has a minimum value F_{min} that is achieved when the real and imaginary parts of the source impedance Z_{sg} are equal to

$$R_{sg} = R_{sg0} \tag{3.97a}$$

$$X_{sg} = X_{sg0} \tag{3.97b}$$

Equation (3.97) can be satisfied by the proper choice of the source impedance Z_{sg}. It is called "noise match" when the condition of Equation (3.97) is satisfied.

The expression for F_{min} given by Equation (3.94) can be simplified by the three-term power series expansion in frequency

$$F_{\min} = 1 + 2\frac{\omega C_{gs}}{g_m}[K_g\{K_r + g_m(R_s + R_g)\}]^{1/2} + 2\left(\frac{\omega Cg_s}{g_m}\right)^2[K_g g_m(R_g + R_s + K_c R_i)] \tag{3.98}$$

Note that $F_{\min}$ decreases with increasing gain–bandwidth product g_m/C_{gs} of the FET. Figure 3.19 shows a typical variation in the minimum noise figure $F_{\min}$ as a function of the drain current I_{ds} normalized to its value I_{DSS} at zero gate bias. The minimum noise figure has a minimum value that usually occurs at drain currents in the range

$$0.1 < \frac{I_{ds}}{I_{DSS}} < 0.2 \tag{3.99}$$

This variation results from two opposing effects. The decrease in diffusion noise with decreasing drain current leads to a decrease in the noise figure with decreasing drain current; the decrease in g_m and the increase in the intrinsic thermal noise with decreasing drain current leads to an increase in the noise figure as pinchoff is approached. At very low value of drain current g_m approaches zero, yielding a sharp increase in $F_{\min}$ near the pinchoff.

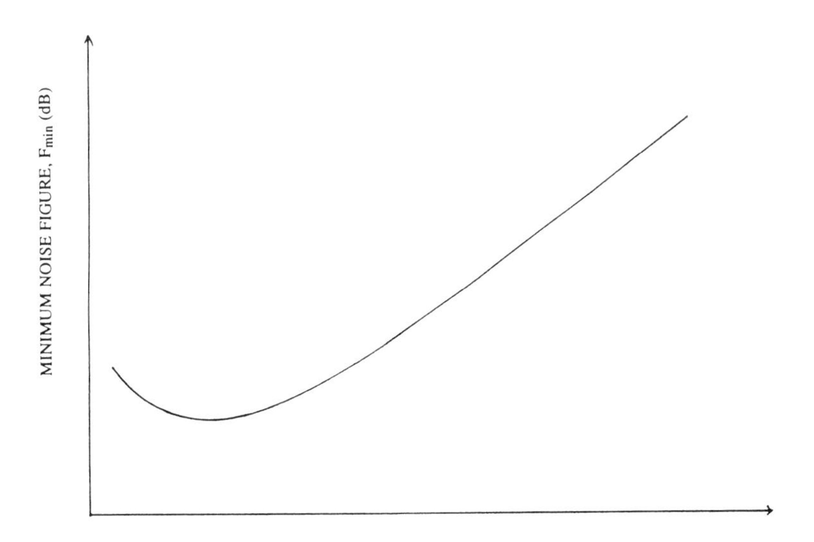

FIGURE 3.19 A typical variation in the minimum noise figure as a function of the drain current normalized to its value at zero gate bias.

REFERENCES

1. R. A. Pucel et al., "Signal and Noise Properties of Gallium Arsenide Microwave Field-Effect Transistors," in *Advances in Electronics and Electron Physics*, Academic Press, New York, 1975, Vol. 38, pp. 195–265.
2. K. Lehovec and R. Zuleeg, "Voltage–Current Characteristics of GaAs JFET's in the Hot Electron Range," *Solid State Electron.*, **13**, 1415–1426 (1970).
3. A. Madjar and F. J. Rosenbaum, "A Large-Signal Model for the GaAs MESFET," *IEEE Trans. MTT*, **MTT-29** (8) (Aug. 1981).
4. L. O. Chua and Y. W. Sing, "Nonlinear Lumped Circuit Model of GaAs MESFET," *IEEE Trans. Electron Devices*, **ED-30** (7) (July 1983).
5. C. M. Snowden et al., "Large-Signal Modeling of GaAs MESFET Operation," *IEEE Trans. Electron Devices*, **ED-30** (12) (Dec. 1983).
6. W. Shockley, "A Unipolar Field Effect Transistor," *Proc. IRE*, **40**, 1365–1376 (Nov. 1952).
7. R. E. Williams and D. W. Shaw, "Guided Channel FETs: Improved Linearity and Noise Figure," *IEEE Trans. Electron Devices*, **ED-25** (6), 600–605 (June 1978).
8. M. S. Shur and L. F. Eastman, "*I–V* Characteristics of GaAs MESFETs with Non-uniform Doping Profile," *IEEE Trans. Electron Devices*, **ED-27** (2), 455–461 (Feb. 1980).
9. N. McIntyre, "Calculation of *I–V* Characteristics for Ion-Implanted GaAs MESFETs," *Electron. Lett.*, **18** (5), 208–210 (March 1982).
10. Pietro de Santis, "Extension of Existing Models to Ion-Implanted MESFET's," *IEEE Trans. MTT*, **MTT-28** (6) (June 1980).
11. M. S. Shur, "Analytical Model of GaAs MESFETs," *IEEE Trans. Electron Devices*, **ED-25**, 612–618 (June 1978).
12. G. W. Taylor et al., "A Device Model for an Ion-Implanted MESFET," *IEEE Trans. Electron Devices*, **ED-26** (3) (March 1979).
13. P. F. Lindquist and W. M. Ford, "Semi-insulating GaAs Substrates," in *GaAs FET Principles and Technology*, J. V. DiLorenzo and D. D. Khandewal, eds., Artech House, Norwood, MA, 1982.
14. J. A. Higgins et al., *IEEE Trans. Electron Devices*, **ED-25**, 587 (1978).
15. R. G. Hunsperger and N. Hirsch, *Electron. Lett.*, **9**, 1 (1973)
16. K. Ohata et al., *IEEE Trans. Electron Devices*, **ED-24**, 1129 (1977).
17. F. H. Doerbeck et al., *Electron. Lett.*, **15**, 577 (1977).
18. R. C. Eden et al., *IEEE J. Solid State Circuits*, **SC-14**, 221 (1979).
19. D. P. Hornbuckle and R. L. Van Tuyl, "Monolithic GaAs Direct-Coupled Amplifiers," *IEEE Trans. Electron Devices*, **ED-28** (2), 175–182 (Feb. 1981).
20. K. Honjo et al., "Ultra-Broad-Band GaAs Monolithic Amplifier," *IEEE Trans. Electron Devices*, **ED-29** (7), 1123–1129 (July 1982).
21. T. Itoh and H. Yanai, "Stability and Performance and Interfacial Problems in GaAs MESFETs," *IEEE Trans. Electron Devices*, **ED-27** (6), 1037–1045 (1980).

22. M. Tanimoto et al., "Anomalous Phenomena of Current–Voltage Characteristics Observed in Gunn-Effect Digital Devices under DC Bias Conditions," *Electron. Commun. Japan*, **60-C** (11), 102–110 (1977).
23. C. Kocot and C. Stolte, "Backgating in GaAs MESFETs," *IEEE Trans. Electron Devices*, **ED-29** (7) (July 1982).
24. H. Goronkin et al., "Backgating and Light Sensitivity in Ion-Implanted GaAs Integrated Circuits," *IEEE Trans. Electron Devices*, **ED-29** (5), 845–850 (May 1982).
25. T. J. Diesel et al., "The Effect of Substrate Chromium Doping on Backgating and Light Sensitivity of GaAs MESFETs," paper presented at the GaAs Integrated Circuit Symposium, 1980.
26. M. S. Birrrittella et al., "The Effect of Backgating on the Design and Performance of GaAs Digital Integrated Circuits," *IEEE Trans. Electron Devices*, **ED-29** (7) (July 1982).
27. R. H. Dawson and J. Frey, "A Simple Model for Determining Effects of Negative Differential Mobility and Magnitude of Saturated Velocity on the Performance of Schottky-Barrier Field Effect Transistors," *Solid State Electron.*, **22**, 343–346 (1979).
28. T. H. Chen and M. S. Shur, "Analytical Models of Ion-Implanted GaAs FETs," *IEEE Trans. Electron Devices* (Jan. 1985).
29. M. S. Shur, "Low Field Mobility, Effective Saturation Velocity and Performance of Submicron GaAs MESFETs," *Electron. Lett.*, **18** (21), 909–911 (Oct. 1982).
30. Himsworth, "A Two-Dimensional Analysis of Gallium-Arsenide Junction Field-effect Transistors with Long and Short Channels," *Solid-State Electron.*, **15**, 1353–1361 (Dec. 1972).
31. M. S. Shur, "Small Signal Nonlinear Circuit Model of a GaAs MESFET," *Solid-State Electron.*, **22**, 723–728 (Aug. 1979).
32. K. Yamaguchi et al., "Two-Dimensional Numerical Analysis of Stability Criteria of GaAs FET's," *IEEE Trans. Electron Devices*, **ED-23** (12), 1283–1289 (Dec. 1976).
33. L. F. Eastman et al., "Design Criteria for GaAs MESFETs Related to Stationary High Field Domains," *Solid-State Electron.*, **23**, 383–389 (1980).
34. A. B. Grebene and S. K. Ghandhi, "General Theory for Pinched Operation of the Junction-Gate FET," *Solid-State Electron.*, **12** 573–589 (1969).
35. W. R. Curtice, "A MESFET Model for Use in the Design of GaAs Integrated Circuits," *IEEE Trans. MTT*, **MTT-29** (5) 448–456 (1980).
36. C. L. Chen and K. D. Wise, "Transconductance Compression in Submicrometer GaAs MESFETs," *IEEE Electron Device Lett.*, **EDL-4** (10), 341–343 (Oct. 1983).
37. A. Materka and T. Kacprzak, "Computer Calculation of Large-Signal GaAs FET Amplifier Characteristics," *IEEE Trans. MTT*, **MTT-33** (2), 129–135 (Feb. 1985).
38. W. R. Curtice and M. Ettenberg, "A Nonlinear GaAs FET Model for Use in the Design of Output Circuits for Power Amplifiers," *IEEE Trans. MTT*, **MTT-33** (12), 1383–1394 (Dec. 1985).

39. H. Statz et al., "GaAs FET Device and Circuit Simulation in SPICE," *IEEE Trans. Electron Devices*, **ED-34** (2), 160–169 (Feb. 1987).
40. H. Fukui, "Determinations of the Basic Parameters of a GaAs MESFET," *Bell Syst. Tech. J.*, **58** (3), (March 1979).
41. Wolf, P., *IBM J. Res. Dev.*, **14** (2), 125–141 (1972).
42. R. L. Vaitus, "Uncertainty in the Values of GaAs MESFET Equivalent Circuit Elements Extracted from Measured Two-port Scattering Parameters," 1983 Cornell Conference on High Speed Semiconductor Devices and Circuits, Cornell University, August 1983.
43. F. Diamond et al., "Measurement of the Extrinsic Series Elements of a Microwave MESFET Under Zero Current Conditions," *Proc. 12th European Microwave Conference* (Finland), September 1982.
44. R. Goyal et al., *Monolithic Microwave Integrated Circuits: Technology and Design*, Chapter 4, Artech House, Norwood, MA, 1989.
45. A. Van der Ziel, *Proc. IRE*, **51**, 461–467 (1963).
46. H. Rothe and W. Dahlke, "Theory of Noisy Fourpoles," *Proc. IRE*, **44**, 811–818 (June 1956).
47. R. A. Pucel et al., "Signal and Noise Properties of Gallium Arsenide Field-Effect Transistors," in *Advances in Electronics and Electron Physics*, Academic Press, New York, 1975, Vol. 38.

CHAPTER FOUR

Schottky Diode and Passive Components

TZU-HUNG CHEN
Hexawave Inc.
1F, 2 Prosperity Road II
Science-Based Industrial Park
Hsinchu, Taiwan, ROC

4.1 INTRODUCTION

Transistors are the heart of high-frequency analog integrated circuits (ICs); however, Schottky diodes and passive devices are also indispensable for high-speed analog circuits. The main applications of Schottky diodes include multipliers, detectors, mixers, varactors, switches, limiters, and level-shifting devices. The passive devices take the form of either lumped or distributed components. Resistors, inductors, and capacitors are the most popular passive lumped components. For example, capacitors are essential for bypass, dc blocking and bias line decoupling applications. Resistors and inductors are often used as loads, impedance-matching elements, and dc biasing elements. At higher frequencies, passive distributed components, such as microstrip lines, coplanar strips, and coplanar waveguides, are also necessary for interconnects and impedance-matching networks.

This chapter is divided into two main sections. Section 4.2 describes the properties, design philosophy, and modeling of Schottky diodes. It also

High-Frequency Analog Integrated-Circuit Design, Edited by Ravender Goyal
ISBN 0-471-53043-3

presents planar Schottky diodes for both epitaxial and ion-implantation fabrication techniques. Section 4.3 deals with the design and modeling of passive devices. Four primary passive devices are discussed in detail, and some secondary passive components are also briefly described.

4.2 SCHOTTKY DIODE

Although rectification is the most obvious property of a p–n junction, it is not the sole device that conducts current in only one direction. Some metal–semiconductor junctions also have the same rectifying property. The device built with a rectifying metal–semiconductor junction and nonrectifying metal–semiconductor junction (or ohmic contact) is called *Schottky barrier diode* or *Schottky diode* for short. In comparison to a p–n junction diode, the Schottky barrier diode usually has a lower forward turnon voltage, lower series resistance and steeper forward *I–V* slope. Especially, it is a majority-carrier device therefore it does not suffer from the minority-carrier charge–storage effects. These properties, together with high cutoff frequency, ruggedness, reproducibility, ease of fabrication, and well-modeled characteristic, have made Schottky diodes very versatile in a variety of applications, such as mixer, detector, multiplier, varactor, and level-shifting devices.

4.2.1 Principle of Operation

Figure 4.1 shows a typical top view and a corresponding cross-sectional view of a Schottky barrier diode to be used in high-speed analog ICs. The diode is built on a three-layer structure: the buffer (intrinsic), active (N), and cap (N^+) layers. The Schottky contacts are located in the recess etched GaAs active layer to reduce reverse saturation and surface leakage currents and to increase reverse breakdown voltage. The ohmic contacts are deposited on the cap layer to reduce the contact resistance. Typically, $N = 10^{17}\ cm^{-3}$ and $N^+ > 10^{18}\ cm^{-3}$. There are two types of Schottky diode fabrication processes, the epitaxial and ion-implantation processes, which are compatible with the GaAs MESFET IC technologies. In the epitaxial process, the buffer layer, active layer, and cap layer are pregrown on top of the GaAs substrate epitaxially. However, in the ion-implantation process, the active layer is performed by implanting the donor ions into the substrate and a follow-up rapid thermal annealing to repair the damage on the crystal structure. For the cap layer, it is formed by either a shallow N^+ implantation or a selective N^+ implantation directly underneath the ohmic contact area.

The Schottky diode is electrically isolated from other IC components by the mesa etch, the selective neutron, oxygen ion bombardment, or both. The mesa etch technique etches away the cap and active layers on the unused area of the wafer. The ion-bombardment technique damages the

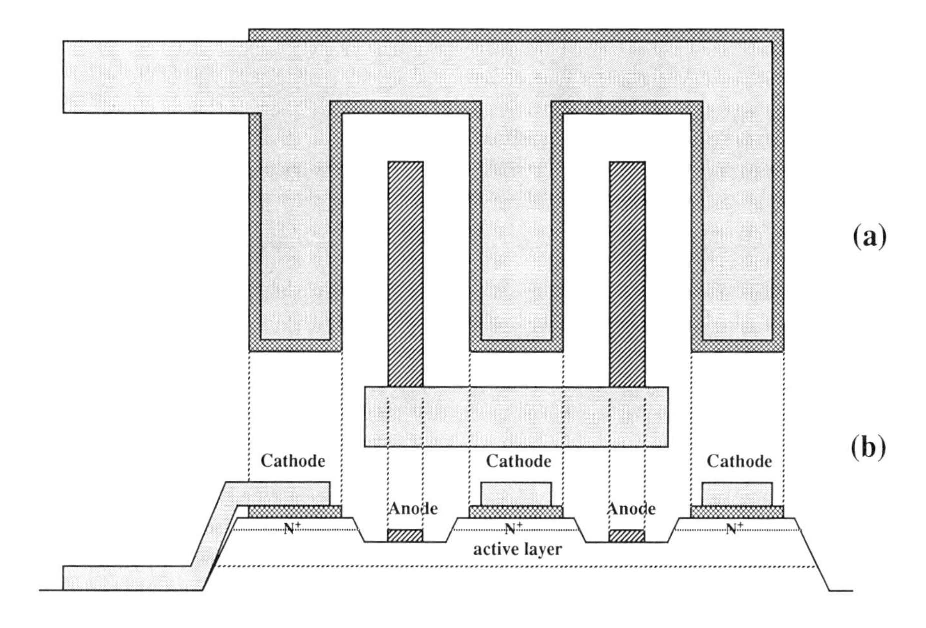

FIGURE 4.1 Top-view (*a*) and cross-sectional view (*b*) of a planar Schottky barrier diode.

crystal structure, thereby destroying the conductivity of the active layer on selective areas. Occasionally, the combination of ion bombardment and shallow mesa etch is also used to achieve better isolation while maintaining flatter surface. In general, the first method is used in the epitaxial process whereas the second and the third methods are used in the ion-implantation process.

Figure 4.2 shows the simplified energy-band diagram of a Schottky contact. Three constraints govern the shape of the energy-band diagram of an ideal metal–semiconductor junction in the absence of surface states:

1. At thermal equilibrium, the Fermi level in the metal and semiconductor must be constant.
2. The vacuum level (the free-space energy level) must change continuously.
3. The semiconductor electron affinity (the energy difference between vacuum level and the bottom of the conduction band) and the metal work function (the energy difference between vacuum level and Fermi level) remain constant throughout each material.

In order to satisfy all three constraints simultaneously, a potential barrier is

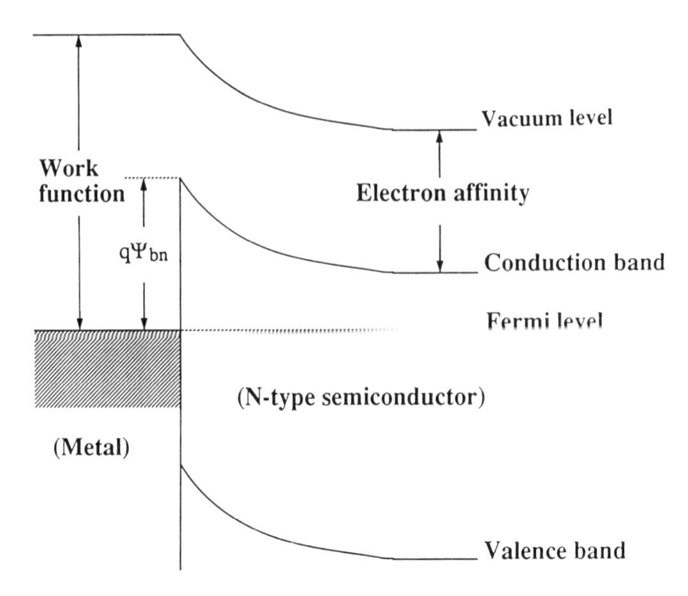

FIGURE 4.2 Simplified energy-band diagram of a Schottky contact.

placed at the metal–semiconductor junction. The barrier height, $q\Psi_{bn}$ (q is the electron charge) is equal to the difference between the metal work function and the semiconductor electron affinity for a metal and n-type semiconductor contact. In the case of a metal and p-type semiconductor contact, the barrier height, $q\Psi_{bp}$ is equal to the difference between the semiconductor energy bandgap and $q\Psi_{bn}$. However, for real metal–semiconductor junctions, a large density of surface states caused by incomplete covalent bonds and other surface imperfections always exist on the semiconductor surface. If the density of surface states is sufficiently large to accommodate any additional surface charges, the Fermi level on the interface will be pinned at a certain level determined solely by the distribution of surface states. As a result, the barrier height is determined by the property of the semiconductor surface and is independent of the metal work function [1].

Another major factor that affects the Schottky barrier height is the image-force-induced barrier lowering. The charge carriers inside the semiconductor experience a force from their image charge in the metal. This force attracts the charge carriers towards the metal surface and effectively lowers the potential barrier. The Schottky barrier lowering is proportional to the square-root of the maximum field at the interface, thereby making the barrier height voltage dependent. Finally, there are other surface imperfections that affect the barrier to current transport through the junction. For example, a thin interfacial layer (~10 Å) that is neither semiconductor nor metal always exists at the junction. Although carriers can tunnel through the

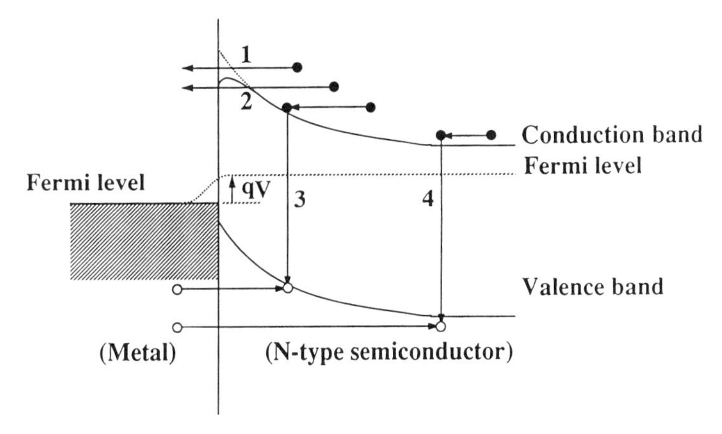

FIGURE 4.3 Energy-band diagram of a Schottky contact incorporating the image-force-induced barrier lowering.

thin interfacial layer, this thin layer does absorb part of the externally applied voltage and causes nonideal behaviors.

4.2.1.1 I–V Characteristics Figure 4.3 shows the energy-band diagram of a Schottky contact incorporating the image-force-induced barrier lowering. The current transport in Schottky contact under forward bias is due mainly to majority carriers. Four basic transport processes account for the current transport of the Schottky contact and are indicated in Figure 4.3 as follows:

1. Transport of electrons from the semiconductor over the potential barrier into the metal
2. Quantum-mechanical tunneling of electrons through the barrier
3. Recombination, via bulk deep levels, in the space-charge region
4. Recombination, via bulk deep levels, in the neutral region

Among these, the first type of transport process is the most important one. Three different theories have been proposed for this process. The thermionic emission theory [2], the diffusion theory [3], and a generalized thermionic emission–diffusion theory [4] can adequately describe the transport of carriers over the barrier. All of these yield a current–voltage characteristic of the Schottky contact given by

$$I = I_s\left(\exp\frac{qV}{nkT} - 1\right) \tag{4.1}$$

where n is the ideality factor with a unity value, I_s is the reverse saturation current, k is the Boltzmann constant, T is the absolute temperature, and V is the external bias voltage.

As to other types of processes, the current transport due to recombination processes leads to the same solution but with an n value of 2. The same type of solution is also obtained from the thermionic emission-tunneling theory [5], which describes the transport of carriers over and through the barrier. However, the n factor in this case varies from unity at low dopings and high temperature to values substantially larger than unity for higher dopings and lower temperature. In addition, there are other minor transport processes, such as interface current due to interface states and edge leakage current due to fringing field at the contact periphery. In practice, the n factor may therefore have any value larger than unity, and the reverse saturation current I_s does not equal the value calculated from the theories due to the presence of the surface leakage current. Furthermore, the series ohmic resistance R_s due to the Schottky and ohmic metallizations, ohmic contact, and neutral semiconductor bulk region can also affect the current–voltage characteristic, especially at high current level. Taking into account the voltage drop across the series resistance and the parallel conduction due to the leakage current, Equation (4.1) becomes

$$I = I_s\left(\exp\left\{\frac{q(V - IR_s)}{nkT}\right\} - 1\right) + G_x V \tag{4.2}$$

where G_x is the parallel conductance. Since the current–voltage characteristic of a Schottky diode deviates from the ideal behavior given in Equation (4.1) at high current level due to the series resistance R_s, it is highly required to minimize the series resistance. The ohmic contact and neutral semiconductor bulk region are the two major contributions to the series resistance. Therefore, the resistance R_s can be reduced by making a high-quality ohmic contact and using a heavily doped buried layer to lower the bulk resistance.

An ohmic contact is a metal–semiconductor contact that has a linear I–V characteristic in both biasing directions and a minimal resistance. In general, low barrier height, high doping concentration, or both can be used to obtain good ohmic contact. For wide-bandgap semiconductors, a metal does not generally exist with a low enough work functions to yield a low barrier. Accordingly, the practical technique to form ohmic contact is by doping the semiconductor heavily in the surface layer. Thus, if a barrier exists at the interface, the depletion region is narrow enough to allow carriers to tunnel through the barrier.

4.2.1.2 C–V Characteristic Since the energy-band diagram of a Schottky junction is similar to that of a one-sided abrupt junction, the depletion layer capacitance of the one-sided abrupt junction can also be applied to the

Schottky junction. The depletion layer capacitance is given by

$$C_j(V) = \frac{C_{j0}}{(1 - V/V_j)^{1/2}} \tag{4.3}$$

where C_{j0} is the junction capacitance at zero bias voltage and V_j is the built-in potential of the barrier height of the junction. The exponent $\frac{1}{2}$ in the denominator of Equation (4.3) comes from the assumption of a uniform doping density. For nonuniform doping profiles, the exponent has different values. In addition, the Schottky diode that is compatible with the IC process has the planar structure shown in Figure 4.1; thus the electric field distribution of the junction becomes highly two-dimensional when the active layer has been completely depleted. Since the depletion-layer capacitance of Equation (4.3) is a one-dimensional solution under low-level injection, it can hold valid for only a small fraction of the external bias voltage range (typically, from -4 to $+1$ V) when applied to real Schottky diodes. Therefore, the simple depletion-layer capacitance equation cannot properly describe the capacitance–voltage characteristic of the real Schottky diode. To closely describe the C–V characteristic of real Schottky diodes, a semiempirical equation has been proposed and widely used. The semiempirical equation is given by

$$C_j(V) = \begin{cases} C_{j0}[1 - V/V_j]^{-m} & \text{for} \quad V < F_c V_j \\ C_{j0}[1 - V/V_j]^{-m-1}[1 - F_c(m+1) + mV/V_j] & \text{for} \quad V > F_c V_j \end{cases} \tag{4.4}$$

where m = grading coefficient
V_j = built-in potential
C_{j0} = zero-bias junction capacitance
F_c = depletion capacitance coefficient

4.2.2 Diode Model

The equivalent circuit of the Schottky diode and origin of the various circuit elements in the structure are shown in Figure 4.4*a*. It consists of a nonlinear current source I_D, junction capacitance C_D, parallel conductance $G_x(G_{x1} + G_{x2})$, series resistance R_s, series inductance L_s, and electrode-to-ground capacitances C_{p1} and C_{p2}. The current source I_D represents the junction current given by Equation (4.2), and the junction capacitance C_D is given by Equation (4.4). As to G_x, R_s, L_s, C_{p1}, and C_{p2}, they are parasitic elements. The existence of parallel conductance G_x is caused by surface, interfacial and substrate leakage currents, and its value is in the order of 10^{-6} to 10^{-9} mhos(℧) for a typical mixer or detector diode. The series resistance R_s includes the ohmic contact resistance, the bulk resistance of the active layer

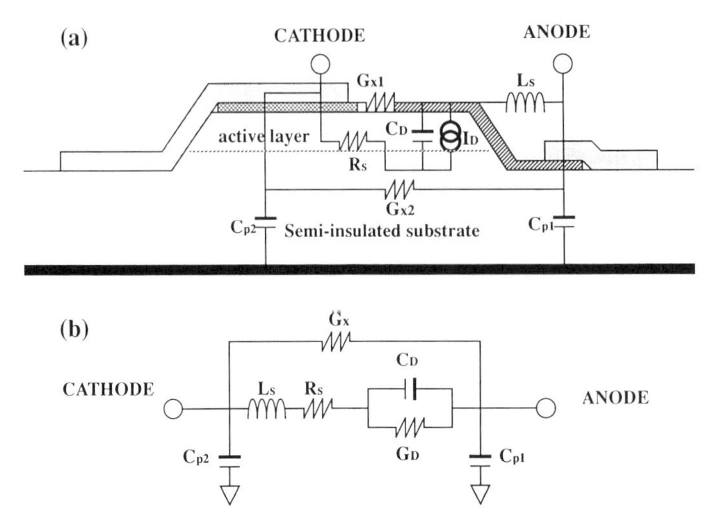

FIGURE 4.4 Nonlinear equivalent circuit and origin of the various circuit elements in the structure (*a*) and linear equivalent circuit (*b*) of the Schottky diode.

between anode and cathode, and the spreading resistance of the neutral semiconductor region underneath the Schottky contact. The series inductance L_s of an interdigitated diode is due primarily to narrow and long metal strips of the Schottky and ohmic contacts. Typically, it has a value in the order of a few pH. Figure 4.4*b* shows the linear equivalent circuit, which is identical to the nonlinear one except the current source I_D is replaced by the junction conductance:

$$G_D = \frac{\partial I_D}{\partial V} \tag{4.5}$$

A figure of merit often used for diodes is the cutoff frequency f_T, which is defined as

$$f_T = \frac{1}{2\pi R_s C_D} \tag{4.6}$$

For most analog IC processes, the cutoff frequency is typically between 100 and 300 GHz. However, for some processes optimized for Schottky diodes, a cutoff frequency as high as 500–800 GHz is not unusual.

Shot noise and thermal noise are two major noise sources in Schottky diodes. The fluctuation in dc current of Schottky contact, as there is a random fluctuation in charge carriers passing through the junction, is termed *shot noise*. The thermal noise is generated by the ohmic resistance of the diode as a result of the random thermal motion of the charge carriers. Typically, noise is specified with respect to a certain noise bandwidth and has the units $V/Hz^{1/2}$ or $A/Hz^{1/2}$. At room or higher temperature, the noise current of shot noise has a mean-square value

$$\langle i \rangle^2 = \frac{2q}{n}(I_D + 2I_s)\,\Delta f$$

$$= 4kTG_D - \frac{2qI_D}{n}\,\Delta f \tag{4.7}$$

with

$$G_D = \frac{q}{nkt}\,(I_D + I_s)$$

where I_D is the diode direct current and I_s and n have the same meaning as given in Equation (4.2). The thermal noise of R_s can also be represented by a shunt current source

$$\langle i \rangle^2 = \frac{4kT}{R_s}\,\Delta f \tag{4.8}$$

In addition, generation–recombination centers in the depletion layer capture and release carriers in a random fashion, and the wide time constant range associated with the process give rise to a noise signal with energy concentrated at low frequencies. This type of noise, called *flicker noise*, is always associated with a flaw of direct current. Its noise current display a spectral density of the form

$$\langle i \rangle^2 = k_f \frac{I_D^a}{f^b}\,\Delta f \tag{4.9}$$

where I_D is the diode direct current, f is the frequency, k_f is a constant for a particular device, a is a constant in the range of 0.5–2, and b is a constant at approximately unity. The current injected by the metallic edge of a diode may also be a source of substantial flicker noise. Experimentally it has been found that any flicker noise present can be represented by a current source in shunt with the shot noise source. Therefore, the noise model for a Schottky diode can be constructed by adding the thermal noise current source i_{R_s} due to R_s and the combined shot noise and flicker noise current source i_{I_D} due to the direct current to the small-signal equivalent circuit in Figure 4.4b. These two noise sources are statistically uncorrelated to each other and given by the equations

$$\langle i_{R_s} \rangle^2 = \frac{4kT}{R_s}\,\Delta F \tag{4.10}$$

$$\langle i_{I_D} \rangle^2 = 4kTG_D - \frac{2qI_D}{n}\,\Delta f + k_f \frac{I_D^a}{f^b}\,\Delta f \tag{4.11}$$

The total noise at a specified output is the rms (root-mean-square) sum of the individual noise contributions.

4.2.3 Diode Characterization and Modeling

The Schottky diode is represented by the nonlinear model shown in Figure 4.5 with the current and capacitance equations given in Equations (4.2) and (4.4) respectively. Therefore, its dc and RF behavior can be characterized by the ideality factor n, the reverse saturation current I_s, the parallel conductance G_x, the zero bias junction capacitance C_{j0}, the built-in potential V_j, the gradient coefficient m, the depletion capacitance coefficient F_c, and the series resistance R_s. These parameters can be determined by the straightforward dc I–V measurement and the curve-fitting to broadband s parameter measurements. In addition, the noise model parameters k_f, a, and b can be estimated from measurement of the diode noise at low frequency.

4.2.3.1 DC I–V Measurement The dc I–V measurement can be used to determine the ideality factor n, reverse saturation current I_s, and series resistance R_s. In this measurement, current should be recorded from the lowest possible value to a level where the voltage drop across R_s is at least a few tens of millivolts. The HP4145 semiconductor parameter analyzer is the most commonly used measuring device. The measured result should be shown on a semilog plot. Figure 4.6 shows a typical Schottky diode I–V curve. At major portions of the current range between the low and high extremes, the measured I–V curve can be approximated by a straight line. That is, when the voltage drop across R_s is insignificant as compared to the bias voltage and the current contributions due to G_x and I_s are much smaller than the total junction current, Equation (4.2) reduces to

$$I = I_s \exp\left(\frac{qV}{nkT}\right) \tag{4.12}$$

or in the logarithmic form

$$\log(I) = \log(e)\frac{q}{nkT}V + \log(I_s) \tag{4.13}$$

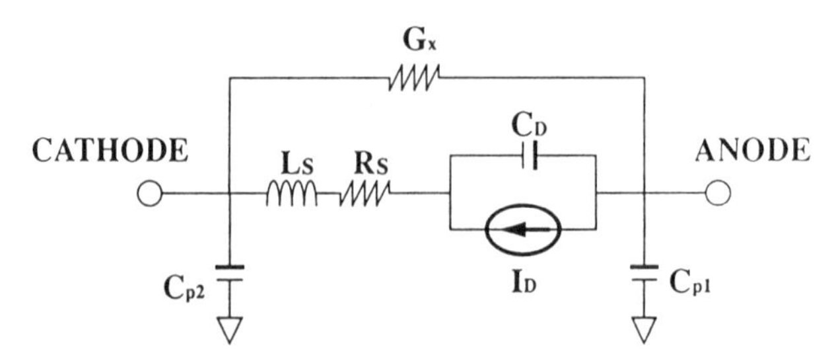

FIGURE 4.5 Nonlinear model of the Schottky diode.

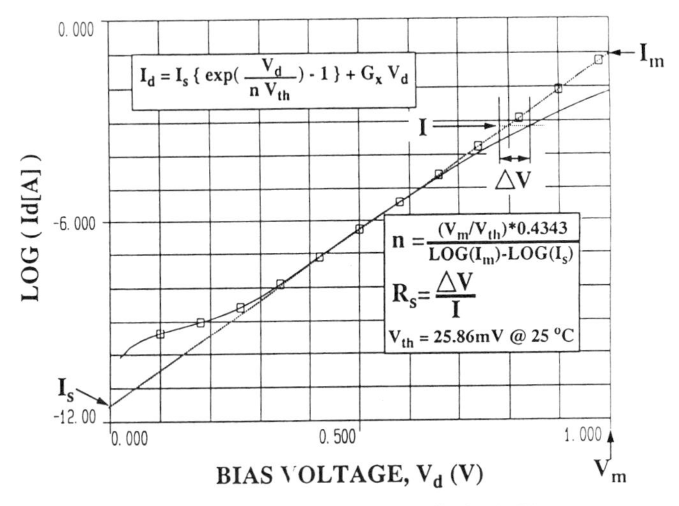

FIGURE 4.6 A typical Schottky diode *I–V* curve.

where $\log(e) = 0.434294\ldots$. By taking the derivative of $\log(I)$ with respect to V in Equation (4.12) and simple manipulation, one can find an expression for n:

$$n = \frac{q \log(e)/kT}{\Delta \log(I)/\Delta V} \tag{4.14}$$

In addition, the intercept point of the closest-fitting straight line at $V = 0$ gives the value of I_s. It can be calculated from any point along the closest-fitting straight line by

$$I_s = I(V) \exp\left(\frac{-qV}{nkT}\right) \tag{4.15}$$

At the low-current extreme, the current contribution due to the effect of G_x is no longer negligible or even becomes a dominant factor. Thus the *I–V* curve deviates from the extended straight line. Knowing n and I_s, one can find G_x by curve-fitting Equation (4.2) to the measured *I–V* curve at the low-current end. At the high-current extreme of the measured *I–V* curve, the curve also deviates from the extended straight line. This deviation is caused by the voltage drop across the series resistance. At high currents, the contributions due to the effect of G_x and I_s are negligible, and Equation (4.2) becomes

$$I = I_s \exp\left\{\frac{q(V - IR_s)}{nkT}\right\} \tag{4.16}$$

or in the logarithmic form

$$\log(I) = \log(e)\frac{q(V - IR_s)}{nkT} + \log(I_s) \tag{4.17}$$

By comparing Equation (4.15) with (4.11), one can find that the additional voltage required in (4.15) to draw the same current I in (4.11) is equal to IR_s. Accordingly, if, at current I, the deviation in voltage between the straight line and the measured I–V curve is ΔV, then R_s is given by

$$R_s = \frac{\Delta V}{I} \tag{4.18}$$

In general, R_s is a weak function of the bias voltage and increases with bias voltage.

4.2.3.2 s-Parameter Measurements Since Schottky diodes are two-terminal devices, their RF behavior can be characterized by either one-port or two-port s parameters. Generally, the s parameters of Schottky diodes are measured on the wafer using GSG (ground–signal–ground) CPW (coplanar waveguide) RF probes and an HP8510 or Wiltron360 vector network analyzer. Figure 4.7 shows the layouts of interdigitated diodes for one-port and two-port RF testings. Typically, four to six sets of s parameters are taken at different dc biases for each diode. The voltages of the dc biases range from ~ -3 to ~ 0.8 V and must at least include three basic points: a zero bias, a forward bias ($\sim 0.7\,V$), and a negative bias (~ -2 V).

The parameter values of the linear model for each dc bias can be determined by curve-fitting the model, including the parasitic elements of

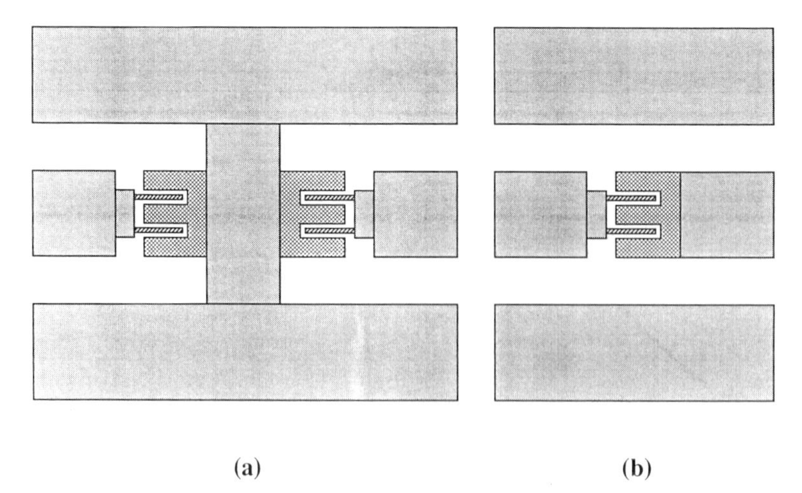

FIGURE 4.7 Layouts of the interdigitated diode for one-port (*a*) and two-port (*b*) on-wafer RF testing.

the pads and grounding paths required for the on-wafer test, to the measured *s* parameters. This optimization process results in the values of the series resistance R_s, the series inductance L_s, the junction capacitance C_j, and the junction conductance G_D. All of them except L_s are bias-dependent. Especially, G_D strongly depends on the dc bias voltage, and its value varies over several orders of magnitude. Typically, the variations of R_s and C_j are less than 2:1 over the measured bias voltage range. However, if R_s has been predetermined from the dc *I–V* measurement mentioned above, the known values of R_s can be used during the optimization. This additional constraint usually gives more accurate values of G_D because it helps to distinguish $1/G_D$ from the ambiguity between R_s and $1/G_D$. Once the C_j–voltage variation is extracted from the measured *s* parameters, another optimization process will be needed to curve-fit the capacitance model given in Equation (4.4) to the extracted *C–V* data. This will give the parameter values of the grading coefficient *m*, the built-in potential V_j, zero-bias junction capacitance C_{j0}, and depletion capacitance coefficient F_c. Figure 4.8 shows the comparison between the extracted and fitted *C–V* characteristics for the 0.25×40-, 0.25×60- and 0.25×80-μm^2 Schottky diodes. Although Equation (4.4) can normally model the extracted *C–V* behavior reasonably well, it often gives values of *m*, V_j, and F_c without physical meaning; namely, Equation (4.4) becomes an empirical formula.

4.2.4 Layout and Fabrication Considerations

The Schottky diode is a very versatile device. It has many applications in high-speed analog circuits. The main applications include mixers, detectors,

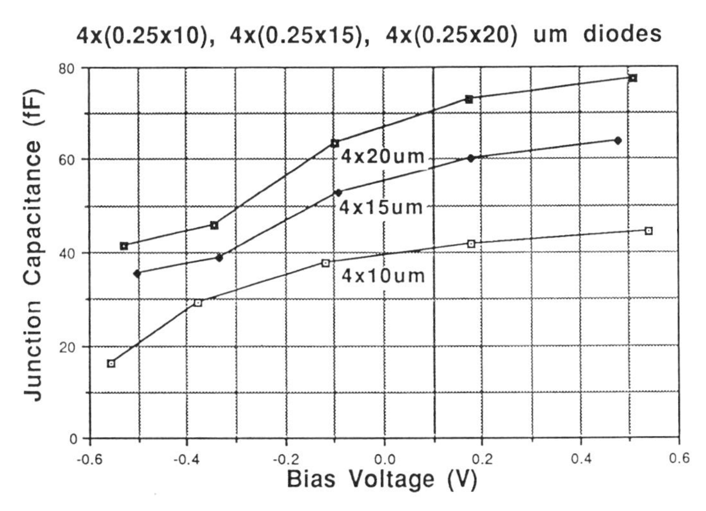

FIGURE 4.8 Comparison between the extracted and fitted *C–V* characteristics for 0.25×40-, 0.25×60-, and 0.25×80-μm^2 Schottky diodes.

multipliers, varactors, switches, limiters, and level-shifting devices. Although all of these applications are related to the rectification property of the Schottky diode one way or another, they do demand different performance requirements of the diode. For example, mixers, detectors, and multipliers require high cutoff frequency, while switches and limiters require high current handling capability and high reverse breakdown voltage. Varactors require large variation range of the junction capacitance and high reverse breakdown voltage. Level-shifting devices require low forward resistance and high current handling capability. Nevertheless, all of these share a common requirement: the minimum series resistance. Therefore, the selection of layer structure and type of layout of Schottky diodes depends on the applications.

There are two types of diode layouts: interdigitated and the overlay diodes. The interdigitated diode has been used as an example in the previous subsections, and its layout is shown in Figure 4.7, whereas the layout of the overlay diode is shown in Figure 4.9. Because of the lateral current flow underneath the Schottky contact caused by the planar structure of the diode, the electric field near the edge of the Schottky contact is much grater than that in the center. Therefore, the current is crowded at the edge of the junction. As a result, the area near the center of the junction contributes capacitance but no significant current. For this reason, the interdigitated structure is used to increase the periphery: area ratio to reduce series resistance and to minimize capacitance. However, the narrow and thin Schottky contact metallization severely limits the current capability of the interdigitated diode. Although the maximum current can be proportionally increased by increasing the number of fingers, its improvement is still very limited. A much more effective way of increasing the current handling capability is to use the overlay structure shown in Figure 4.9, where the Schottky contact area is much larger, allowing the thicker air-bridge metal to be deposited on top of the contact area, thereby increasing the current carrying capability by several orders.

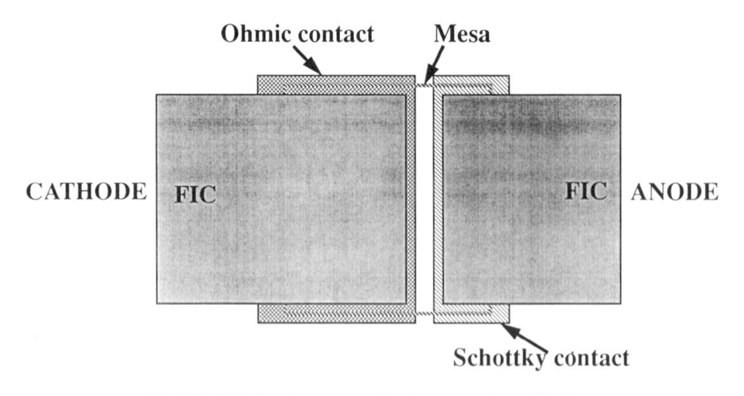

FIGURE 4.9 Layout of the overlay diode.

For applications requiring high cutoff frequency, a low active-layer doping concentration ($5\text{–}9 \times 10^{16}\,\text{cm}^{-3}$) is used to reduce the junction capacitance, and an interdigitated structure is used to reduce both the junction capacitance and the series resistance. For varactor applications, a very low doping density ($\sim 1 \times 10^{16}\,\text{cm}^{-3}$) and thick active layer are used to increase the variation range of the junction capacitance and reverse breakdown voltage. For level-shifting applications, a high active-layer doping concentration ($1 - 3 \times 10^{17}\,\text{cm}^{-3}$) is normally used to minimize the "on-state" resistance of the diode, and an overlay structure is used to increase the current capability.

4.3 PASSIVE COMPONENTS

The design philosophy for analog ICs is to replace passive components by active devices wherever possible because active devices occupy less area. However, there are exceptions, such as low-noise, high-power, and high-linearity requirements. Accordingly, analog ICs employ not only transistors and diodes but also a large number of passive lumped elements, such as resistors, inductors, and capacitors. For example, capacitors are essential for bypass and dc blocking applications. Resistors and inductors are often used as loads and impedance-matching elements. At higher frequencies, passive distributed elements, such as microstrip lines, coplanar waveguides, and coupled transmission line components, are also necessary.

Because of existing parasitics associated with the real lumped element, it is unrealistic to regard the lumped element as an ideal element for applications in high-speed analog ICs. Thus the properties and accurate modeling of the lumped elements are essential to accomplish high-speed analog IC designs. The following subsections will discuss the properties, models, and applications of the most frequently used lumped elements.

4.3.1 MIM Capacitors

MIM (metal–insulator–metal) capacitors, also called *overlay* or *parallel-plate* capacitors, use a thin film of dielectric, such as SiO, SiO_2, or Si_3N_4, sandwiched between two metal layers. Film thickness is usually in the range of 1000–6000 Å. MIM capacitors possess the advantage of wide range of realizable capacitance values (between 0.1 and 100 pF) yet need only a small area (between 20×20 and $500 \times 500\,\mu\text{m}^2$). Their main applications are bypass, dc blocking, and impedance matching. Table 4.1 summarizes the properties of some dielectric films commonly used for MIM capacitors. The top view and a corresponding cross-sectional view of the MIM capacitor are shown in Figure 4.10. Usually, the bottom metal plate is realized with the first-level interconnection (FIC) metallization while the top metal plate is realized with the air-bridge (second-level or top-level) metallization. The

TABLE 4.1 Properties of Some Dielectric Films

Dielectric	Relative Dielectric Constant, ε_r	Thickness, t (Å)	Capacitance per Unit Area (pF/mm^2)	Thermal Coefficient (ppm/°C)
SiO	4.5–6.8	1,500–4,000	100–400	100–500
SiO_2	4.0–5.0	1,500–3,500	100–280	40–80
Si_3N_4	6.0–7.0	1,500–2,500	210–410	25–500
Al_2O_3	6.0–9.0	1,500–2,500	210–500	300–500
Polyimide	3.0–4.5	8,000–10,000	25–50	−450 to −550

thin dielectric layer of thickness t is sandwiched between the two metal plates. The area of the MIM capacitor is defined by the dimensions L and W in the top view of the MIM capacitor shown in Figure 4.10*a*. The capacitance C is simply given by

$$C = \varepsilon_r \varepsilon_0 \frac{LW}{t} \tag{4.19}$$

where ε_r is the relative permittivity (relative dielectric constant) of the dielectric and ε_0 is the permittivity of free space. Normally, the MIM capacitor value is specified in pF/mm^2 (picofarads per square millimeter) or $fF/\mu m^2$ (femtofarads per square micrometer) and is in the range of 100–

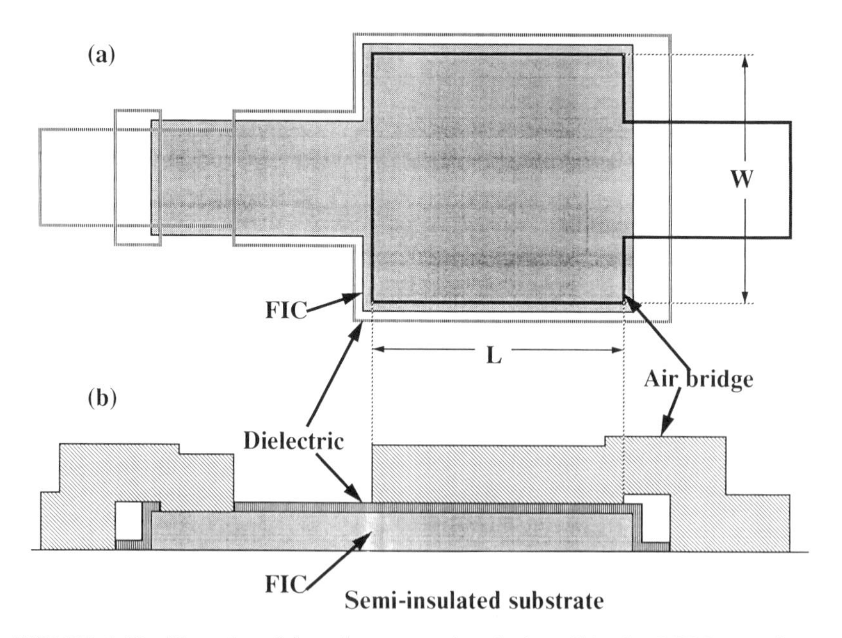

FIGURE 4.10 Top view (*a*) and cross-sectional view (*b*) of a MIM capacitor.

400 pF/mm^2. The breakdown voltage of MIM capacitors is typically higher than 20 V and, generally, is not a concern for analog IC applications.

For high-speed analog IC applications, the MIM capacitor can no longer be treated as an ideal lumped element because of its finite dimension and conductor loss. Figure 4.11 shows an equivalent circuit of the MIM capacitor. The capacitor is modeled as a four-port element with two from the top and two from the bottom plate on the opposite sides. In Figure 4.11, C represents the plate capacitance, C_b represents the parasitic capacitance of the bottom plate to ground, R_p represents the dielectric loss, R_t and R_b represent the conductor losses of the top and bottom plates, and L_t and L_b represent the parasitic inductances due to the top and bottom metal plates. Usually, the series resistance R_b is 3–7 times larger than R_t because the top plate made of air-bridge metal is 3–7 times thicker than the bottom plate made of FIC metal. The sheet resistivity of FIC metal is 30–40 mΩ per square as constructed with 6–9 mΩ per square of air-bridge metal. The parasitic inductances L_t and L_b and the shunt capacitance C_b can be estimated by assuming the MIM capacitor structure to be a microstrip section of length L and width W Then L_t, L_b, and C_b can be calculated approximately by

$$L_b = 2L\left(\ln\frac{2\pi L}{W} - 1 + \frac{W}{L\pi}\right) = L_t \tag{4.20}$$

$$C_b = \varepsilon_r\varepsilon_0\frac{LW}{h} \tag{4.21}$$

where L_t and L_b are in nH (nanohenries), L and W are in cm and h is the substrate thickness in cm. The shunt resistance R_p is, normally, in the order of MΩ (megohms) and does not have any significant effect on the RF behavior of the MIM capacitor.

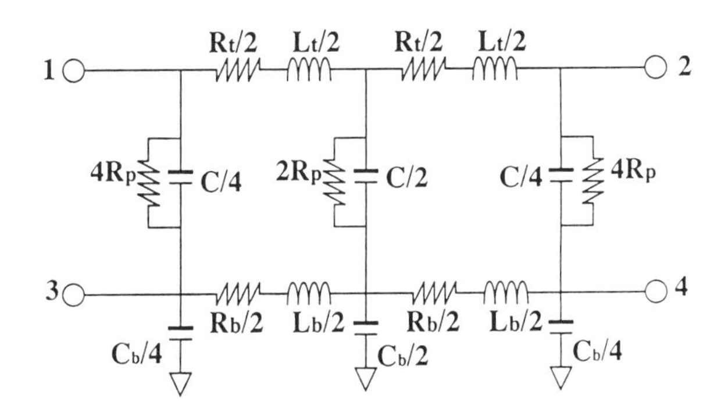

FIGURE 4.11 Equivalent circuit of the MIM capacitor.

4.3.2 Planar Spiral Inductors

Planar spiral inductors can be realized with landed and air-bridged metal lines and laid out in rectangular, octagonal, or circular shape. Figure 4.12 illustrates these three shapes of spiral layouts in two types of metal lines. Although the space between two adjacent lines and the line width are identical for the examples, a combination of different line-to-line space and line width can also be used. Typical values for both dimensions are between

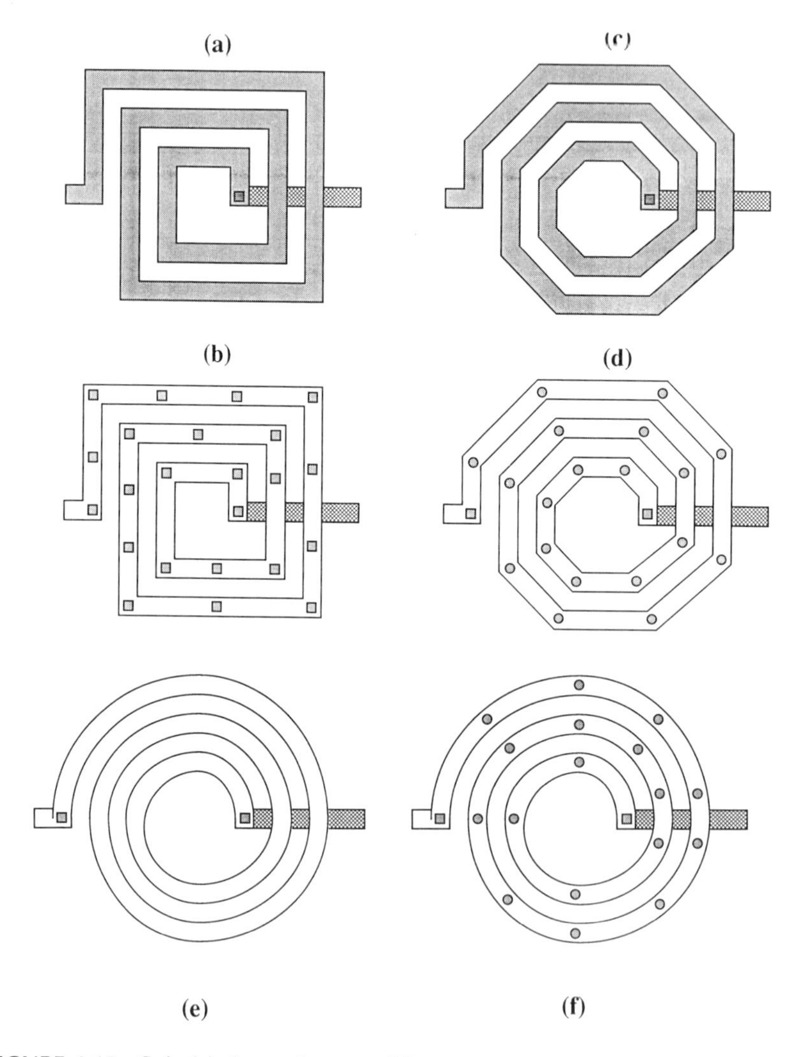

FIGURE 4.12 Spiral inductor layouts: (*a*) rectangular shape and landed metal line, (*b*) rectangular shape and air-bridged metal line, (*c*) octagonal shape and landed metal line, (*d*) octagonal shape and air-bridged metal line, (*e*) circular shape and landed metal line, and (*f*) circular shape and air-bridged metal line.

7 and 10 μm, but a minimum of 4 μm and a maximum of up to 30 μm have also been used. However, when the spiral inductor needs to carry direct current, the minimum line width should be determined by the maximum allowable current density of the metal. Typically, the maximum ratings of the current density are 8–15 mA/μm for the top-level metal and 2–3 mA/μm for the FIC metal. Usually, the spiral is made of top-level metal and an underpath made of FIC metal is used to extend the inner node of the spiral to outside. Since the maximum current density of the FIC metal is much lower than that of the top-level metal, the current carrying capability of the spiral inductor is often set by the underpath width.

The structure of a spiral inductor allows the strong coupling amount individual turns and thus provides high values of inductance on a small area. Typically, the range of realizable inductance values is between 0.2 and 20 nH, with the number of turns ranging from 1 to 10. The physical size of spiral inductors varies from 100×100 to $500 \times 500\ \mu m^2$ depending on the number of turns and inner dimension.

The equivalent circuit of the spiral inductor is shown in Figure 4.13, where L is the inductance of the inductor, C_1 and C_2 are the parasitic capacitances of the metal line to ground, R_1 and R_2 represent the losses associated with C_1 and C_2 due to the conductor loss, and R_s represents the loss associated L due to the conductor loss. The values of R_s, R_1, and R_2 increase with frequency as a result of the skin effect of the conductor. Their frequency dependence can be given approximately by

$$R_s = R_0\left[1 + \left(\frac{f}{f_0}\right)^2\right]^{1/4} \tag{4.22}$$

$$R_1 = R_{10}\left[1 + \left(\frac{f}{f_0}\right)^2\right]^{1/4} \tag{4.23}$$

$$R_2 = R_{20}\left[1 + \left(\frac{f}{f_0}\right)^2\right]^{1/4} \tag{4.24}$$

where R_0, R_{10}, and R_{20} are the dc values of R_s, R_1, and R_2; f is the operating frequency; and f_0 is the skin effect critical frequency of the metal line. The critical frequency f_0 depends on the property of the conductor and the geometry of conductor cross section. Its value is, typically, between 1 and 4 GHz. Among all model parameters, the calculation of the dc

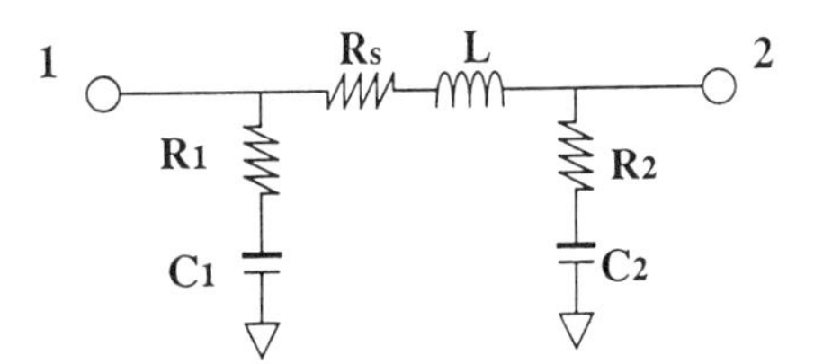

FIGURE 4.13 Equivalent circuit of the spiral inductor.

resistance of spiral inductor R_0 is the simplest. It is the product of the sheet resistance and length (in number of squares) of the metal line. It is difficult to accurately calculate the values of the others because of the complex coupling among individual segments of the spiral. The most accurate formulas for the inductance value of a rectangular spiral are given by Greenhouse [6]. They are derived from Grover's formula [7] by taking into account the negative mutual inductance between turns. Greenhouse has shown that the disagreement between experimental measurements and theoretical calculations is less than 10%. On the other hand, Remke and Burdick [8] have also derived formulas for the inductance value of a circular spiral based on Clivei's formula [9]. Their method takes into account the mutual inductance between loops and uses the Neumann form for mutual inductance and the method of current images for the ground plane effect. The disagreement between Remke's formulas and measurements is less than 8%. Figure 4.14 shows the calculated inductance value as a function of the inner radius in (*a*) and a function of the number of turns in (*b*) for circular spirals having 10-μm line weight and 10-μm space on a 19-mil-thick GaAs substrate.

Currently, there is no accurate method to calculate the parasitic capacitances C_1 and C_2. A rough estimation for C_1 and C_2 is half of the conductor-to-ground capacitance for the total spiral area. In addition, an extra underpath-to-ground capacitance needs to be added to the capacitance C_2 associated with the inner node. Fortunately, several commercial simulators based on full-wave electromagnetic theory [10], such as Em [11], Explorer [12], and HFSS [13], are currently available for the simulation of 2.5- or 3-dimensional structures, such as spiral inductors. Figure 4.15 shows a comparison between the EM simulated and measured *s* parameters for a 2.5-turn rectangular spiral constructed with landed line having a 10-μm width and 10-μm line-to-line spacing on a 100-μm-thick GaAs substrate. The agreement between them is reasonably good up to 20 GHz. However, the EM simulation can be applied only to relatively simple and/or regular structures, such as rectangular shape, due to the limitation of computer's memory size and computing time. Therefore, the most practical way of modeling spiral inductors is to measure the inductors and curve-fit the element values of the equivalent circuit to the measured *s* parameters. Figure 4.16 shows the extracted parameter values of circular spirals as a function of the number of turns. The inductors are constructed with landed lines on 100-μm-thick substrate. The inner radius of the circular spirals is 60 μm, and both conductor width and spacing are 10 μm. Note that the theoretical value of the inductances calculated by using Remke and Burdick's data [8] is also shown in Figure 4.16 for comparison. Figure 4.17 shows the self-resonant frequency of the inductors as a function of the number of turns. The self-resonant frequency decreases when the number of turns increases because the electrical length or the line capacitance of the spiral inductors increases with the number of turns. Since the inductors are

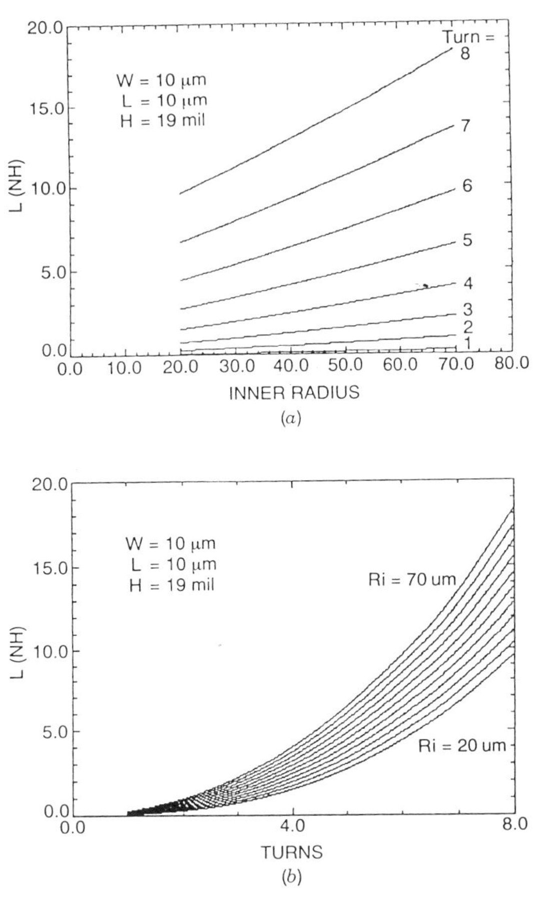

FIGURE 4.14 Calculated inductance as a function of the inner radius (*a*) and the number of turns (*b*) for circular spirals having 10-μm line width and 10-μm space on a 19-mil-thick GaAs substrate.

applicable only for frequencies below the self-resonant frequency, it is highly desired to make the self-resonant frequency: inductance ratio is high as possible. In words, we need to minimize the electrical length of the spiral while maintaining the same inductance or to maximize the inductance of the spiral inductor while maintaining the same spiral length. The electrical length of the spiral inductor can be reduced by using air-bridge metallizations to lower line capacitance. The inductance can be increased by maximizing the spiral area: length ratio or using thicker substrate, which reduces the image current effect of the backside ground plane. Since the circular shape has the largest area: length ratio, a circular spiral inductor is preferable. However, it is more difficult to lay out a compact circuit with circular spirals than rectangular spirals. Perhaps the simplest way to increase

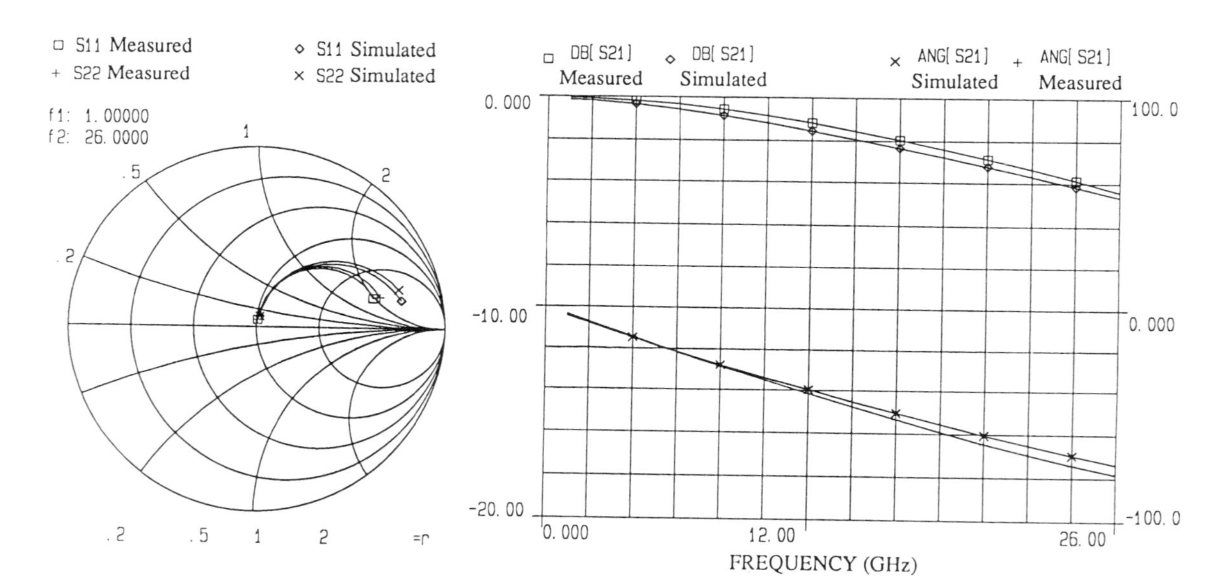

FIGURE 4.15 Comparison between the EM simulated and measured *s* parameters for a 2.5-turn rectangular spiral constructed with landed line having 10-μm width and 10-μm space on a 100-μm-thick GaAs substrate.

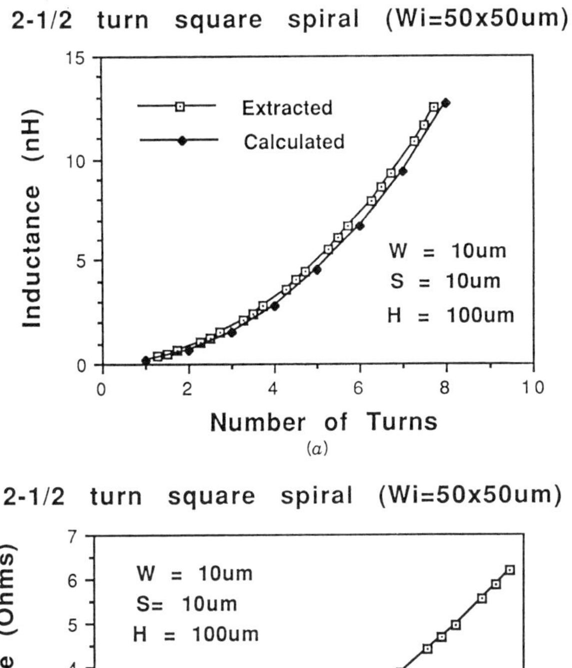

(*a*)

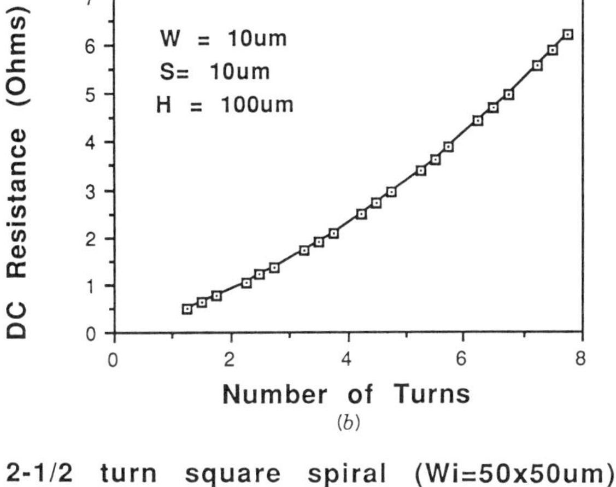

(*b*)

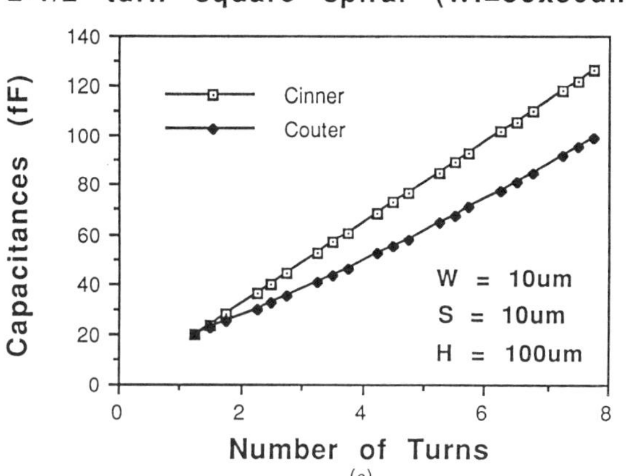

(*c*)

FIGURE 4.16 Extracted parameter values of circular spirals: (*a*) inductance, (*b*) series dc resistance, and (*c*) shunt capacitance versus the number of turns.

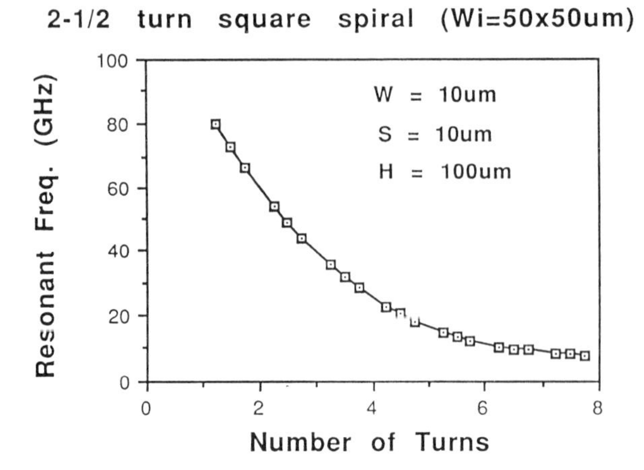

FIGURE 4.17 Self-resonant frequency versus the number of turns.

spiral area: length ratio is to use the largest possible inner dimension (inner diameter for circular spirals or inner side length for rectangular spirals).

For applications requiring only small inductance values, single-loop inductors or meandered narrow line are often used. Figure 4.18 shows the single-loop inductor structure, which is described by the mean radius a and conductor width W. The inductance can be calculated approximately [7] by

$$\mathscr{L} = 12.57a[\ln(8\pi a/W) - 2] \tag{4.25}$$

where $\mathscr{L}$ is in nanohenries and a and W are in centimeters. When a backside ground plane is present, the inductance is reduced due to the image current by the factor (for a 200-μm-thick substrate) [14]

$$1 + (5.2 \pm 1.5)a \tag{4.26}$$

For a meandered narrow line, if the coupling between adjacent segments is negligible, the inductance can be calculated by regarding the meandered line as a straight line and is given by

$$\mathscr{L} = 2L\left[\ln\frac{L}{W+t} + 1.193 + \frac{0.2235(W+t)}{L}\right] \tag{4.27}$$

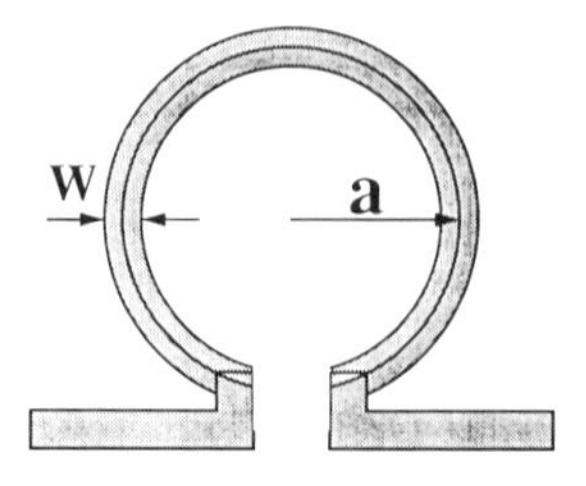

FIGURE 4.18 Single-loop inductor structure.

where $\mathscr{L}$ = inductance (nH)
L = conductor length (cm)
W = conductor width (cm)
t = conductor thickness (cm)

The equivalent circuit of the spiral inductor shown in Figure 4.13 can also be applied to the single-loop inductor and meandered narrow metal line. The method to estimate parasitic element values for spiral inductors is also applicable here.

4.3.3 Planar Transformer Baluns

Baluns are required in a variety of important analog circuits such as balanced mixers, push–pull amplifiers, and push–pull multipliers. Although active baluns are commonly used in analog IC designs because of their very small size and wide frequency bandwidth, they have many disadvantages, such as dc power consumption, low power handling capability, high noise figure, and high intermodulation distortions. For some applications, active baluns can not meet the requirements; therefore, small-size and broad-bandwidth passive baluns are preferable. The transformer balun can fulfill this requirement because it has a small size ($\sim 400 \times 800\ \mu m^2$) and operates over multioctave bandwidth for frequency above 1 GHz. The planar transformer balun consists of two spiral transformers. Each transformer is made up of two interleaved spiral inductors. These two transformers are connected in series and have one of their four outer nodes grounded. The rest of the outer nodes serve as the unbalanced and balanced ports, while one of their two inner common nodes serves as the center tap. Figure 4.19 shows the simplified circuit diagram and the photograph of a rectangular spiral transformer balun. Here, the center tap of the transformer balun is short-circuited to the ground.

The resonant frequency of the spiral coil divides the operating frequencies of the transformer balun into two regions: the magnetic coupling region (frequencies below the resonant frequency) and magnetic–electric coupling region (frequencies above the resonant frequency) [15]. The magnetic coupling region is usually more useful because of its wider relative bandwidth. In this region, the inductance value and the resonant frequency of the spiral coil are the two bandwidth-limiting factors. The coil inductance sets the lower limit of the frequency band, while the resonant frequency sets the upper limit. Therefore, the bandwidth can be increased by either increasing the resonant frequency while maintaining the same inductance or by increasing the inductance of the coil while maintaining the same resonant frequency. To increase the resonant frequency without lowering the inductance, one can reduce the electrical length of the spiral coil at high frequencies and/or coupling capacitance between the primary and secondary coils while maintaining the same physical length of the coil. For example,

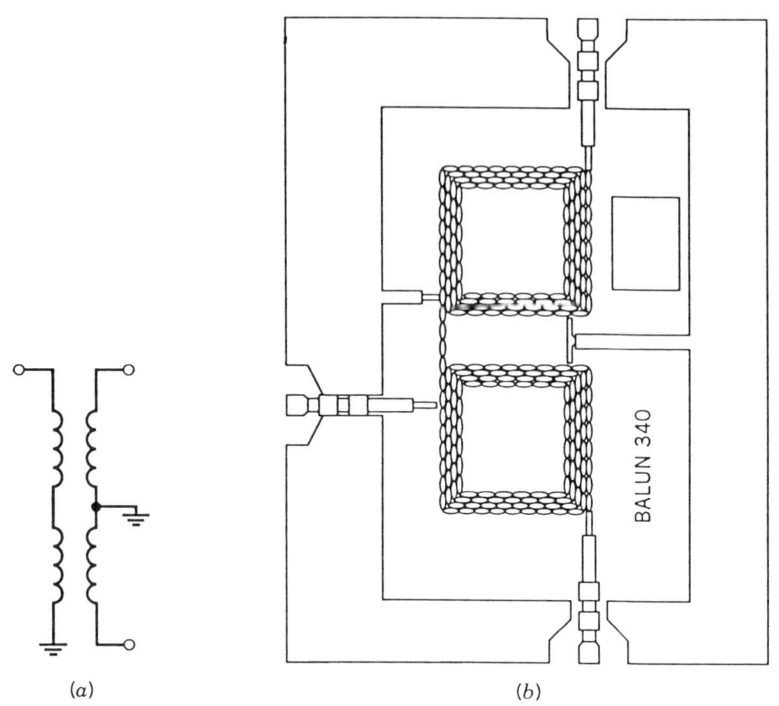

FIGURE 4.19 Simplified circuit diagram (*a*) and the photograph (*b*) of a rectangular spiral transformer balun.

the electrical length of the spiral coil can be reduced by using air-bridged metallizations to lower line capacitance, whereas the coupling capacitance between coils can be reduced by using thinner substrate. On the other hand, the coil inductance can be increased without lowering the resonant frequency by maximizing the spiral area : length ratio.

Figure 4.20 shows the equivalent circuit of a transformer balun consisting of two transformers and some parasitic elements. The parameters associated with the transformers are the primary inductance L_1, secondary inductance L_2, and mutual inductance M. The series resistances R_1 and R_2 represent the conductor losses of the primary and secondary coils and the shunt capacitances, C_1, C_2, and C_p, represent the coil-to-ground capacitances. The capacitance C_{12} represents the capacitive coupling between the primary and secondary coils.

4.3.4 Planar Resistors

Planar resistors are perhaps the second most frequently used elements in analog ICs. Their applications include loads for transistors, feedback resistors, lossy impedance matchings, damping or stabilizing resistors, terminations for other components, RC decouplings for supply lines, and

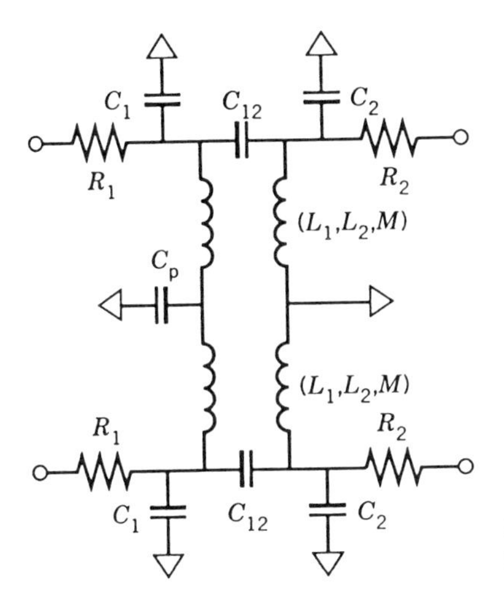

FIGURE 4.20 Equivalent circuit of the transformer balun.

biasing networks for active devices. Planar resistors for analog ICs may be fabricated from either deposited resistive films or from the semiconductor layers. Some factors that determine the choice of planar resistor type are (1) sheet resistivity, (2) current density handling capability, (3) power density capability, (4) resistance accuracy or reproducibility, (5) temperature coefficient of resistance, (6) thermal stability and reliability, and (7) frequency response.

Thin-film resistors are usually formed by evaporating a thin resistive film on the substrate and defining the desired pattern by photolithography. Two types of materials are commonly used for the resistive films: the composite thin-film material, such as TaN and Ta_2N; and the thin metal film, such as Ta, Ti, Cr, and NiCr. Because of the electromigration, which is a phenomenon associated with crystalline grain boundaries [16], at high current density of the resistive film, current densities should be kept within a certain limit to prevent the device from a catastrophic failure over a sufficiently long period. Typical maximum allowable value is about 10^4 A/cm^2 for composite thin-film materials and about 5×10^5 A/cm^2 for thin metal films. Therefore, current handling of the thin-film resistor is determined by the specified current density rating and cross-sectional are of the resistive film. Since the deposited films are very thin, typically 100–400 Å, the current density per unit width is quite low, about 0.01–0.2 mA/μm.

GaAs or bulk resistors that use the active layer in the GaAs substrate can be fabricated by forming an isolated region of active layer on the substrate by mesa etching or by isolation implant of the surrounding conducting layer. Another way is by selective implantation of a conducting region within the semiinsulated substrate. The sheet resistivity of active-layer GaAs resistors is typically in the range between 100 and 1500 Ω per square. Because bulk resistors are fabricated from single-crystal GaAs, they do not suffer from electromigration. However, they have three major disadvantages. First, the surface potential may change over long periods, even on a surface that has been covered by dielectric. This results in the depth of surface depletion changes, giving rise to changes in the sheet resistivity. Second, the temperature coefficient of GaAs resistors is about 10 times higher than the values of deposited thin-film resistors. Third, bulk resistors exhibit a nonlinear behavior at high electric field as a result of carrier velocity saturation. The current saturation characteristic depends not only on the carrier concentration but also on the geometry of the resistor. To avoid nonlinearity, which is undesirable for linear circuits, the resistor should be designed so that the maximum field strength within it does not exceed about 1 kV/cm. Consequently, the minimum length of the resistor L is determined by the ratio between the maximum voltage across it and the critical field strength, 1 kV/cm. The width of the resistor is then calculated by

$$W = L \frac{R_{sh}}{R} \tag{4.28}$$

where R_{sh} is the sheet resistivity and R is the resistance value. However, the current density rating also sets the minimum width of the resistor to ensure sufficient current handling capability. Therefore, if the calculated width is less than the minimum width, then minimum width should be used and the length should be scaled up accordingly. On the other hand, the current saturation characteristic can be used to advantage in circuits where constant-current sources are required. For example, the use of the saturated resistor loads in amplifier stages [17] rather than linear resistor loads results in higher signal gain. This is because the ac resistance of the saturated resistor is substantially higher than the dc resistance, which is equal to the value of the linear resistor if it were used instead.

The properties of some candidate resistive films and bulk GaAs for use in planar resistors are summarized in Table 4.2. The planar resistor can be modeled very accurately as an ideal resistance R connected in series

TABLE 4.2 Properties of Resistive Films and GaAs Layer

Material	Sheet Resistivity (Ω/square)	Current Density Rating (mA/μm)	Stability	Thermal Coefficient (ppm/°C)
Cr	4–15	0.05–0.2	Good	3000
Ti	5–150	0.1–0.5	Good	2500
Ta	20–200	0.1–0.4	Excellent	−100–500
NiCr	30–200	0.05–0.3	Good	−50–−200
Ta_2N	50–200	0.02–0.1	Excellent	−50–−120
TaN	60–220	0.02–0.1	Moderate	−160–−300
GaAs (linear)	100–1500	0.1–0.5	Good	3000
GaAs (saturated)	>10,000	0.15–0.7	Good	3000

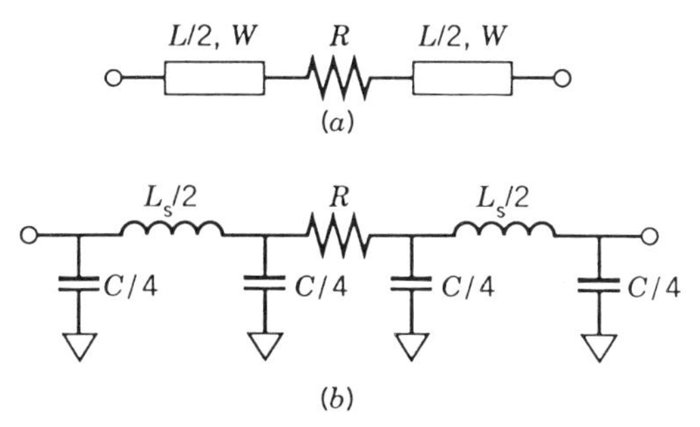

FIGURE 4.21 Equivalent circuit (*a*) and lumped-element equivalent circuit (*b*) of a planar resistor.

between two sections of microstrips having half of the physical size of the resistor. It can also be modeled equally well with only lumped elements. Both equivalent circuits are shown in Figure 4.21, where C and L_s represent shunt capacitance and series inductance of the metal strip having the same dimension as the resistor; L_s and C can be estimated by using Equations (4.20) and (4.21), respectively.

4.3.5 Other Passive Components

Besides the capacitors, inductors, transformer baluns, and resistors discussed in the previous subsections, other passive components are either used frequently but are less critical or are used occasionally in high-speed analog ICs. These secondary passive components include microstrips, coplanar strips, coplanar waveguides, through-substrate vias, air-bridge crossovers, and interdigitated capacitors.

Basically, all interconnections between elements on a substrate can be regarded as some sort of transmission line. The most commonly used transmission lines in analog ICs are, in the order of frequent use, microstrips, coplanar strips, and coplanar waveguides. The microstrip line is a narrow metal strip on a substrate with backside ground plane. The coplanar strip line consists of two parallel metal strips separated by a slot on a substrate. Here, one of the two conductors is a ground plane and the other is the signal line. The coplanar waveguides consists of a central "hot" conductor separated by a slot from two adjacent ground planes. In fact, most interconnections in analog ICs can be modeled with the simple equivalent circuit shown in Figure 4.22, where L, R, and C are respectively the line inductance, line resistance, and lie capacitance. The only exception is when interconnections are longer than one-tenth of the wavelength of the signal. In such a circumstance, the proper transmission line models need to be used for the extreme long interconnections to obtain better accuracy of simulation. Since transmission line models have already been described in

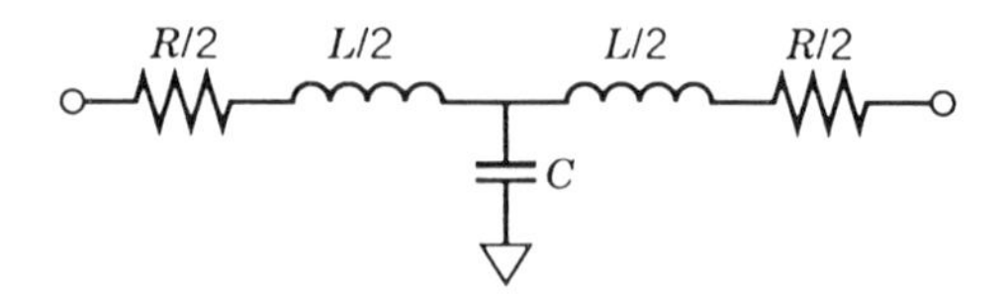

FIGURE 4.22 Equivalent circuit of the interconnection metal line.

details in many microwave books [18–20], we will not repeat this subject here.

Through-substrate vias are sometimes used to provide low-inductance paths to the backside ground for certain circuit elements on the front side of the substrate. However, they are available only for thin substrates with thickness less than or equal to 7 mils. Since the sue of through-substrate via requires additional processing steps including wafer thining, via etching, and backside metallization, it increases fabrication time and cost and lowers circuit yield. On the other hand, it provides several advantages, such as denser circuit layout, better heat-dissipation capability, and backside ground plane for microstrips. There are two different shapes of vias: the conical shape produced by chemical etch (wet etch) and the cylindrical shape produced by reactive-ion etch (RIE, or dry etch). Figure 4.23 shows the top view and a corresponding cross-sectional view of the through-substrate via for both types. The equivalent circuit of the through-substrate via is shown in Figure 4.24, where L_p and L_v represent the inductances of the metal pad and plated metal wall of the via hole; L_p can be roughly estimated from Equation (4.20) by substituting L_{pad} for W and 0.35^*L_{pad} for L, whereas L_v can be calculated approximately by using the formula for a tubular cylinder

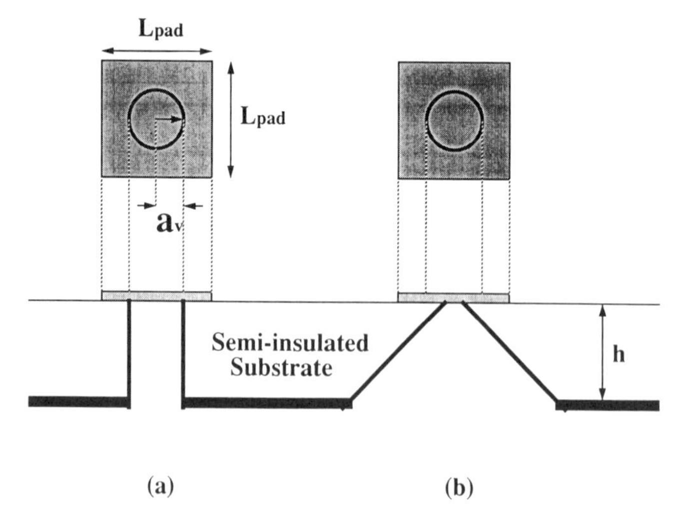

FIGURE 4.23 Top view and cross-sectional view of the cylindrical shape (*a*) and conical shape (*b*) through-substrate vias.

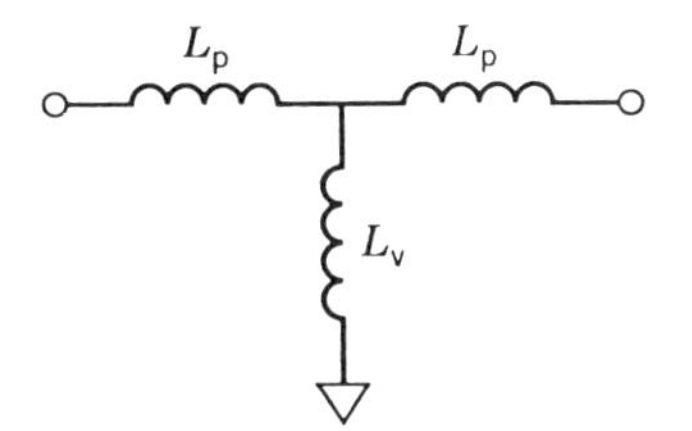

FIGURE 4.24 Equivalent circuit of the through-substrate via.

given by

$$L_v = 2d\left(\ln\frac{2d}{a} - 1\right)\text{ nH} \tag{4.29}$$

where a and d are the radius and length of the tube in centimeters. For a cylindrical shape via hole, d is equal to the substrate thickness h and a is equal to the via-hole radius a_v, while for a conical shape via hole, approximate values for d and a are assumed to be 1.414^*h and $a_v + h/3$. Thus, for a via hole of 60-μm diameter mounted perpendicular to a $100 \times 100\text{-}\mu\text{m}^2$ metal pad on a 100-μm-thick substrate, L_p and L_v are approximately 14 and 18 pH, respectively.

The so-called air-bridge crossovers are generally used whenever conductors must cross over each other. The crossover consists of a deposited air-bridge metal strap that crosses over one or more FIC metal strips with an air gap in between for low capacitive coupling. The air gap is typically 1.5–3.0 μm, and a thin dielectric layer of 0.1–0.2-μm thickness is also often present in the gap for surface protection. Figure 4.25 shows the graphical illustration of a crossover. The equivalent circuit of the crossover is shown in

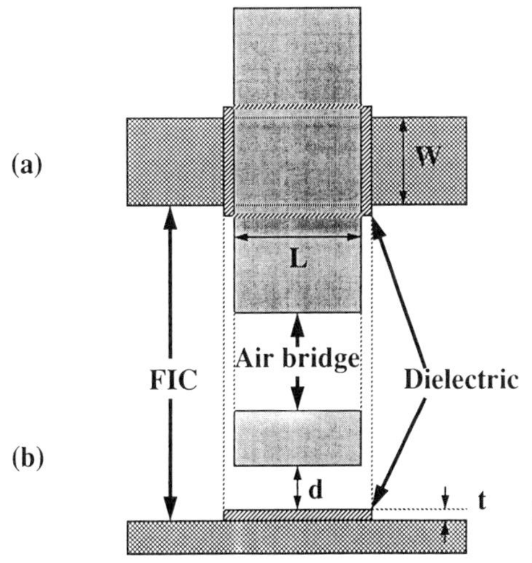

FIGURE 4.25 Top view (*a*) and cross-sectional view (*b*) of a crossover.

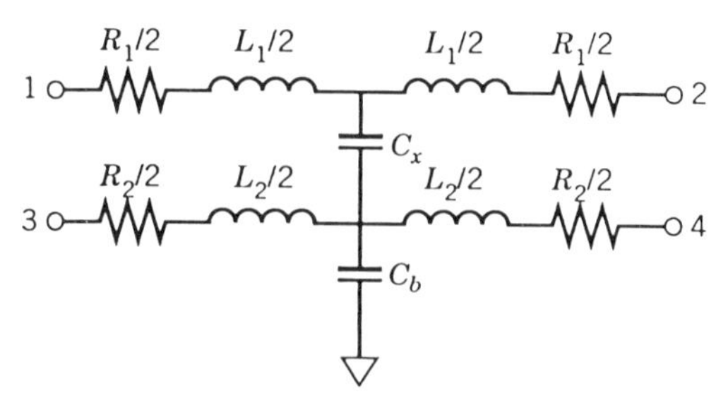

FIGURE 4.26 Equivalent circuit of the crossover.

Figure 4.26, where L_1 and R_1 as well as L_2 and R_2 are the inductances and resistances of the air bridge and FIC metal strips, C_x is the coupling capacitance of the crossover, and C_b is the FIC-to-ground capacitance. As an approximation, L_1 and L_2 can be estimated from Equation (4.27) by regarding the air bridge and FIC metals as isolated metal strips of the same length and width. The coupling capacitance is estimated as

$$C_x = \varepsilon_0 \frac{LW}{d + t/\varepsilon_r} \tag{4.30}$$

where ε_r = relative dielectric constant of dielectric layer
L = length of crossover area
W = width of crossover area
d = vertical height of air gap
t = thickness of dielectric layer

Interdigitated capacitors are used where high precision and small values of capacitance are required. Since the maximum realizable capacitance of the interdigitated capacitance is less than 1 pF, its application is limited to impedance matching networks at high frequency. The capacitance results from the fringing fields between two sets of interleaved metal fingers as shown in Figure 4.27. The equivalent circuit of the interdigitated capacitor is

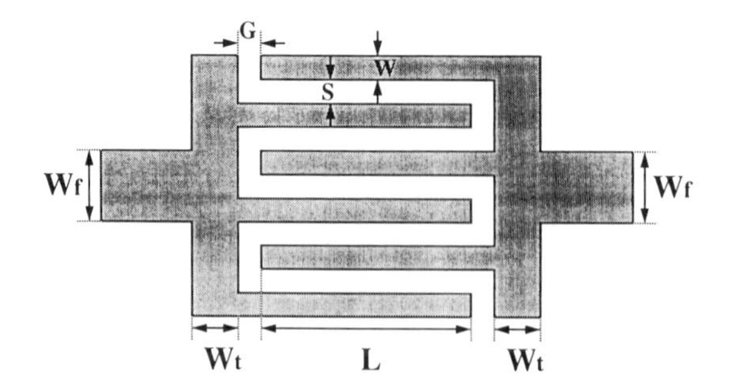

FIGURE 4.27 Interdigitated capacitor structure.

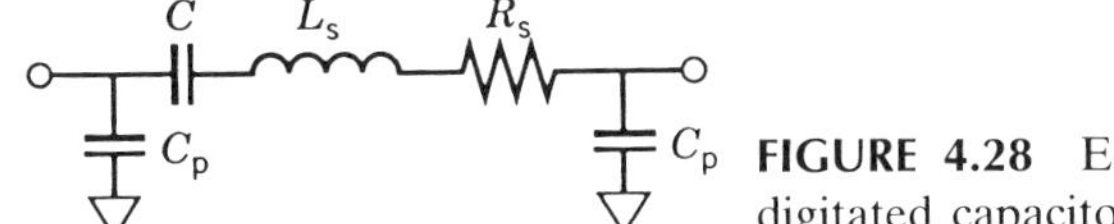

FIGURE 4.28 Equivalent circuit of the interdigitated capacitor.

shown in Figure 4.28, where C is the coupling capacitance, R_s and L_s are the parasitic series resistance and capacitance of the fingers, and C_p is the capacitance of the fingers to ground. Hobdell has introduced a generalized method [21] for the derivation of the various circuit elements shown in Figure 4.28. This method is based on the results given by Simith [22] to calculate the even- and odd-mode fringing capacitances for coupled lines in a periodic array of lines. Wilson [16] has also published a set of design curves for the capacitance resulting from calculation in terms of elliptic integrals without taking into account the end effects.

REFERENCES

1. A. M. Cowley and S. M. Sze, "Surface States and Barrier Height of Metal–Semiconductor Systems," *J. Appl. Phys.*, **36**, 312, 1965.
2. H. A. Bethe, "Theory of the Boundary Layer of Crystal Rectifiers," *MIT Radiat. Lab.*, Rep. 43–12, 1942.
3. W. Schottky, "Halbleitertheorie der Sperrschicht," *Naturwissenschaften*, **26**, 843, 1938.
4. C. R. Cowley and S. M. Sze, "Current Transport in Metal–Semiconductor Barriers," *Solid State Electron.*, **9**, 1035 (1966).
5. C. Y. Chang and S. M. Sze, "Carrier Transport Across Metal–Semiconductor Barriers," *Solid State Electron.*, **13**, 727 (1970).
6. H. M. Greenhouse, "Design of Planar Rectangular Microelectronic Inductors," *IEEE Trans. Parts, Hybrids, and Packaging*, **PHP-110** (2), 101–109 (June 1974).
7. F. W. Grover, *Inductance Calculations*, Van Nostrand, Princeton, NJ, 1946.
8. R. L. Remke and G. A. Burdick, "Spiral Inductors for Hybrid and Microwave Applications," *Proceedings of 24th Electronic Components Conference*, May, 1974, pp. 152–161.
9. A. Clivei, "Optimized Minature Thin-Film Planar Inductors Compatible with Integrated Circuits," *IEEE Trans. Parts, Materials, and Packaging*, **II**, 71–88 (June 1969).
10. J. C. Rautio et al., "An Electromagnetic Time-Harmonic Analysis of Shield Microstrip Circuits," *IEEE Trans. MTT*, **35**, 726–730 (1987).
11. Em, Sonnet Software, Inc., Liverpool, NY.
12. Explorer, Compact Software, Inc., Paterson, NJ.
13. HFSS, Hewlett-Packard Company, Santa Clara, CA.
14. H. Thomas et al., *Gallium Arsenide for Devices and Integrated Circuits*, Peter Peregrinus, London, 1986, pp. 265–269.

15. G. E. Howard et al., "The Power Transfer Mechanism of MMIC Spiral Transformers and Adjacent Spiral Inductors," *IEEE MTT-S International Microwave Symposium Digest*, 1989, pp. 1251–1254.
16. K. Wilson, "Other Circuit Elements for MMICs," *GEC J. Res.*, **4** (2), 126–133 (1986).
17. C. P. Lee et al., "Saturated Resistor Load for GaAs Integrated Circuits," *IEEE Trans. Electron Devices*, **29** (7), 1103–1109 (1982).
18. R. Goyal, *Monolithic Microwave Integrated Circuits:Technology and Design*, Artech House, Norwood, MA, 1989, pp. 351–375.
19. K. C. Gupta et al., *Microstrip Lines and Slotlines*, Artech House, Norwood, MA, 1984.
20. K. C. Gupta et al., *Computer-Aided Design of Microwave Circuits*, Artech House, Norwood, MA, 1981, pp. 57–85.
21. J. L. Hobdell, "Optimization of Interdigital Capacitors," *IEEE Trans. MTT*, **9** (9), 788–791 (1979).
22. J. I. Simith, "The Even- and Odd-Mode Capacitance Parameters for Coupled Lines in Suspended Substrate," *IEEE Trans. MTT*, **19** (5), 424–431 (1971).

CHAPTER FIVE

Basic Building Blocks

DONALD ESTREICH
Microwave Technology Division
Hewlett-Packard Company
1412 Fountaingrove Parkway
Santa Rosa, CA 95403

5.1 INTRODUCTION

The earliest GaAs IC analog applications (i.e., 1976 through perhaps 1983) were primarily microwave amplifiers [1]. Most of these microwave amplifiers were translations of hybrid (thin-film dielectric substrates and chip active components connected with wire bonds) microwave integrated circuits (MICs). In hybrid MIC technology FETs are expensive, and passive components such as transmission lines, spiral inductors, and capacitors are inexpensive. On GaAs substrates, as used in GaAs IC technology, the opposite is true—FETs are relatively inexpensive and passive components, which consume large areas, are expensive.

As a technology matures engineers learn how to leverage the strengths of that technology. Circuit design will evolve to fit the strengths of the technology used to build them. This chapter presents some of the building blocks that have evolved and proved their utility for GaAs IC analog circuit design. The building blocks discussed herein by no means form a complete set, but most of the important ones are presented. We begin with a discussion of MESFET (or FET) biasing considerations before discussing the circuit building blocks.

High-Frequency Analog Integrated-Circuit Design, Edited by Ravender Goyal
ISBN 0-471-53043-3

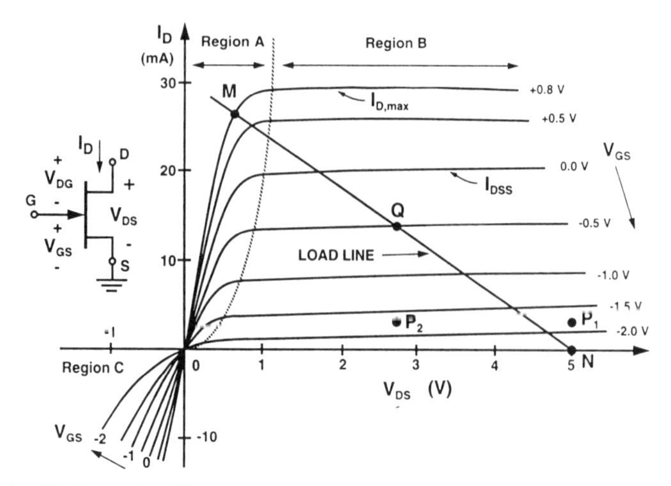

FIGURE 5.1 *N*-channel gallium arsenide MESFET *I–V* characteristic. The device's gate dimensions are 1 μm × 100 μm.

5.2 BIASING CIRCUITS

Field-effect transistors are active devices only under appropriate bias conditions. Consider Figure 5.1, showing the common-source *I–V* characteristic of an N-channel (1-μm × 100-μm)[1] gallium arsenide FET [2]. In addition, Figure 5.1 shows the commonly used N-channel FET symbol with the practice used herein for voltage polarities and positive drain current direction. Although a 100-μm FET geometry will be used in the discussion in this section, the reader should be aware that in GaAs IC design one advantage is freedom to create any size FET to satisfy a variety of requirements.

In the first quadrant of the *I–V* characteristic, the dashed line separates the *I–V* characteristic into two distinct operating regions. Region A is the *linear* region of operation, and region B is the *saturated* region of operation. FETs used in switch and attenuator applications are usually biased in the linear region (used as passive devices). When the GaAs FET is required to be active, the saturated region is used. In the saturated region the FET has its largest usable gain; hence amplifiers almost exclusively use the saturated region.

In the saturated region the highest operative drain current available from an FET, denoted here by $I_{D,max}$, occurs under strong forward bias applied to the gate–channel junction. Parameter $I_{D,max}$ is important for FET power applications because it is the maximum current an FET can draw from a load [where the maximum output power is approximately equal to

[1] This notation is gate length by total gate width.

$R_L(I_{D,max}/2\sqrt{2})^2]$. Another parameter even more often referred to is I_{DSS}, the drain current in the saturated mode under zero gate bias ($V_{GS} = 0$ V).

Two other curves that can be useful for biasing purposes or understanding FET behavior are (1) the *transfer curve* (I_D vs. V_{GS}) and (2) the *transconductance scan* (g_m vs. V_{GS}). Figure 5.2 shows the transfer curve for the FET *I–V* characteristic in Figure 5.1. The slope of the curve $\partial I_D/\partial V_{GS}$, is the transconductance of the FET, commonly denoted by the symbol g_m. The transconductance concept is useful because the FET is generally viewed as a voltage-controlled device because of its high input impedance (small gate input capacitance). Inspection of the transfer curve in Figure 5.2 reveals that for this FET g_m is approximately constant over the range of $-1.0\text{ V} < V_{GS} < 0.7$ V. The transconductance scan shown in Figure 5.3 verifies this conclusion. For many FETs g_m is at or near its maximum value at $I_D = I_{DSS}$. In applications requiring low distortion it is highly desirable for the window of nearly constant g_m to be as wide as practical without greatly compromising the value of g_m. For values of V_{GS} less than -1.0 V, the transfer curve deviates from a straight line (compare to the dashed line extrapolation in Fig. 5.2). In this region the FET behaves approximately with a *square-law characteristic*, and g_m falls in value as the FET approaches the *pinchoff* condition.

Region C, the third quadrant in Figure 5.1, is the common-source *inverse* operation of the FET. The *I–V* characteristics in the first and third quadrants are clearly not symmetric about $V_{DS} = 0$. For symmetric operation, such as might be required for transmission gates, the drain and source must be reversed as the drain voltage changes polarity.

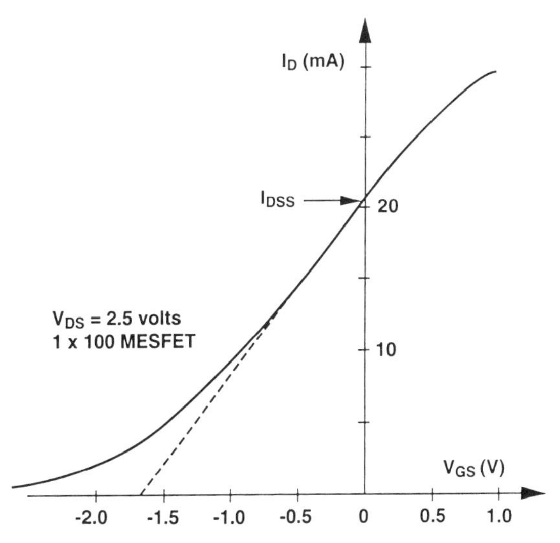

FIGURE 5.2 Transfer curve of drain current versus gate-to-source voltage for the 100-μm FET *I–V* characteristic in Figure 5.1.

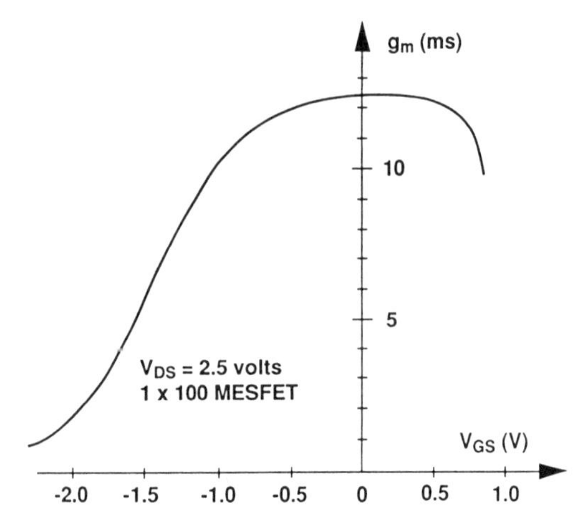

FIGURE 5.3 Transconductance scan showing the behavior of the transconductance versus the gate-to-source voltage. The 100-μm FET's *I–V* characteristic is shown in Figure 5.1.

The *safe operating region* of an FET establishes bias limits. All FETs exhibit drain-to-gate and drain-to-source breakdown. The FET's breakdown voltage places a limit on the maximum voltages that may be applied to the drain and gate terminals. This obviously has important consequences for power applications. The FET *I–V* characteristic in Figure 5.4 illustrates drain-to-source breakdown. In addition, a maximum gate current (under

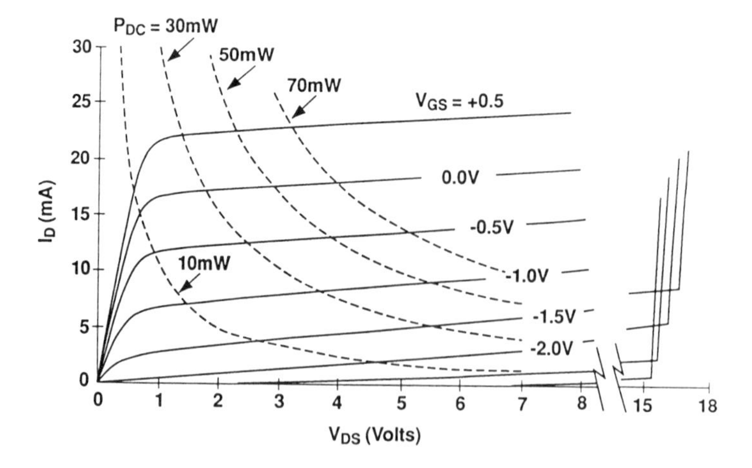

FIGURE 5.4 *I–V* characteristic of a 100-μm FET with contours of constant dc power dissipation shown.

forward bias) is set by the metallization[2] used for the gate. A maximum dc power dissipation is determined by the thermal resistance of the FET and the maximum junction temperature (approximately 175°C for GaAs FETs presently). Contours of constant power are included in Figure 5.4 and are an aid in determining safe bias points if the thermal properties of the FET are known.

5.2.1 Biasing a Single FET

Selecting the bias point (setting V_{GS} and V_{DS}, or I_D and V_{DS}) of an FET within a circuit depends on how the FET is being used (i.e., its circuit function). For example, for the amplification function biasing for high g_m values is important because the voltage gain of an FET is proportional to g_m. The transconductance scan (Fig. 5.3) clearly identifies the maximum small-signal g_m value near $V_{GS} = 0$ for this FET. However, for an amplifier g_m alone is not sufficient to set the bias point; other considerations come into play. DC power or thermal limitations, low-noise requirements, output power requirements, load impedance, and harmonic distortion are examples of other factors that often influence the bias point selection in amplifiers.

As an illustration, consider the FET of Figure 5.1 with a pure resistive load presented to the drain. A load line appears in Figure 5.1 corresponding to a resistive load of 167 Ω and a 5-V dc power supply. Typically the FET would be biased at point Q along the load line [this corresponds to ($V_{GS} = -0.5$ V, $V_{DS} = 2.75$ V, and $I_D = 14$ mA) in Fig. 5.1]. Point Q is approximately the midpoint between point N (maximum drain voltage, minimum drain current) and point M (minimum drain voltage, maximum drain current). This allows for equal voltage swings along the load line about point Q (small-signal or class A amplifier operation). Note that point M takes the FET out of saturation—this results in distortion of the output waveform because g_m falls. However, a more severe cause of distortion is the sharp drop in g_m over the swing between points Q and N (Fig. 5.3 clearly shows the g_m collapse as point N is approached).

The bias point Q and load line shown in Figure 5.1 are more representative of a power amplifier than a low-noise amplifier. For power amplifiers the load line is usually chosen such that point N is close to the drain-to-source breakdown of the FET. This allows for the maximum voltage swing possible with the FET. For higher-efficiency class AB or B operation, bias points closer to point P_1 in Figure 5.1 are selected.

For best noise figure it is well known [3] that lower drain currents (set at approximately $0.15 \cdot I_{DSS}$) and moderate drain voltages (e.g., $V_{DS} = 2$–4 V)

[2] The cross section of the gate is small because of the short gate length; hence small gate currents produce very high current densities, leading to reliability problems (e.g., electromigration).

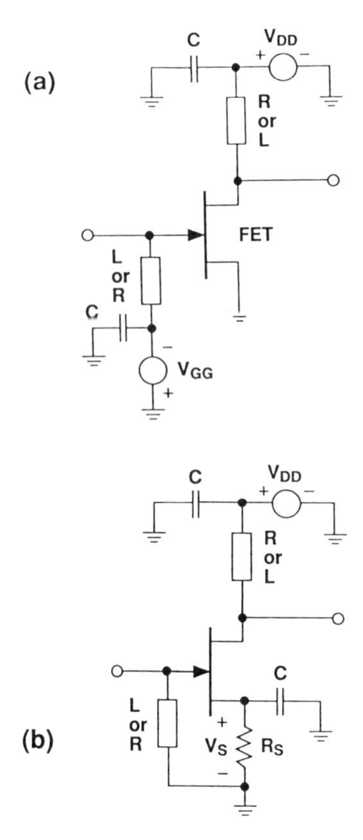

FIGURE 5.5 (*a*) FET biased using two voltage supplies to set V_{DS} and V_{GS} independently; and (*b*) FET biased with only a single voltage supply and a series source resistor to set the bias operating point.

give best noise performance in MESFETs. This corresponds to $I_D \cong 3$ mA for the 100-μm FET (shown as bias point P_2 in Fig. 5.1). This compares to $I_D = 14$ mA, or $0.7 \cdot I_{DSS}$ at point Q. Under low-noise bias conditions the lower g_m is partially compensated for by using a larger load resistance.

Bias schemes for FETs generally fall into two categories: (1) the two-supply bias configuration and (2) the single-supply configuration. Both configurations are illustrated in Figure 5.5. The two-supply approach shown in Figure 5.5*a* allows V_{DS} and V_{GS} to be independently specified, whereas the single-supply scheme of Figure 5.5*b* determines only V_{DS} and relies on the voltage V_S developed across R_S to determine V_{GS}. The capacitors are for bypassing at sufficiently high frequencies.[3]

5.2.2 Biasing Multiple FET Configurations

In principle, knowing the *I–V* characteristic of each FET in multiple FET configurations allows the bias point of each FET to be uniquely determined

[3] Capacitive bypassing results in a finite source-to-ground impedance that can degrade the gain of the amplifier. Bypassing requires $|1/\omega C| \ll R_S$ to be effective.

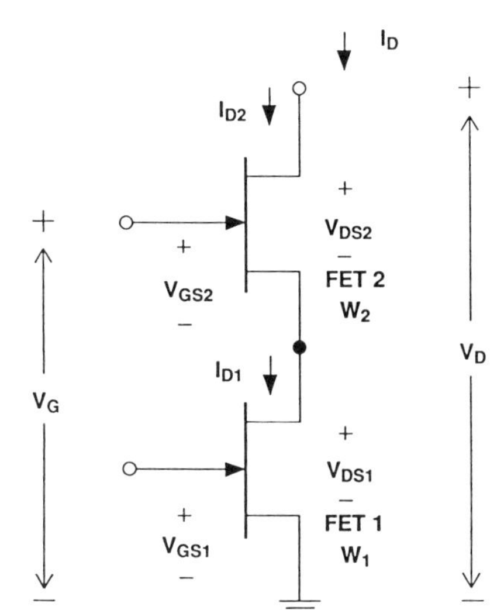

FIGURE 5.6 Stack of two FETs with voltage polarities and current directions defined.

by either computational techniques [e.g., CAD (computer-assisted design) nonlinear simulator] or graphical means. Consider the two-FET stack shown in Figure 5.6, where current and voltage notations and polarities are defined. This stack forms several commonly used circuit cells. For example, it would be a *cascode* cell when FET1 serves as a common-source gain stage driving a common-gate stage (FET2)—see Section 5.3.2.1 for a discussion of the cascode cell.

Figure 5.7 shows the *I–V* characteristic of FET1 with solid lines and the *I–V* characteristic of FET2 with dashed lines. The graphical approach requires knowledge of the *I–V* characteristics of the FETs. The constraints

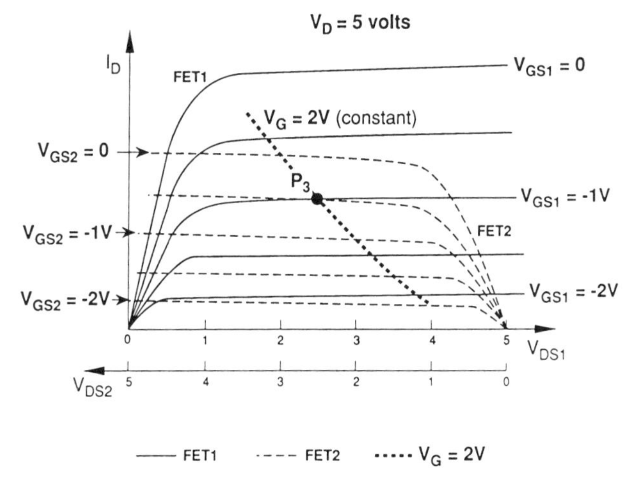

FIGURE 5.7 *I–V* characteristic of FET1 (as configured in Fig. 5.4) with FET2's *I–V* characteristic superimposed as a family of load lines.

on the FET1 and FET2 in this configuration are

$$I_D = I_{D1} = I_{D2} \tag{5.1}$$

assuming that the gate currents are negligible, and

$$V_D = V_{DS1} + V_{DS2} \tag{5.2}$$

With these constraints the I–V characteristic of FET2 is superimposed, with its drain voltage inverted, over the I–V characteristic of FET1. Hence, FET2 is drawn as a family of load lines to FET1. FET2 does not have to be the same-size device as FET1 as assumed in Figure 5.7. In this example the total voltage supplied across both FETs is 5 V ($V_D = 5$ V). With $V_{GS1} = -1.0$ V and $V_{GS2} = -0.5$ V, the FET2 gate potential V_G must be 2.0 V from the relationship

$$V_G = V_{DS1} + V_{GS2} \tag{5.3}$$

given that V_{DS1} is graphically found to be 2.5 V at bias point P_3. As V_{GS1} is varied, but with V_G constant, bias point P_3 varies as shown in the dotted line in Figure 5.7.

5.2.3 Temperature and Backgating Effects in Biasing

The I–V characteristics of FETs are temperature-dependent; hence, stability in the bias point of an FET over temperature must be considered at the design step. In many cases the most important temperature dependence to be considered in biasing is the drain current variation. Consider Figure 5.8, showing the change in drain current with ambient temperature for an

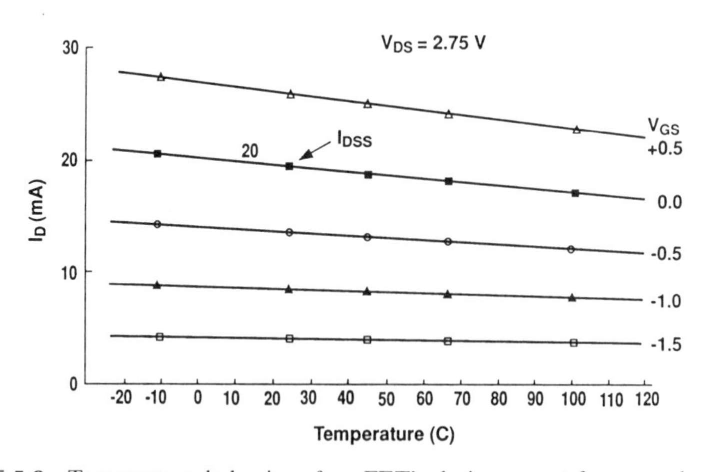

FIGURE 5.8 Temperature behavior of an FET's drain current for several values of gate-to-source voltage.

N-channel (1-μm × 100-μm) GaAs FET. The magnitude of the change in the drain current per unit change in temperature is clearly a function of V_{GS}—typical temperature coefficients for GaAs FETs fall in the range of −0.12 to −0.18%/°C.

Backgating (or *sidegating*) is analogous to the backgate bias effect in MOSFET transistors [4] where the backside substrate-to-channel junction acts as a second gate in modulating the drain current. Backgate bias modulation in MOSFETs is well behaved and predictable; in GaAs MESFETs backgating is much more variable and the backgate thresholds can vary substantially from wafer to wafer and even among devices on a single wafer. Backgating [5,6] occurs when the drain current of an FET decreases because an ohmic region adjacent to the FET is negatively biased in relation to the FET's channel. Figure 5.9 shows the general behavior of an FET drain current as the negatively biased backgating electrode is swept in voltage (V_{BG} denotes the potential of the backgating electrode referenced to the source of the FET). The characteristic feature of backgating is that a threshold voltage exists, denoted here by V_{th}, defining the onset of the drain

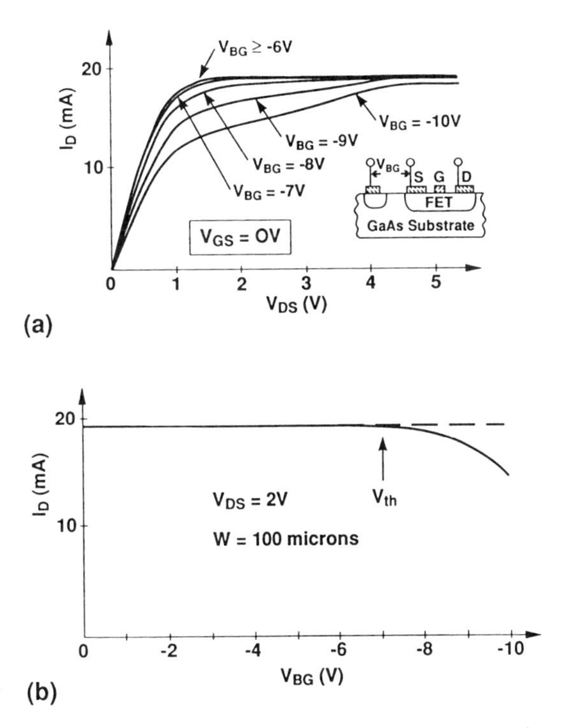

FIGURE 5.9 (*a*) Drain current versus drain-to-source voltage characteristic of a 100-μm FET with backgating voltage applied; (*b*) behavior of drain current as backgating voltage is increased showing decrease of the drain current beyond the backgating threshold voltage.

current reduction. In Figure 5.9 the threshold of backgating is approximately -7 V. There is a corresponding increase in substrate current flowing between the body of the FET and the region responsible for the backgate effect with the onset of backgating [5]. The backgating threshold voltage decreases with decreasing temperature; hence, GaAs ICs are most susceptible to backgating problems at lower temperatures. Backgating is often more severe at lower drain-to-source voltages. Furthermore, as the spacing between the FET and backgating electrode increases, the magnitude of the backgating threshold voltage increases. The layout of an IC can be an important consideration when backgating is a potential problem in a circuit.

5.3 BASIC GAIN STAGES

In GaAs IC design it is convenient to have a family of well-characterized gain blocks to choose from during the design phase. This section presents the most common gain blocks.

5.3.1 Single-Transistor Configurations

The simplest gain block is a single device. There are three configurations possible with a three terminal device. For the GaAs FET these are referred to as the *common-source FET* (CSF), *common-gate FET* (CGF), and *common-drain FET* (CDF) configurations (see Fig. 5.10). Table 5.1 compares some important parameters of the three configurations.

5.3.1.1 Common-Source FET The most widely used FET configuration is the common-source connection. This is schematically shown in Figure 5.10*a*. The advantages of the CSF connection include high gain, good noise performance, and high input and output impedances. Most GaAs IC amplifiers use the CSF configuration—it gives the most flexibility to the

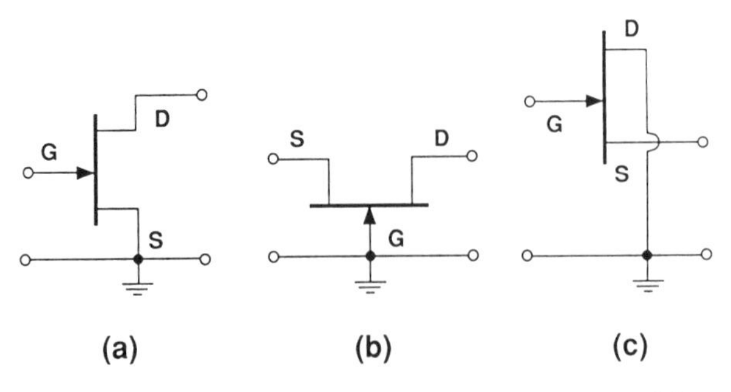

FIGURE 5.10 The three connections of a three-terminal FET: (*a*) common-source FET configuration; (*b*) common-gate FET configuration; (*c*) common-drain FET configuration. Bias components are not shown.

TABLE 5.1 Comparison of CSF, CGF, and CDF Configurations[a]

Parameter	Common-Source	Common-Gate	Common-Drain
Voltage gain[b]	$-g_m r_{ds}$	$(g_m r_{ds} + 1)$	$\frac{g_m r_{ds}}{g_m r_{ds} + 1} < 1$
Current gain	Very high	≈ 1	Very high
Input resistance	High	$\frac{r_{ds} + R_L}{g_m r_{ds} + 1}$	High
Output resistance[c]	r_{ds}	$r_{ds} + (g_m r_{ds} + 1)R_S$	$\frac{r_{ds}}{g_m r_{ds} + 1} \cong \frac{1}{g_m}$
Reverse isolation	Best	Good	Poorest

[a] Low-frequency to midfrequency approximations.
[b] Voltage gain computed assuming load resistance R_L is large (i.e., $R_L \to \infty$).
[c] For the CGF, R_S is the resistance presented to the source.

designer in choosing matching elements and/or feedback elements. As a simple gain cell the FET is usually biased in the saturated region of operation.

The *intrinsic voltage gain* of an FET is $-g_m r_{ds}$, where r_{ds} is the small-signal drain-to-source resistance of the FET. For a typical GaAs IC foundry process $|g_m r_{ds}|$ is of the order of 15 (but can range from a low of 10 to a high of approximately 20). The intrinsic gain is the highest voltage gain achievable from a single FET (excluding the possibility of positive feedback). Loading the drain of the CSF stage with a finite resistance R_L reduces the FET's voltage gain to $-g_m R_{eq}$, where R_{eq} is the parallel combination of r_{ds} and R_L. The minus sign indicates signal inversion (i.e., the output is 180° out of phase with the input).

The input impedance of the CSF is large at low frequencies because the gate is a reverse-biased junction. The junction capacitance C_{gs} is typically small (e.g., for GaAs FETs the gate capacitance is of the order of 1 pF/mm of gate width). Hence, at high frequencies the input impedance goes as $1/j\omega C_{gs}$. The current gain is extraordinarily high at low frequencies because of the high input impedance, but falls to unity at the frequency $f_T = [g_m / 2\pi(C_{gs} + C_{gd})]$. The value of f_T is often approximated by $g_m / 2\pi C_{gs}$ because typically C_{gd} is about one-tenth the value of C_{gs}. The output impedance is simply the channel drain-to-source resistance r_{ds} of the FET (e.g., assuming g_m to be about 125 mS/mm as from Fig. 5.3; then r_{ds} is of the order of 100 Ω-mm).

5.3.1.2 Common-Gate FET The common-gate FET (CGF) stage has the property that its broadband input resistance is primarily determined by its transconductance, which in turn is proportional to the width of the FET. For this reason the CGF is often used as an impedance transformer (this is commonly called *active matching*). From Table 5.1, if $r_{ds} \gg R_L$, which is often a fair approximation, the input resistance is approximately equal to $1/g_m$, with g_m directly proportional to the width of the FET [7,8]. For a 50-Ω impedance, a g_m of 125 mS/mm dictates an FET width of about 160 μm.[4] When compared to matching with passive elements, active matching has the advantages of giving better broadband match and smaller chip size, and in some cases producing a design more tolerant to process variations. The disadvantages include higher power dissipation if impedance matching is the sole function of the CGF stage, lack of flexibility in choosing the width of the FET, and has higher noise figure than the CSF stage.

In contrast to the CSF stage discussed in Section 5.3.1.1, the CGF stage has a noninverting voltage gain. Its voltage gain is approximately equal in magnitude to that of the CSF (see Table 5.1). However, the CGF has a small amount of positive feedback [7], which accounts for the factor of 1 being added to the $g_m r_{ds}$ product. The CGF has unity current gain and high output resistance, which is dependent on the resistance R_S presented to the source terminal of the FET (see Table 5.1).

5.3.1.3 Common-Drain FET The major feature of the common-drain FET (CDF) is its capability to source large currents because of its low output impedance. The CDF is often referred to as the *source follower* because its source voltage "follows" the gate or input voltage. It is often used as an output stage to drive large capacitive loads (although this can cause instability) or low impedance loads and as an impedance transformer to extend bandwidths (one would want $R_{out}C_L$ to be as small as possible). It has slightly less than unity voltage gain. The CDF has low input capacitance because the source node voltage tracks the input gate node voltage, thereby reducing the gate-to-source capacitance. Again, the current gain is very large because of high input impedance, but this feature alone is seldom useful because it is strongly frequency-dependent (which is ignored in the discussion here). The poorest reverse isolation occurs with the CDF stage compared to the CSF and CGF stages. The CDF stage is often used for level shifting in ICs as discussed in Section 5.6.

5.3.2 Multiple-Transistor Configurations

There are numerous ways in which two or more FETs can be combined to perform useful functions such as voltage gain. This section describes some of the more commonly used multiple-transistor gain blocks.

5.3.2.1 Cascode Cell The *cascode* cell is formed by a cascade of a CSF stage driving a CGF stage (see Fig. 5.11). Historically, the cascode gain cell

[4] When resistors are included for biasing purposes, the FET width must to modified to maintain a 50-Ω input resistance.

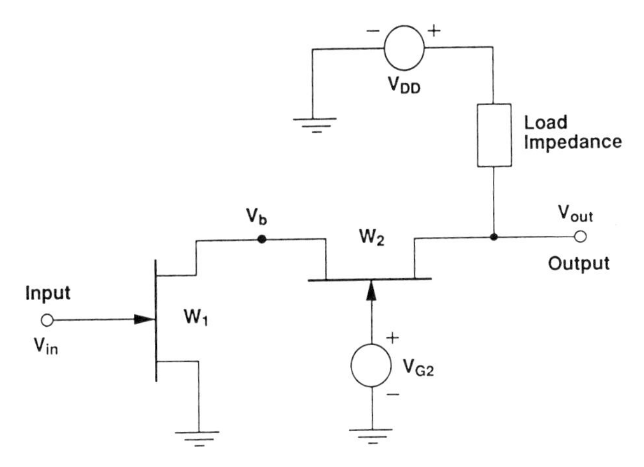

FIGURE 5.11 The cascode cell consists of a CSF stage (FET1) driving a CGF stage (FET2). Simplified bias circuitry is shown—voltage source V_{G_2} is an ac low impedance.

has been used for wideband amplifier and low-noise amplifier applications [9]. The cascode cell has two advantages over the simpler CSF stage: (1) the output-to-input feedback capacitance is reduced—therefore, the Miller effect capacitance is smaller thereby allowing wider bandwidths; and (2) the output impedance is higher. Not only is the reverse isolation very high, but the cascode cell is less sensitive to changes in r_{ds}. The transconductance of the cascode cell is approximately the same as the CSF stage (of course, given equal-size FETs in the comparison).

The lower feedback capacitance of the cascode configuration can be understood by the following argument. The Miller capacitance referred to the input of a CSF stage is equal to the gate-to-drain capacitance C_{gd} multiplied by the voltage gain of the stage, that is, $C_{miller} = C_{gd}(1 + g_m R_L)$, where $g_m R_L$ is the voltage gain. For the simple CSF stage the voltage gain can be large; thus the Miller capacitance can be large and substantially reduce the bandwidth of the stage. For the cascode cell the voltage gain of the input (common-source) FET is low because it is presented with a low impedance from the CGF. In fact, the voltage gain of the CSF stage is approximately unity because with a load impedance of approximately $1/g_{m2}$, where FET2 is the CGF stage, the voltage gain is approximately $g_{m1}(1/g_{m2}) \cong 1$ if $g_{m1} \cong g_{m2}$. The nearly unilateral nature of the cascode cell is desirable for achieving wider bandwidths and allowing for circuit techniques such as shunt peaking to be used (reduced interaction between the input and the output, which is a major source of stability problems). Of course, the price for this advantage is the use of two FETs without an increase in overall transconductance compared to the CSF stage, and double the dc power dissipation compared with a single CSF stage. With two FETs a larger voltage is required and more area is consumed in IC layout, although using dual-gate FETs [10] will reduce the IC area consumption.

The output resistance of the cascode cell is equal to the output resistance of the CSF multiplied by the CGF intrinsic voltage gain

$$r_{\rm ds}^{\rm cascode} = (g_{\rm m2} r_{\rm ds1} + 1) r_{\rm ds2} + r_{\rm ds1} \tag{5.4}$$

where FET1 ($g_{\rm m1}$ and $r_{\rm rd1}$) is the CSF stage and FET2 ($g_{\rm m2}$ and $r_{\rm ds2}$) is the CGF stage. If both FETs are the same width, then $g_{\rm m1} = g_{\rm m2} = g_{\rm m}$ and $r_{\rm ds1} = r_{\rm ds2} = r_{\rm ds}$, and

$$r_{\rm ds}^{\rm cascode} = (g_{\rm m} r_{\rm ds} + 2) r_{\rm ds} \tag{5.5}$$

One way to explain the result expressed in Equations (5.4) and (5.5) is that a change in the output voltage $V_{\rm out}$ produces a change in the internal node voltage $V_{\rm b}$ that is $(1/g_{\rm m} r_{\rm ds})$ smaller. Hence, the output resistance of the CSF ($r_{\rm ds1}$ or $r_{\rm ds}$) is multiplied by the intrinsic voltage gain of the CGF because the incremental output current is reduced by the reciprocal of the gain of the CGF. This is an order of magnitude increase (assuming $g_{\rm m} r_{\rm ds} \cong 10$) in output resistance over the single CSF stage.

Another possible advantage of the cascode configuration is higher voltage swings at the output before FET breakdown is reached. With GaAs FETs the drain-to-gate breakdown voltage is the limiting voltage. This is more of a problem with a single CSF stage because it is inverting. In the CSF stage, as the gate swings negative in voltage, the drain (output node) goes positive. Therefore, the CSF stage's drain-to-gate breakdown is reached at lower output node voltages than with the cascode cell with its noninverting CGF output. This is only an advantage under large-signal operation.

The cascode configuration can be applied equally well to both single-ended and differential amplifiers (refer to the discussion on differential amplifiers below). It has been applied to distributed amplifiers with great success [11,12]. The principle used in the cascode configuration can be applied to circuit functions other than amplification. For example, Section 5.4 describes cascoded active loads producing larger load resistances than achievable with single devices.

5.3.2.2 Double-Cascode Cell The *double-cascode* cell is an extension of the cascode cell discussed above. It is shown in Figure 5.12. A still higher $V_{\rm DD}$ voltage is required to bias all three FETs into saturation. The double-cascode cell would be used where an even higher output resistance is needed beyond that of the cascode cell. An analysis of the output resistance at low

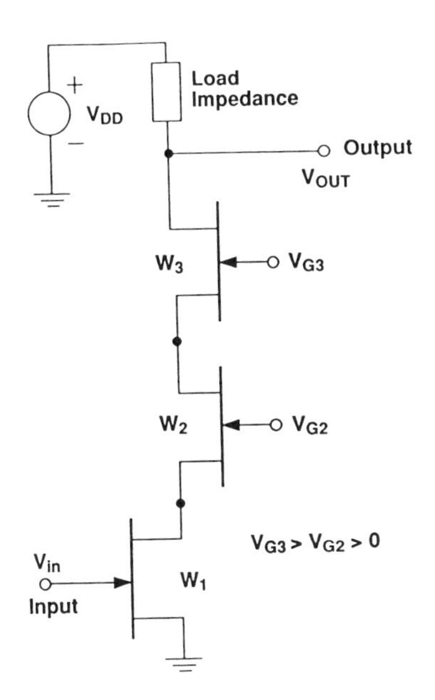

FIGURE 5.12 The double-cascode cell consists of a CSF stage driving a cascade of two CGF stages.

to midband frequencies gives

$$r_{\mathrm{ds}}^{2,\mathrm{cascode}} = [g_{\mathrm{m}} r_{\mathrm{ds}} (g_{\mathrm{m}} r_{\mathrm{ds}} + 3) + 3] r_{\mathrm{ds}} \tag{5.6}$$

where all three FETs are assumed to be of the same width (i.e., identical electrical parameters). Of course, parasitic capacitances act to reduce the impedance levels achieved in cascode cells at higher frequencies.

5.3.2.3 Differential Cell The single device and cascode gain cells discussed above form *single-ended* amplifiers. There is a special class of circuits that are *double-ended*—the *differential amplifier* is in this class. The differential cell [13] performs the function of amplifying the difference between two signals, but signals common to both inputs are not amplified (such signals, called *common-mode* signals, are completely rejected in a perfect differential cell). The differential amplifier uses the CSF configuration with symmetry applied to select the difference between two signals.

Figure 5.13 shows the differential-gain block in its simplest form with double-ended input and double-ended output. The two amplifying FETs (W_1 and W_2) are source-coupled with the tail current set by a current source. The Norton equivalent of this current source is shown in Figure 5.13—an ideal current source I_{SS} paralleled with resistor R_{SS}. For simplicity resistive drain loading is assumed.

For analytical purposes the input terminal signals, v_{in1} and v_{in2}, can be

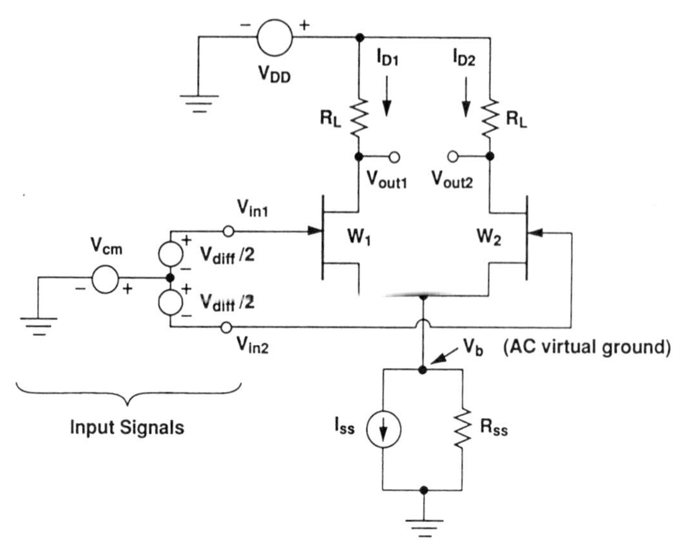

FIGURE 5.13 The differential cell with identical load resistors and Norton equivalent circuit of current source setting the tail current. The input signal is a superposition of common-mode and differential signals.

considered to be a superposition of a *common-mode* signal v_{cm} and a *differential-mode* signal v_{diff}. using the representation of the input signals in Figure 5.13, we obtain

$$v_{\mathrm{in1}} = v_{\mathrm{cm}} + \tfrac{1}{2} v_{\mathrm{diff}} \tag{5.7}$$

$$v_{\mathrm{in2}} = v_{\mathrm{cm}} - \tfrac{1}{2} v_{\mathrm{diff}} \tag{5.8}$$

The differential and common-mode signals in terms of v_{in1} and v_{in2} are

$$v_{\mathrm{diff}} = v_{\mathrm{in1}} - v_{\mathrm{in2}} \tag{5.9}$$

$$v_{\mathrm{cm}} = \tfrac{1}{2}(v_{\mathrm{in1}} + v_{\mathrm{in2}}) \tag{5.10}$$

Likewise, the output terminal signals v_{out1} and v_{out2} can be expressed in terms of differential and common-mode signals $v_{\mathrm{diff,out}}$ and $v_{\mathrm{cm,out}}$, respectively.

The differential gain A_{diff} is defined to be

$$A_{\mathrm{diff}} = \frac{v_{\mathrm{out1}} - v_{\mathrm{out2}}}{v_{\mathrm{in1}} - v_{\mathrm{in2}}} = \frac{v_{\mathrm{diff,out}}}{v_{\mathrm{diff}}} = -g_{\mathrm{m}}(r_{\mathrm{ds}} \,\|\, R_{\mathrm{L}}) \tag{5.11}$$

where $(r_{\mathrm{ds}} \,\|\, R_{\mathrm{L}})$ is the parallel combination of r_{ds} and R_{L}. Equation (5.11) assumes a double-ended output. Thus, in the differential cell the amplification of a differential signal is equal to the gain of one side of the differential circuit (e.g., CSF stage). Sometimes a single-ended output is used, in which case the differential gain is reduced by a factor of 2 (or 6 dB). The common-mode gain A_{cm} is defined to be

$$A_{cm} = \frac{v_{out1} + v_{out2}}{v_{in1} + v_{in2}}$$

$$= \frac{v_{cm,out}}{v_{cm}} = \frac{-g_m r_{ds} R_L}{2(g_m r_{ds} + 1)R_{SS} + r_{ds} + R_L} \tag{5.12}$$

The common-mode gain A_{cm} depends strongly on the Norton equivalent resistance R_{SS} of the current source. As R_{SS} approaches infinity (as it would for an ideal current source), A_{cm} goes to zero. This never happens in practice. In the other extreme, as R_{SS} approaches zero (current source removed and FET sources grounded), then $A_{cm} \rightarrow A_{diff}$.

The preceding expressions for the differential and common-mode gains assumed both FETs to be identical. Small mismatches in FET parameters (e.g., transconductance) results in perturbations to the preceding equations. Two of the advantages in fabricating a differential cell in an integrated circuit process is the superior parameter matching inherent in simultaneous fabrication of active devices and components and the thermal tracking between active devices by being in close proximity to each other.

The FET differential cell has high input impedance (as do both the CSF stage and the cascode cell). The output impedance is high, but is generally dominated by the load impedance (e.g., R_L in Fig. 5.13). The single-ended output resistance is given by

$$Z_{out} = [r_{ds} + 2(g_m r_{ds} + 1)R_{SS}] \parallel R_L \tag{5.13}$$

The double-ended (differential) output resistance is $2Z_{out}$.

One of the most powerful concepts in engineering is the use of symmetry. The differential cell is one of the best examples of symmetry in circuit design. Symmetry in the differential cell's transfer curve results in cancellation of the even harmonics. Low second harmonic generation (which is generally the dominant harmonic generated) is one of the principal advantages of the differential cell. The symmetric operation in a differential cell reduces power supply current variations, thereby simplifying power supply bypassing.

It is, of course, possible to apply the cascode principle to the differential cell as shown in Figure 5.14. To take advantage of the cascode differential cell's higher resistance, very high impedance loads are required such as with cascoded active loads (See Section 5.4).

5.3.2.4 Push–Pull Cell The *push-pull* is another gain block using symmetry to achieve certain advantages (e.g., lower even harmonic distortion) [14].

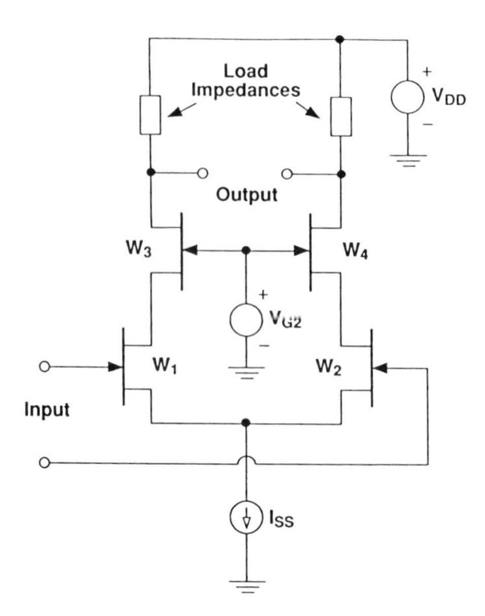

FIGURE 5.14 A differential cell using cascode arrangement for higher output resistance (and higher gain).

In fact, when the push–pull cell is used in class A operation, it is similar to the differential cell discussed above. Figure 5.15 shows the push–pull cell. The two input terminals are driven 180° out of phase. In Figure 5.15 transformers are shown driving the input terminals with antiphase signals and combining the output signals. Jamison et al. [15] have demonstrated the application of planar spiral transformers (formed using a standard MMIC technology) for interfacing push–pull amplifiers. A differential stage connected with single-ended input, but double-ended output (*paraphase am-*

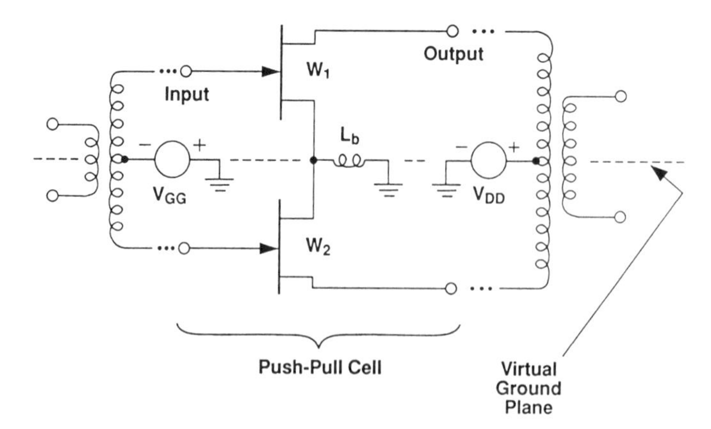

FIGURE 5.15 A push–pull cell with transformer interface at input and output.

plifier in Sokolov et al. [16]), can be used to also accomplish the same function as the input transformer.

Just as with the differential cell, even harmonic distortion is reduced by cancellation of the nonlinearity in the transfer curves due to the symmetric positioning of the transistors. Of course, this requires selection of the bias points of both transistors so that cancellation is facilitated.

The push–pull cell has two important advantages at high frequencies. First, the input and output impedances are easier to match to as compared to the case where both transistors are simply connected in parallel. Given that Z_{in} and Z_{out} are the input and output impedances of a single FET, the push–pull configuration results in input and output impedances of $2Z_{in}$ and $2Z_{out}$, respectively, whereas for two FETs in parallel they would be $\frac{1}{2}Z_{in}$ and $\frac{1}{2}Z_{out}$, respectively. In power amplifiers larger transistors are hard to impedance-match because of their low input impedances. The second advantage is the minimization of grounding impedance problems because common lead impedances, such as L_b in Figure 5.15, are not in the fundamental signal path. This can have the added benefits of improved gain and stability. Also, this reduces the constraints on the bonding and packaging requirements for the high frequency push–pull amplifier [17].

The push–pull cell is often used in power applications where the transistors are operated in the class B mode. An advantage in this case is better RF power to dc power efficiency because the bias point currents are greatly reduced.

5.3.2.5 f_T Multiplier The f_T *multiplier*[5] consists of transistors arranged with their drains summed together to increase the ac output signal current [18,19]. The single-ended version is shown in Figure 5.16. With this arrangement, assuming both FETs to be of equal gate width and identical transconductances, the output current is doubled. FET1 acts as a source follower with the source resistor chosen to be $1/g_m$ so that the input signal v_{in} evenly splits between the two FETs. The effective input capacitance is halved, while the transconductance for the f_T multiplier itself is g_m. The output resistance Z_{out} is

$$Z_{out} = \frac{(2g_m r_{ds} + 1)}{2(2g_m r_{ds} + 1) - 1} r_{ds} \cong \frac{r_{ds}}{2} \tag{5.14}$$

as would be expected from the theory that Z_{out} is simply the output resistance of two FETs connected in parallel. A disadvantage of the f_T multiplier is the more complicated biasing scheme needed. At higher

[5] The f_T of an active device or cell is defined as the frequency for which the current gain becomes unity. For an FET f_T is calculated from $g_m/2\pi(C_{gs} + C_{gd})$.

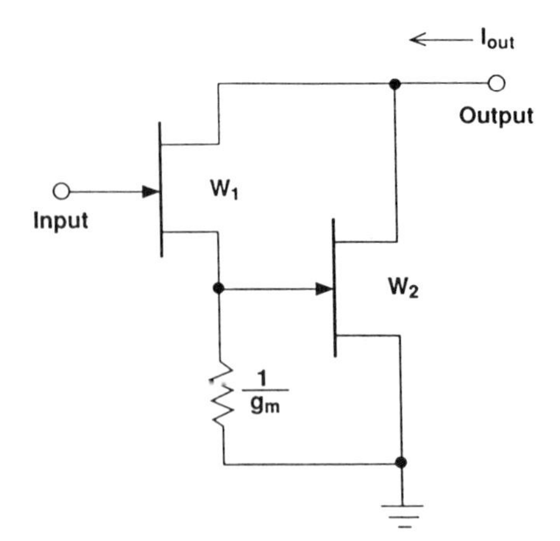

FIGURE 5.16 Single-ended f_T multiplier cell. Bias circuitry is not shown.

frequencies FET2 (designated as W_2) has phase lag problems that can result in distortion.

A practical implementation of the f_T multiplier appears in Figure 5.17 [20]. Here a differential cell uses the basic idea behind the f_T multiplier to increase the output signal current. The biasing of the differential implementation is less demanding than for the single-ended version of the f_T multiplier. The voltage gain for this cell is the same as the simple differential cell, but the useful bandwidth is approximately doubled (in practice it is found to increase about 1.7). An integrated circuit implementation of the differential f_T multiplier version (even a cascoded version) is practical because IC technologies favor FET-intensive circuits rather than circuits

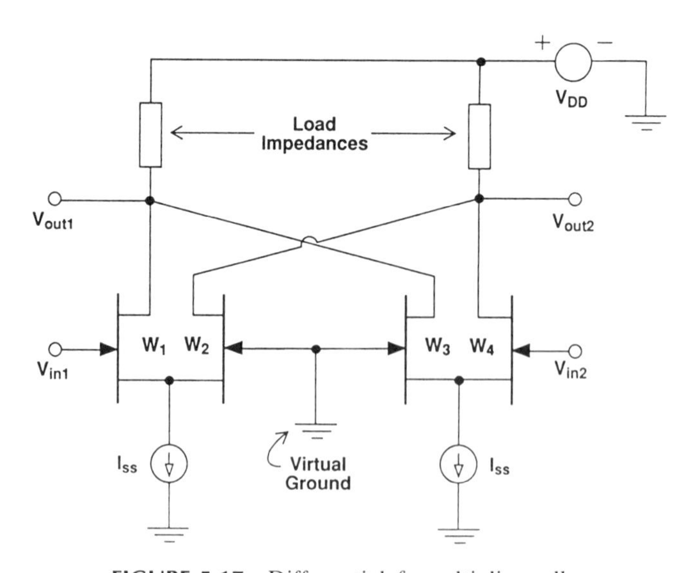

FIGURE 5.17 Differential f_T multiplier cell.

using mostly passive components (because passive components often consume more area than FETs).

5.4 ACTIVE LOADS

Several of the most commonly used gain cells were discussed in Section 5.3. One of the variables in IC design is the selection of the loading component used with a gain cell. FET active loads are described below. In addition, the active load is compared to standard passive loads.

5.4.1 Single-FET Active Load

The *single-FET active load* (SFAL) is formed by connecting the gate and source together so that the gate-to-source voltage is forced to zero (see Fig. 5.18*a* for schematic). With the drain-to-source voltage greater than the knee voltage V_{knee} required for saturation, the drain current I_{DSS} is approximately constant with drain voltage. Looking into the source of FET1 the resistance is r_{ds}; this is the load resistance (denoted by Z_{load}) of the active load.

The major design parameter of the active load is the FET's gate width. The load resistance varies inversely with the gate width. Figure 5.19 shows the load line of an active load superposed on the common-source FET's *I–V* characteristic. In this example the active load width is two-thirds of the width of the common-source FET and the supply voltage is $V_{\text{DD}} = 5$ V. Therefore, the width of the active load relative to the width of the amplifying FET limits the common-source FET's drain current. The load resistance corresponds to the slope of the SFAL's *I–V* characteristic as

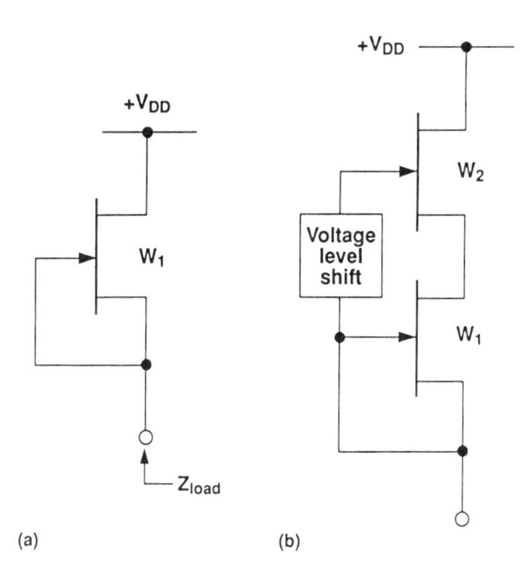

FIGURE 5.18 (*a*) Simple FET active load; (*b*) the FET cascode active load.

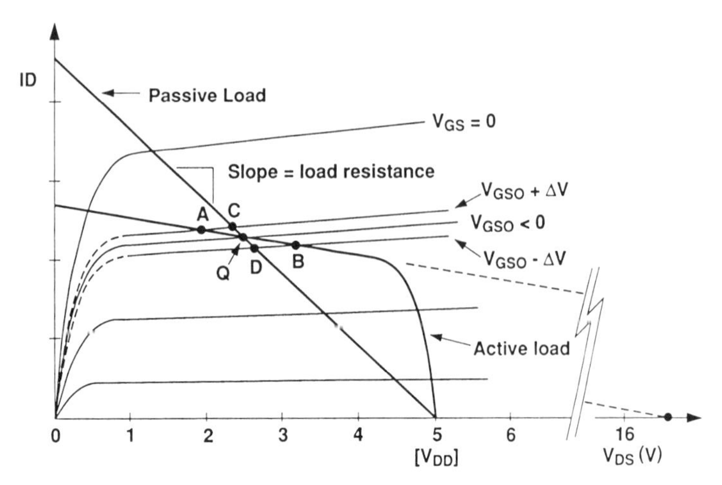

FIGURE 5.19 The load line of the single FET active load superimposed on the I–V characteristic of a common-source FET. For comparison a passive resistor load line is included at point Q.

indicated in Figure 5.19. Active loads have the advantage of providing large values of load resistance without the need for large supply voltages that would be required for a large-valued passive resistor. They are usually smaller in size than passive resistors, which has area and cost advantages in IC design. However, active loads are generally noisier than passive loads.

The active load sets the current for the amplifying cell. For example, a common-source FET gain cell operating at one-half of I_{DSS} requires an active-load FET with one-half the gate width. A wide range of currents is possible by scaling the width of the active load. The resistance of the active load varies inversely with the width; for example, halving the width doubles the resistance (r_{ds}). Of course, the common-source FET must have the appropriate gate-to-source voltage V_{GS} set to establish the required drain–source voltage V_{DS}.

5.4.2 Multiple-FET Active Load

If the load resistance attainable from a single-FET active load is not large enough, a *cascoded FET active load* (CFAL) can be constructed. Consider the cascoded active load shown in Figure 5.18*b*. If both FETs (W_1 and W_2) are identical, a voltage level shift is required to elevate the gate potential of FET2 so that FET1 remains in saturation during operation (see Section 5.6 for a discussion of level-shift networks). The load resistance becomes

$$Z_{load} = (g_m r_{ds} + 2) r_{ds} \tag{5.15}$$

which is approximately an order of magnitude larger compared to the

SFAL. The increase in Z_{load} is obtained by forcing FET2 to absorb most of the voltage swing across the CFAL.

It is possible to keep applying the cascode idea to raise the load resistance to still higher values. For example, for the *double-cascode FET active-load* cell

$$Z_{load} = [(g_m r_{ds} + 1)(g_m r_{ds} + 2) + 1] r_{ds} \tag{5.16}$$

another order of magnitude increase in Z_{load} is obtained (see Fig. 5.12 for the double-cascode arrangement). However, biasing becomes more complicated because three FETs are used and diminishing returns are soon reached beyond the double-cascode active load.

5.4.3 Comparing Active and Passive Loads

The simplest load is the passive resistor. An active load approximates the linear passive resistor over part of its operating range. Consider again Figure 5.19 showing the load lines of both a passive resistor and an active load positioned on the I–V characteristic of a common-source FET. In this example the active-load FET is two-thirds the width of the common-source FET; with the gate of the common-source FET set at V_{GS0} the bias point Q results from current continuity considerations.

The load line of the active load is linear over a smaller range compared to an ideal resistor. The largest voltage supportable across the active load is $(V_{DD} - V_{knee})$, where practically V_{knee} is approximately 1 V, before significant distortion sets in as a result of the active load's nonlinearity. If the common-source FET's gate voltage is varied between $V_{GSO} + \Delta v$ and $V_{GSO} - \Delta v$, the resulting voltage swing at the drain node is from point A to point B along the active load's load line. A pure resistive load line is also drawn through V_{DD} and point Q. For the same gate voltage swing the drain node's voltage swing would correspond to points C and D along the passive load line. For the values given in Figure 5.19, the voltage gain with the active load is greater by a factor of almost 4 compared with this passive load. If a linear resistor were chosen to have an identical gain and also pass through point Q, corresponding to the dashed line in Figure 5.19, the required value for V_{DD} would be 16.5 V. Hence, active loads allow higher voltage gain to be achieved with lower dc power dissipation. In comparison to passive resistors, active loads are not as linear (lead to greater distortion) and lead to greater degradation of the noise figure because they generate more noise than would a passive resistor.

All gain cells must drive load capacitance—the capacitive loading of the following stage or wiring (parasitic) capacitance. Generally, the drive resistance in combination with the load capacitance sets the upper frequency of operation. The greater voltage gain available with active loads has the penalty of smaller bandwidths. The 3-dB bandwidth for a simple RC load is given by $1/(2\pi RC_L)$, where R is the equivalent drive resistance through which the loading capacitance C_L must be charged.

Much work has gone into finding ways to extend the upper frequency of operation of amplifiers beyond the RC limit. One compensation technique known as *shunt peaking* is relatively easy to implement because it requires the addition of only a single inductive component. Rather than use a resistor as a load in parallel with the loading capacitance C_L, a series inductor is included with the resistor. The inductive reactance cancels in part the capacitive reactance over a band of frequencies; hence the gain is improved at the high end of the band. If the common source FET is modeled as shown in Figure 5.20*a*, the forward voltage transmittance A_v of the shunt-peaked stage is

$$A_v = -\frac{g_m}{C_L} \frac{s + R/L}{s^2 + (R/L)s + 1/LC_L} \tag{5.17}$$

where $s = j\omega$. This response function has the general form of two complex poles and a single zero—the pole-zero diagram is shown in Figure 5.20*b*. Defining the quantities

$$2\alpha = \frac{R}{L} \quad \text{and} \quad \frac{1}{LC_L} = 2\alpha\omega_H \equiv \omega_0^2 \tag{5.18}$$

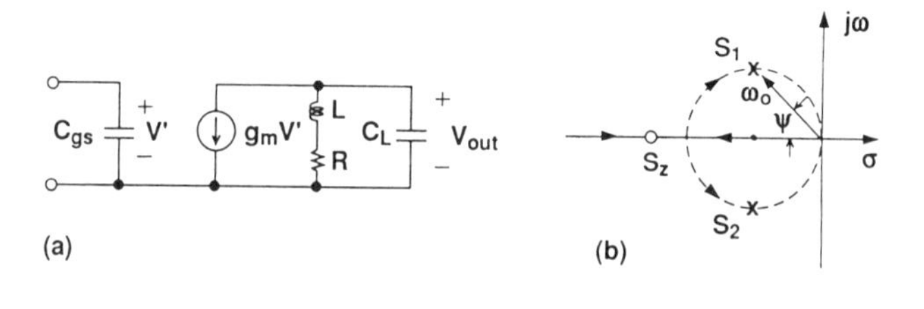

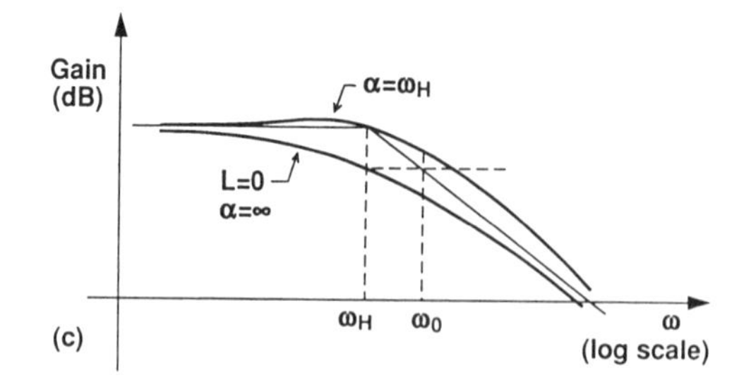

FIGURE 5.20 (*a*) Small-signal equivalent circuit of FET (g_m and C_{gs} elements only) connected to a shunt-peaked load with parasitic capacitance C_L; (*b*) pole-zero diagram for shunt-peaked case; (*c*) plot of gains (in decibels) versus radian frequency both with shunt peaking ($\alpha = \omega_H$) and without shunt peaking ($\alpha = \infty$).

allows Equation (5.17) to be written as

$$A_v = -\frac{g_m}{C_L}\frac{s + 2\alpha}{s^2 + 2\alpha s + 2\alpha\omega_H} \tag{5.19}$$

which factors into the form

$$A_v = -\frac{g_m}{C_L}\frac{(s - s_z)}{(s - s_1)(s - s_2)} \tag{5.20}$$

The complex poles are located in the complex s plane at

$$s_1, s_2 = -\alpha \pm j\sqrt{2\alpha\omega_H - \alpha^2} \tag{5.21}$$

and the zero is on the real axis at

$$s_z = -2\alpha = -\frac{R}{L} \tag{5.22}$$

When R and C_L are held constant, as L is varied (with $\alpha < 2\omega_H$, which is the shunt-peaked case) the poles moves on a circle of radius ω_H and center located at $-\omega_H$. If $\alpha = 2\omega_H$, the poles are coincident on the negative real axis. For values of α greater than $2\omega_H$, the poles are real (one pole asymptotically approaching $-\omega_H$ and the other going to infinity as $\alpha \to \infty$).

Figure 5.20*c* demonstrates the bandwidth improvement from shunt peaking with $\alpha = \omega_H = R/2L$. The case with no compensation ($L = 0$ and $\alpha = \infty$) is plotted for comparison—the half-power bandwidth is extended by almost a factor of 2 with shunt peaking (Fig. 5.20*c* illustrates this).

The shunt-peaked stage can be adjusted for either of two commonly used responses: (1) the *maximally flat magnitude* (MFM) characteristic or (2) the *maximally flat delay* (MFD) characteristic, which is sometimes called the most linear phase-shift characteristic. For the MFM case the pole angle $\psi = 39°$ and $Q_0 = \omega_0/2\alpha = \{\sqrt{L/C_L}\}/R = (\sqrt{2} - 1)^{1/2} = 0.645$. The angle ψ is defined from the origin as shown in Figure 5.20*b*. With the MFD choice the pole angle $\psi = 32°$ and $Q_0 = 0.59$.

Another technique is to use *series peaking*, in which an inductor is placed in series with the load capacitance. A pair of complex poles still result, but the zero is removed from the expression for A_v. The conditions for the MFM and MFD cases are changed with the removal of the zero. Using series peaking the MFM case (Butterworth) requires a pole angle $\psi = 45°$ and the MFD case (Thomson) requires $\psi = 30°$.

Practical spiral inductors can be fabricated in MMIC processes to realize inductances from 0.5 nH up to approximately 10 nH [21,22]. The semi-insulating property of GaAs allows for low loss transmission lines and moderate Q (e.g., 20–50) spiral inductors to be fabricated on GaAs wafers. Spiral inductors are widely used in MMIC designs for matching and bandwidth enhancement. However, one penalty associated with the use of spiral inductors is the large amount of (expensive) area consumed by these elements. For this reason some MMIC designers have attempted to find alternatives to the spiral inductors such as the synthetic or active inductor discussed next.

5.4.4 Synthetic Inductor Loads

In some cases an effective alternative to the passive spiral inductor is a *synthetic inductor*, which can be formed with a single FET and resistor combination. Small size is one of the primary advantages of the synthetic inductor. This configuration is shown in Figure 5.21 [23].

The impedance looking into the FET's source terminal is that of a resistor in series with an inductor. To see how an inductive reactance arises, refer to the simplified ac equivalent circuit shown in Figure 5.22. The gate-to-source capacitance plays the central role in establishing the inductive reactance. The signal v_s applied to the source terminal is divided between the resistor R_G and the gate-to-source capacitance C_{gs}. Voltage v' across C_{gs} is the control voltage for the dependent drain-to-source current generator of magnitude $g_m v'$. The phase of voltage v' lags that of the input voltage v_s. Hence, current $g_m v'$ must also have the same phase lag relative to v_s. By definition, when the current lags the voltage by 90°, the reactance is inductive.

For the reactance to be inductive $g_m R_G$ must be greater than unity. The reactance $X(\omega)$ is easily shown to be

$$X(\omega) = \frac{\omega(g_m R_G - 1)C_{gs} r_{ds}^2}{(g_m r_{ds} + 1)^2 + \omega^2 (R_G + r_{ds})^2 C_{gs}^2} \tag{5.23}$$

The behavior of $X(\omega)$ is such that it increases linearly with ω at low frequencies with an inductance value of $[(g_m R_G - 1)C_{gs} r_{ds}^2]/(g_m r_{ds} + 1)^2$.

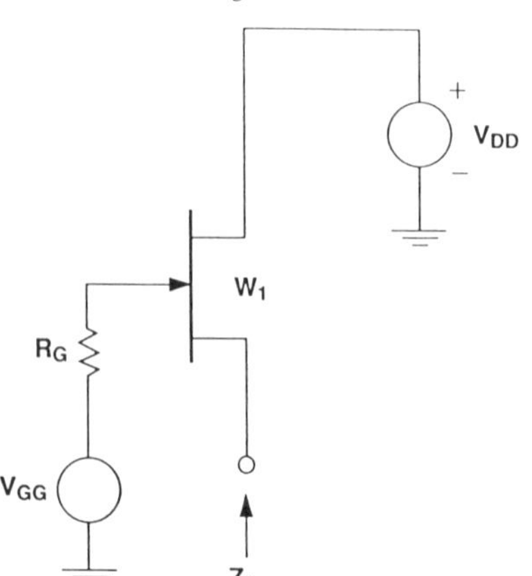

FIGURE 5.21 Single FET synthetic inductor configuration. The bias supply V_{GG} is an ac short circuit over the useful inductance range.

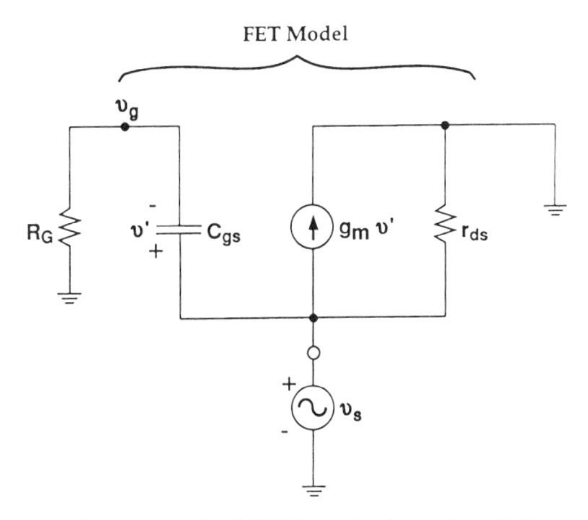

FIGURE 5.22 Simple three-terminal FET equivalent circuit for computation of the impedance looking into the source terminal of the FET.

The maximum value of $X(\omega)$ is reached at ω equal to $(g_m r_{ds} + 1)/(R_G + r_{ds})C_{gs}$; beyond this frequency $X(\omega)$ monotonically decreases.

If the drain-to-source resistance is assumed to be very large, Equation (5.23) can be simplified. The inductance $L(\omega)$ is [23]

$$L(\omega) = \frac{(g_m R_G - 1)C_{gs}}{g_m^2 + \omega^2 C_{gs}^2} \tag{5.24}$$

and the series resistance $R(\omega)$ is

$$R(\omega) = \frac{g_m + \omega^2 R_G C_{gs}^2}{g_m^2 + \omega^2 C_{gs}^2} \tag{5.25}$$

where $Z(\omega) = R(\omega) + j\omega L(\omega)$. Choosing larger values of R_G allows for larger values of inductance; however, the useful frequency range is lowered. Obviously the Q of the synthetic inductor is quite low.

Another active inductor is shown in Figure 5.23 consisting of two FETs in a cascode configuration along with a feedback resistor [24]. When both FETs

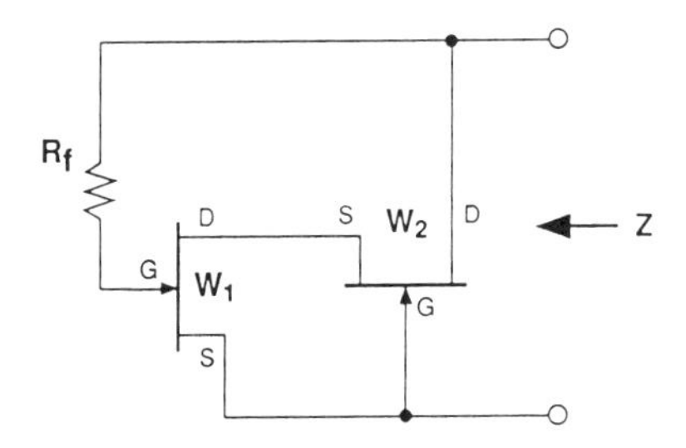

FIGURE 5.23 Active inductor formed with a cascoded FET pair and a feedback resistor R_f (after Hara et al. [24], © IEEE; reproduced by permission).

are identical and each FET is modeled only as a transconductance g_m and a gate-to-source capacitor C_{gs}, the impedance Z is

$$Z(\omega) = \frac{1 + j\omega C_{gs} R_f}{g_m + j\omega C_{gs}(\omega C_{gs}/g_m)} \tag{5.26}$$

The practical condition for inductive reactance is $g_m \gg \omega C_{gs}(\omega C_{gs}/g_m)^2$. This configuration is useful up to frequencies of approximately $f_T/2$.

5.5 CURRENT SOURCES AND CURRENT MIRRORS

The need often arises in analog design to set currents at predetermined levels. Although off-chip components can do this, there are many advantages to doing it on-chip (e.g., less system complexity, less overall expense generally possible with an on-chip solution, on-chip component matching, and temperature tracking). The discussion that follows focuses on using depletion-mode FETs as current sources and current mirrors.

5.5.1 Single-FET Current Source

In the saturated region of operation the drain current of an FET is approximately constant with drain-to-source voltage for a constant gate-to-source voltage (refer to Fig. 5.1). This behavior is the principle behind the FET current source. The single-FET current source is identical in nature to the single-FET active load presented in Section 5.4.1 (compare Fig. 5.24*a* with Fig. 5.18*a*). Generally FET current sources operate with zero gate-to-source voltage achieved by short-circuiting the gate and source terminals together. The width of the FET sets the current as long as the FET remains

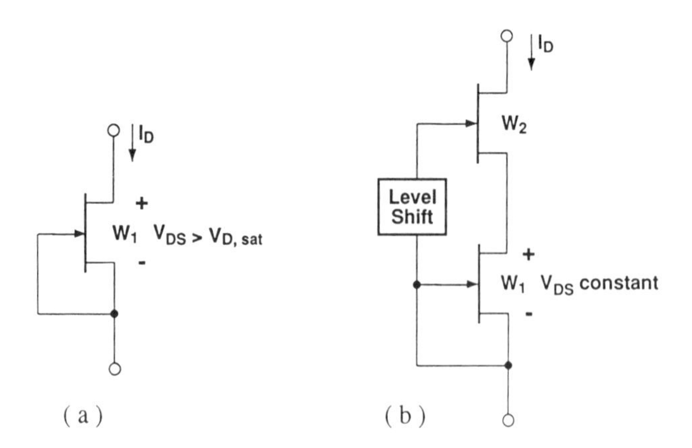

FIGURE 5.24 (*a*) Single-FET current source; (*b*) cascoded FET current source. Bias circuitry for level-shift network not shown.

in saturation. The resistance associated with the single-FET current source is r_{ds} (this would be infinite for an ideal current source). One advantage of the single-FET current source is its relatively small size. Unfortunately, second-order effects in GaAs ICs can cause problems in the single-FET current source.

5.5.2 Second-Order Effects in FET Current Sources

The finite output resistance r_{ds} of the single-FET current source means that the current will vary as the drain-to-source voltage is changed. Accurately modeling the current source using a single value for r_{ds} is not possible. One source of error arises from changes in temperature due to ambient change and/or FET internal heating. As the ambient-temperature changes we expect the current to change due to conductivity variations as discussed in Section 5.2.4 (see Fig. 5.8). Furthermore, as V_{DS} varies, the self-heating of the FET will result in changes in the channel temperature of the FET beyond ambient-temperature changes, leading to additional current shifts. AC thermal effects are present only at low frequencies because of long thermal time constants (perhaps of the order of a microsecond or longer). A partial solution to this problem is to prevent V_{DS} from changing across the current source FET (Section 5.5.3 discusses how to achieve this using the cascode configuration).

As also discussed in Section 5.2.4, backgating or sidegating results in unwanted current variations in the drain current of an FET (refer to Fig. 5.9 for example). Negative potentials applied to conductive regions positioned close to a current source can reduce the current below its design value (which is I_{DSS} when $V_{GS} = 0$). The problem is further complicated by the fact that backgating thresholds decrease (i.e., a smaller negative backgate voltage is required for the onset of backgating) as the substrate temperature decreases. Geometric layout is usually the key to controlling backgating by either designing larger spacings between the current source FET and negative-bias regions or placing fixed positive-potential conductive regions between them for shielding. Of course, some GaAs IC processes are more susceptible to backgating than others.

5.5.3 Cascoded Current Sources

Cascoded configurations have already been discussed in detail in Sections 5.3 and 5.4. Figure 5.24*b* shows a cascoded current source where FET2 is usually designed to be the same width as FET1. When the level-shift network maintains a constant voltage (e.g., typically 2 to 3 V), the drain-to-source voltage V_{DS} across FET1 remains fixed. Voltage changes across the cascoded current source are absorbed by FET2. The cascode configuration of Figure 5.24*b* raises the resistance of the current source to $(g_m r_{ds} + 2)r_{ds}$ from r_{ds}. This is approximately an order of magnitude higher than for a single-FET current source, although thermal and backgating effects can degrade it.

A doubly cascoded current source can be formed by adding still another FET and level-shift network. While higher-order cascoded structures are possible, they require higher supply voltages (i.e., higher dc power dissipation) and add complexity that is not warranted for many circuits.

A simplified implementation of the cascoded current source, which omits the level-shift network (i.e., both gate terminals are connected to the source of FET1), can be formed by choosing FET2 larger than FET1. Two requirements must be met for this implementation. First, the FET pinchoff voltage must be larger than the drain voltage required for the onset of saturation ($V_{DS} > V_{knee} \cong 1$ V). Second, FET2 must be sufficiently larger in gate width than FET1 such that the gate-to-source voltage of FET2 is greater than the drain voltage required for FET1 to be in saturation[6] (i.e., $|V_{GS}| > V_{knee}$). For example, suppose a current source is designed using the process represented by the FET characteristics in Figures 5.1 and 5.2. For this process the pinchoff voltage V_p is approximately -1.7 V (as defined by the dashed-line extrapolation in Figure 5.2) and V_{knee} is approximately 1 V. Assuming the drain current scales directly with gate width, choosing FET2 to be 4 times larger than FET1 requires V_{GS} of FET2 to be -1.45 V to maintain equal drain currents (determined from Fig. 5.2 with V_{GS} corresponding to $I_D = 0.25 \cdot I_{DSS}$). Although the requirement $|V_{GS}| > V_{knee} \cong 1$ V is met, the margin is quite small. The impedance of this form of cascoded current source is $(g_{m1} r_{ds1} + \frac{4}{3}) \cdot r_{ds1}$, where g_{m1} and r_{ds1} are FET1's parameters. Although this implementation of the current source is simpler, it generally does not reduce the area consumed and is less tolerant to process variations because FET1 is biased close to the onset of saturation.

5.5.4 Current Mirrors

Current mirrors are another valuable building block for analog circuits. Current mirrors fall into two categories—*positive* if they function as a current "sink" (quiescent output current is "positive") and *negative* if they function as a current "source" (quiescent output current is "negative"). Current mirrors can be *inverting* or *noninverting*. Not only can current mirrors be used as current sources; they can also be used as active loads to differential pairs and as current amplifiers.

A *positive current mirror* as described by Scheinberg [25] is shown in Figure 5.25*a*. Current I_1 is the input (or reference) current, and I_2 is the mirrored (output) current. The feedback path formed by source follower

[6] The drain-to-source voltage where the FET transitions from the ohmic region of operation to the saturated region of operation is sometimes called the *knee voltage*.

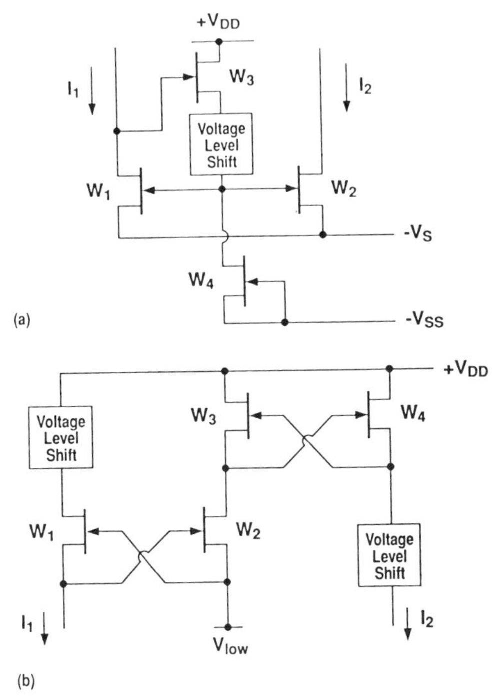

FIGURE 5.25 (*a*) A positive current mirror (after Scheinberg [25], © IEEE; reproduced by permission); (*b*) a negative current mirror (after Toumazou and Haigh [26], © IEEE; reproduced by permission).

FET3 and the level-shift network adjust the gate potential of FET1 to sink current I_1. Since the gates of FET1 and FET2 are tied together, the current in FET2 will be the same as in FET1 if they have equal gate widths (or scale by the ratio of the gate widths if unequal). All four FETs must be in saturation, and supply voltages V_s and V_{ss} are chosen to accomplish this. This current mirror is very fast, but its ac transfer ratio (I_2/I_1) peaks above unity near $f = f_T$ because it has a zero located at f_T. The output conductances of FET1 and FET2 can be reduced by using a cascode configuration.

An example of a *negative current mirror* [26] is shown in Figure 5.25*b*. Current I_1 is the input (or reference) current. The gate-to-source voltage of FET1 is inverted and becomes the gate-to-source voltage applied to FET2. The drain currents in FET3 and FET2 must be equal (usually FET2 and FET3 are identical in width). As with FET1 and FET2, the gate-to-source voltage of FET3 is inverted and applied to FET4. Hence, the mirrored current I_2 is equal to I_1 multiplied by the ratio of the width of FET4 to the width of FET1. The dc current transfer ratio is $\{1/[1 + (3/g_m r_{ds})]\}$ and the pole frequency $\omega = \frac{1}{2} g_m / C_{gs}$ (which is approximately $\frac{1}{2} f_T$). The magnitude of the dc transfer function can be brought closer to unity by cascoding techniques. Furthermore, additional FETs can be paralleled with FET4 if more than one current must mirror I_1.

5.6 VOLTAGE LEVEL-SHIFT NETWORKS

The level-shift function is important to many analog circuits because the threshold of the depletion-mode FET is negative, thereby necessitating adjusting dc levels in biasing direct-coupled, multi-FET circuits. The diode stack is by far the most commonly used approach to the level-shift function.

5.6.1 Schottky Barrier Diode Stacks

Schottky barrier diodes are readily available in IC processes as described in Chapter 2. The diode has an exponential forward I–V characteristic of the form

$$I_f = I_s\left(\exp\frac{qV_f}{kT} - 1\right) \tag{5.27}$$

where kT/q is the *thermal voltage* (0.026 V at $T = 25°C$) and I_s is the diode *saturation current* [27]. The exponential I–V relationship is useful for level shifting because the forward voltage is approximately constant with moderate changes in the current. For GaAs Schottky diodes forward voltages ranging from 0.65 to 0.85 V are common.

In circuits diodes are connected in series to develop the voltage shift desired. For example, a three-diode stack would be a logical choice for a 2-V shift. Manipulation of Equation (5.27) gives the forward voltage of the diode

$$V_f \cong \frac{kT}{q}\ln\frac{I_f}{I_s} \tag{5.28}$$

This is the voltage across the junction as established by the thermodynamics of the barrier. There is an additional contribution to the forward voltage from the parasitic series resistance associated with the structure of the diode. The series resistance is a function of current [28] for planar diode structures such as those used in GaAs ICs. However, for the level-shift application the diode stacks are operated at a fixed current. With constant current the series resistance can be taken as constant, denoted here by R_s^d, which simplifies the modeling of the diode. When the series resistance is included, the total forward voltage of the diode is

$$V_{total} = V_f + I_f R_s^d \tag{5.29}$$

The forward voltage of a solid-state diode is also temperature-dependent. When Equation (5.28) is examined, the forward voltage is clearly seen as a function of temperature—the thermal voltage is proportional to temperature T and the saturation current I_s varies with temperature in a complicated way [28]. The temperature coefficient of V_f is negative in sign. In addition, the series resistance R_s^d is temperature-dependent because semiconductor resistivity increases with temperature as a result of increased lattice scattering. For this reason it has a positive temperature coefficient. Although these temperature coefficients are opposite in sign, the negative temperature coefficient of the barrier usually dominates. Typical values of V_{total} range from -1 to -2 mV/°C. Low diode forward currents have temperature coefficients close to -2 mV/°C, while high diode forward currents yield smaller values closer to -1 mV/°C. The temperature behavior of the current source setting the diode stack's operating current is also important in determining the temperature stability of the diode stack.

5.6.2 Diode Level-Shift Circuits

In the design of a level-shift network the design variables includes diode geometry, the number of diodes, and the current flowing through the diode stack. Figure 5.26*a* shows a commonly used *step-down level shifter*. The function of this circuit is to pass the signal from input to output with the output level shifted negatively by an amount equal to the voltage across the stack of diodes D_1, D_2, and D_3. FET2 is a current source establishing the current flowing in the diodes (and the drain current of the source follower

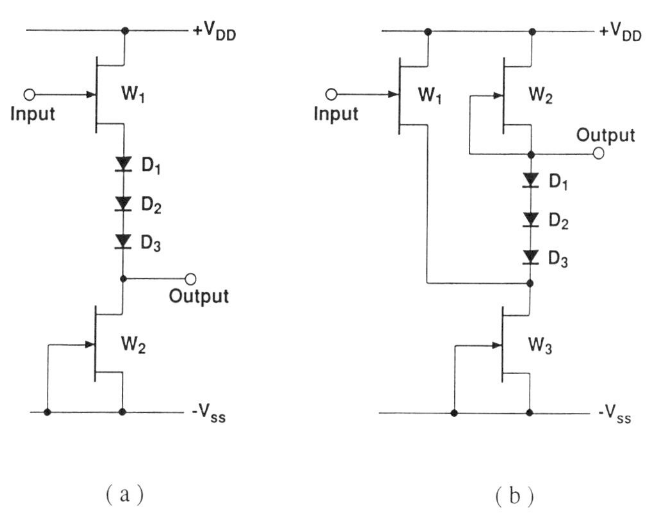

FIGURE 5.26 DC level-shift networks: (*a*) step-down level-shift network; (*b*) step-up level-shift network.

FET1). It is common practice to design both FET1 and FET2 to be identical in width; under these circumstances the nominal gate-to-source voltage of FET1 is zero. The gain of the circuit in Figure 5.26*a* is less than unity because of the series resistance of the diode stack and the gain of the source follower. The resistance of the diode stack ($3R_s^d$ for a triple stack) and the output resistance of the single-FET current source (r_{ds}) form a voltage divider. It is good design practice to minimize the resistance of the diode stack. From practical limitations in diode layout, typical gain values for this level-shift network are decreased by −0.25 to −0.65 dB (assuming three diodes). For example, the gain–attenuation of the level-shift network (Fig. 5.26*a*) is given by

$$G_{lsn} \cong \left(\frac{g_m r_{ds}}{g_m r_{ds} + 1}\right)\left(\frac{r_{ds}}{r_{ds} + R_{diode}}\right) \tag{5.30}$$

where R_{diode} is the total resistance of the diode stack. The first bracketed term on the right-hand side (RHS) is the source follower's gain (from Table 5.1). Assuming $g_m r_{ds} = 12$ and $R_{diode} = 0.05 r_{ds}$, $G_{lsn} = (0.923)(0.952) = 0.879$ [or -0.70 dB + (-0.42 dB) = -1.12 dB].

A variation of the step-down level shifter allows a *step-up level-shifter* to be constructed. This is shown schematically in Figure 5.26*b*. The diode stack current is set by FET2. Current source FET3 determines the current in FET1 and also sinks the current through the diode stack. FET1 is a source follower with its source connected to the bottom of the diode stack instead of the top. This allows the output to follow the input with the output level shifted positive by an amount equal to the voltage across diodes D_1, D_2, and D_3. As with the step-down level shifter, gain is less than unity.

5.7 OUTPUT–BUFFER STAGES

Special demands are placed on the output circuitry of an IC. For example, output stages often must drive large capacitive loads or meet stringent distortion requirements. They may be required to transform impedances or provide power matching to a load impedance. In this section three useful output stages are described—the common-source FET, the source follower, and the stacked push–pull.

5.7.1 Common-Source FET

The common-source FET (CSF) as a gain stage was discussed in Section 5.3.1.1. It features high gain, high input impedance, and a large output resistance. Furthermore, the CSF's output capacitance C_{ds} is small where C_{ds} is typically less than one-tenth of the input capacitance C_{gs}.

The CSF is well suited for amplifier output stages. It can perform the transition from a high internal node impedance to a low output impedance (such as 50 Ω). A wide range of output impedances can be achieved with the CSF output stage by varying the FET width simultaneously with any impedance loading the FET. For example, for a 1000-μm wide FET r_{ds} is approximately 100 Ω. If paralleled with a 100-Ω load resistor (denoted here by R_L) the equivalent output resistance becomes $R_L \parallel r_{ds} = 50\ \Omega$. For a 1000-μm FET, g_m is approximately 125 mS (following the FET example from Section 5.2), giving a voltage gain of $g_m(R_L \parallel r_{ds}) \cong 0.125 \times 50 = 6.25$ (or 16 dB). The I_{DSS} of a 1000-μm FET is about 200 mA, and its maximum drain current, denoted by $I_{D,max}$, is approximately 300 mA. The maximum deliverable RF output power from the CSF stage is $R_L(I_{D,max}/2\sqrt{2})^2 \cong 1$ W (rms) under the best of conditions. [*Note*: An often encountered case is a 50-Ω reverse termination driving a 50-Ω load giving a deliverable output power of approximately 0.5 Watt (rms).] If instead a 500-μm FET is selected, r_{ds} is doubled (i.e., 200 Ω), and R_L must be 66.67 Ω to form an equivalent output resistance of 50 Ω. This smaller FET gives a smaller voltage gain [i.e., $g_m[R_L \parallel r_{ds}] \cong 0.0625 \cdot 50 = 3.125$ (or 9.9 dB)]. In general, smaller FETs have lower voltage gain and less RF output power (assuming a fixed output resistance), but have the advantages of smaller dc power consumption, smaller IC layout area, and smaller input capacitance. Driving a large output stage FET from a high-impedance node reduces the bandwidth because of the large input capacitance C_{gs} of the FET (however, shunt or series peaking can recover some of the bandwidth shrinkage).

There are three primary FET parameters [29] that affect distortion in FET stages: transconductance g_m, output resistance r_{ds}, and input capacitance C_{gs}. For low-gain, single-stage CSF amplifier stages the transconductance nonlinearity usually dominates. Referring to Figure 5.3, it is clear that with large variations in gate-to-source voltage V_{GS} the transconductance g_m will vary considerably over a complete cycle. Hence, the nonlinearity in g_m produces distortion in the FET's output current. However, for high-gain CSF amplifiers, the input swing is smaller and nonlinearity in r_{ds} becomes more important because the output signal amplitude is greater than the input signal amplitude. The output resistance r_{ds} is both a function of V_{GS} (r_{ds} increases with V_{GS} more negative) and V_{DS} (increases as V_{DS} increases in the saturated region of operation). Strong distortion results when the FET swings out of the saturated region of operation (i.e., into cutoff or breakdown).

5.7.2 Source Followers

The *source follower* uses an FET in the common-drain configuration. It is often used as an output driver when a low output impedance is required for driving either low-impedance loads or high-capacitance loads. The input capacitance of the source follower is smaller than C_{gs} because the source potential tracks the gate potential. The input capacitance C_{in} is given by

$$C_{in} \cong C_{gd} + C_{gs}(1 - A_v) \tag{5.31}$$

where A_v is the voltage gain of the source follower and is slightly less than unity. Hence, C_{in} is usually much smaller than C_{gs} alone, so a source follower will present less loading to the circuit driving it. The source follower's excellent isolation property is probably the most important reason why it is so widely used as an output stage.

A commonly used implementation of the source follower is shown in Figure 5.27. FET1 is the follower and FET2 is the current source (both FETs are typically the same width by design). Identical devices are required to avoid a deliberate offset voltage between output and input, although device mismatches from process fluctuations produce small unwanted offsets. The ideal voltage gain of the source follower is unity—for this to hold the FET's output resistance must be infinite. With finite output resistance, the voltage gain is

$$G_v = \frac{g_m r_{ds}}{g_m r_{ds} + 2} < 1 \tag{5.32}$$

With $g_m r_{ds}$ typically being 10–15 for GaAs FETs, values for G_v range from 0.83 (−1.6 dB) to 0.88 (−1.1 dB).

With the simple configuration in Figure 5.27, the output voltage V_{out} can effectively swing over the range, $-V_{SS} + V_{knee} < V_{out} < V_{DD} - V_{knee}$, where the knee voltage V_{knee} is the voltage corresponding to the onset of saturation ($V_{knee} \cong 1$ V). The output resistance of this connection is

$$R_{out} = \frac{r_{ds}}{g_m r_{ds} + 2} \tag{5.33}$$

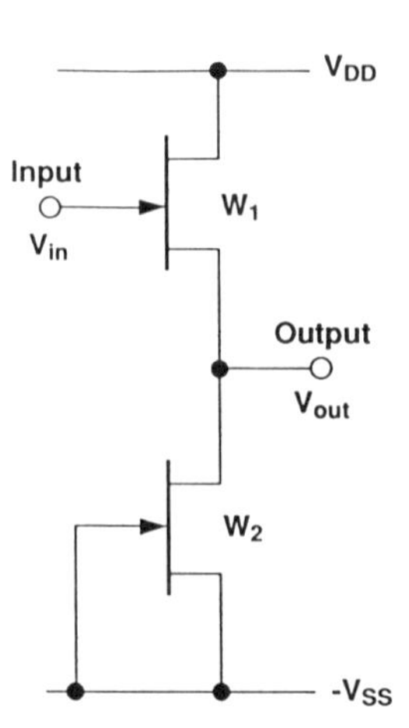

FIGURE 5.27 Source follower output–buffer stage with current source FET2 setting source follower's quiescent current.

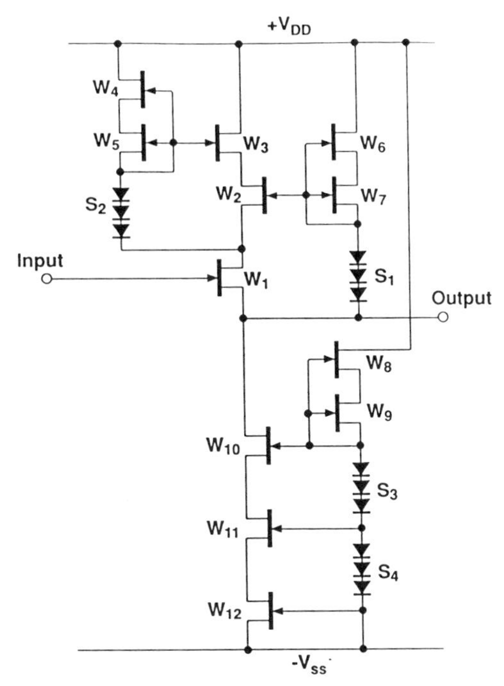

FIGURE 5.28 Doubly cascoded source follower with all bias circuitry shown (after Poulton et al. [30], © IEEE; reproduced by permission).

An important cause of distortion in the source follower comes from slew-rate-limiting [29]. For example, if the source is not able to follow a fast increasing input (gate) signal, large gate-to-source voltages can develop across the input of the FET1, thereby forward-biasing the gate junction.[7] A forward-biased gate junction results in a low input impedance, causing distortion in the output waveform. The forward biasing of the gate junction compromises the source follower's ability to isolate a high-impedance internal node from the output node.

In some applications a voltage gain of 0.8–0.9 is not adequate. The cascode arrangement can be used to come closer to a unity transfer ratio. Consider the *doubly cascoded source follower* [30] shown in Figure 5.28. This source follower was designed to operate with a 6-bit sample-and-hold GaAs circuit; hence accurate replication of the signal was of great importance. FET1 is the source follower, and FET12 is the primary current source. To keep the drain-to-source voltage constant across FET1, its drain is "bootstrapped" by FET2. That is, FET1's drain potential is forced to track its source potential but is voltage-shifted up by the voltage across

[7] A disadvantage of the Schottky barrier FET is that under forward bias the junction conducts appreciable current (i.e., low impedance), whereas the metal–oxide–semiconductor junction of a MOSFET does not.

diode stack S1. FET2 is, in turn, bootstrapped by FET3. Note that the gate of FET3 tracks the drain potential of FET1 rather than the output node. This improves the high-frequency performance by lowering the parasitics between the output node and supply V_{DD}.

Current source FET12 is doubly cascoded (i.e., FET10 and FET11 perform the cascode function). Diode stacks S3 and S4 (consisting of three diodes each with about 2 V across each stack) provide voltage translations to FET10 and FET11. The diode stacks are biased using the cascoded current source constructed with FET8 and FET9 (the width of FET8 is larger than the width of FET9 by typically a factor of 3). FET4 through FET7 also serve to bias diode stacks in an analogous manner. With both FET1 and FET12 doubly cascoded, the source follower's output resistance is greatly improved; therefore by Equation (5.32) it is obvious that the voltage gain will be much closer to unity. However, imperfect substrate isolation, backgating, and output conductance dispersion [30] prevent realization of the full benefit of multiple cascoding.

5.7.3 Stacked Push–Pull

The *stacked push–pull* output–buffer stage [31] described below has the advantages of providing voltage gain, produces a broadband match to a constant resistance, and can partially cancel second harmonic distortion. The cell consists of two FETs as shown schematically in Figure 5.29. FET1 is an inverting amplifier, and FET2 is a source follower; together they form the push–pull structure. These FETs must be driven out of phase in order to sum their respective output signals at the output node. FET3 in Figure 5.29 serves to invert the phase and establish the magnitude of the drive signal v_2 to FET2.

There are two considerations in choosing the sizes of FET1 and FET2. First, FET size determines the slew rate available to drive the load impedance. Current $I_{D,max}$ is the maximum current an FET can source or sink and is directly proportional to the width of the FET (each GaAs IC

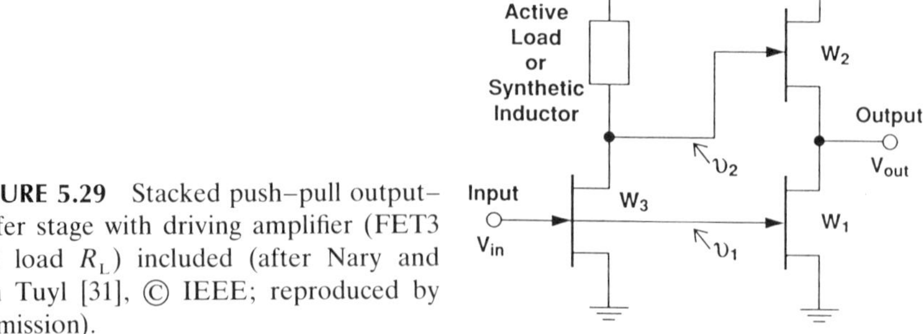

FIGURE 5.29 Stacked push–pull output–buffer stage with driving amplifier (FET3 and load R_L) included (after Nary and Van Tuyl [31], © IEEE; reproduced by permission).

process has a characteristic $I_{D,max}$ per unit width). Second, the sizes of FET1 and FET2 determine the output resistance R_{out}; R_{out} is the parallel combination of the resistance looking into the source node of the common-drain FET and the drain node of the common-source FET (Table 5.1 conveniently gives the resistance expressions). The output resistance R_{out} is

$$R_{out} = \frac{r_{ds1} \cdot r_{ds2}}{g_{m2} r_{ds1} r_{ds2} + r_{ds1} + r_{ds2}} \tag{5.34}$$

where the numeric parts of the subscripts correspond to FET1 and FET2, respectively. It is convenient to choose FET1 and FET2 to be the same size, although this is not necessary. If identical FETs are selected, using parameters from the process used to illustrate Section 5.2, 150-μm FETs ($g_{m1} = g_{m2} = 0.0165$ S, $r_{ds1} = r_{ds2} = 725\ \Omega$, and $g_m r_{ds} = 12$) give an R_{out} of approximately 52 Ω (which matches well to an external 50-Ω load $R_L = 50\ \Omega$).

The signal v_2 to FET2 should be larger in amplitude than the signal v_{in} to FET1. An inverting amplifier is required to drive the source follower FET2 for proper operation. Let the voltage gain of the driver stage (FET3 and its load) be A_i as indicated in Figure 5.30. While the voltage gain of the source follower is less than unity, the inverting amplifier driving the source follower added to the inverting amplifier of the push–pull cell has an overall voltage gain of

$$G_{pp} = \frac{-(g_{m1} + A_i g_{m2}) R_{eq}}{g_{m2} R_{eq} + 1} \tag{5.35}$$

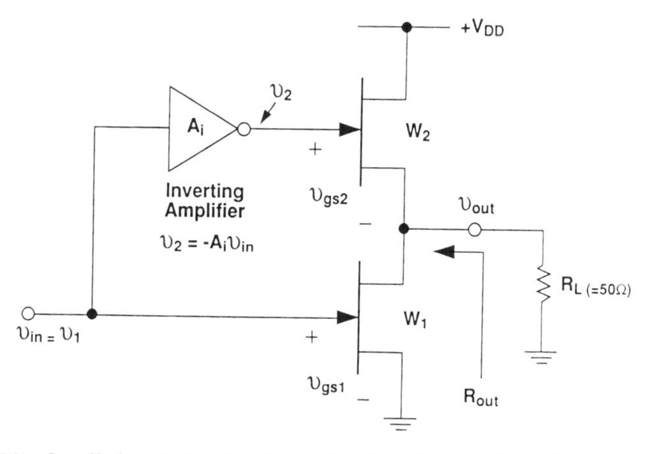

FIGURE 5.30 Small-signal circuit schematic of stacked push–pull output–buffer for computation of output resistance and voltage gain.

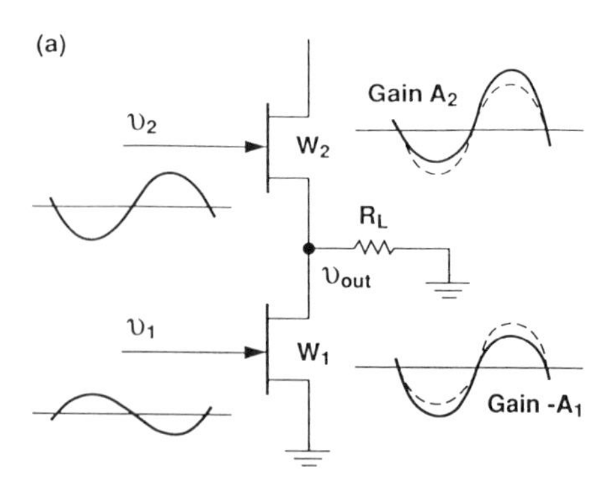

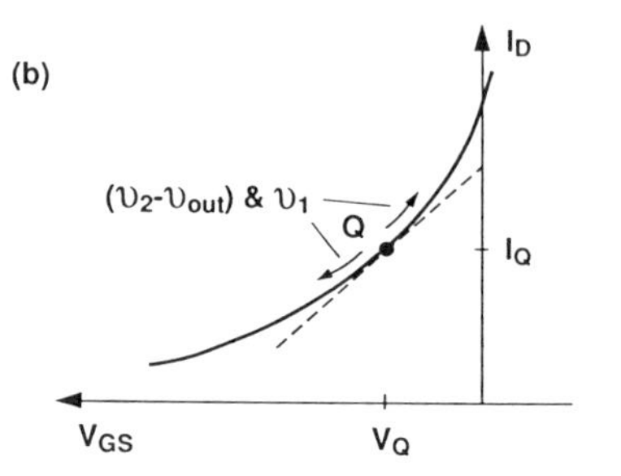

FIGURE 5.31 (*a*) Output waveforms for FET1 and FET2 assuming two different transfer curves [dashed waveform is for constant g_m and solid waveform assumes transfer curve shown in part (*b*)]; (*b*) FET transfer curve showing relation between V_{GS} and I_D used to construct part (*a*).

where the equivalent resistance R_{eq} is formed by the parallel combination

$$\frac{1}{R_{eq}} = \frac{1}{r_{ds1}} + \frac{1}{r_{ds2}} + \frac{1}{R_L} \tag{5.36}$$

Note that the voltage gain G_{pp} is inverting. In general, gain A_i can be chosen to fall within a range of values, say, from 2 to 3, which are practical design values. Other requirements, such as minimizing the second harmonic distortion as discussed below, can set the value for gain A_i.

The principle behind minimizing second harmonic distortion can be explained by using Figure 5.31. The key assumption in the argument below is that the load resistance R_L at the output node of the push–pull stage is the dominant impedance (i.e., $R_L \ll r_{ds}$ and $R_L \ll 1/g_m$). This is equivalent to assuming that the transconductance nonlinearity dominates because the load impedance is linear.[8] In Figure 5.30*a* the output waveforms of the source follower and the inverting amplifier are shown separately. If the transconductance nonlinearities of both FETs are identical, the second harmonic signals from both FETs will cancel and the fundamental signals will add. This can be seen in Figure 5.31 by examining the output signals of both

[8] The drain of FET1 looks into the source node of FET2 and is presented with a resistance of $1/g_{m2}$ that is nonlinear. Unless R_L is much smaller than $1/g_{m2}$, the second harmonic distortion from FET1 is accentuated and will not fully cancel that of FET2. This is a major reason why the stacked push–pull stage in practice only partially cancels even-order distortion products.

FETs. The output waveforms represent two different cases. The dashed output waveforms assume a constant transconductance corresponding to the dashed I_D–V_{GS} transfer curve in Figure 5.31*b* with no distortion. For comparison, the solid output signal waveforms include the transconductance nonlinearity corresponding to the solid I_D–V_{GS} transfer curve shown in Figure 5.31*b*. The difference between the solid and dashed signal lines is the even order distortion. When both output signals are added at the output node the distortion cancels. The condition for cancellation is

$$\frac{v_2}{v_1} = -\frac{1 + A_1}{1 - A_2} = A_i \tag{5.37}$$

where A_1 is the inverting voltage gain of FET1 and A_2 is the voltage gain of the source follower. With a linear output impedance R_L, voltage gain A_1 is $-[g_{m1}R_L/(g_{m2}R_L + 1)]$ and voltage gain A_2 is $[g_{m2}R_L/(g_{m2}R_L + 1)]$, and the inverting voltage gain A_i is

$$A_i = -(g_{m1} + g_{m2})R_L - 1 \tag{5.38}$$

Continuing the example calculation using 150-μm FETs, $|A_i|$ is found to be 2.45 (or 7.78 dB). When this value for A_i is substituted into Equation (5.35), $|G_{pp}|$ is found to be 1.45 (or 3.23 dB).

The preceding analysis is valid only at low to midrange frequencies. At higher frequencies voltage v_2 experiences excess phase delay relative to v_1 ($= v_{in}$) because of the delay introduced from the inverting driver stage (FET3 with gain A_i). This results in an additional source of distortion for this output–buffer stage. Other sources of distortion result from nonlinearity in the output impedance of the push–pull stage (i.e., r_{ds} is not linear and $1/g_{m2}$ is often important) and transfer curve mismatches between FET1 and FET2.

REFERENCES

1. R. S. Pengelly, *Microwave Field-Effect Transistors—Theory, Design and Applications*, 2nd ed., Wiley, New York, 1986.
2. R. L. Van Tuyl, V. Kumar, D. C. D'Avanzo, T. W. Taylor, V. E. Peterson, D. P. Hornbuckle, R. A. Fisher, and D. B. Estreich, "A Manufacturing Process for Analog and Digital Gallium Arsenide Integrated Circuits," *IEEE Trans. Microwave Theory Tech.*, **MTT-30** (7), July 1982, pp. 935–942.
3. R. A. Pucel, H. A. Haus, and H. Statz, "Signal and Noise Properties of Gallium Arsenide Field-Effect Transistors," in *Advances in Electronics and Electron Physics*, Academic Press, New York, 1975, Vol. 38.
4. Y. Tsividis, *Operation and Modeling of the MOS Transistor*, McGraw-Hill, New York, 1987.

5. C. Kocot and C.A. Stolte, "Backgating in GaAs MESFETs," *IEEE Trans. Electron Devices*, **ED-29** (7), 1059–1064 (July 1982).
6. M. Rocchi, "Status of the Surface and Bulk Parasitic Effects Limiting the Performance of GaAs IC's," *Physica*, **128B**, 119–138 (1985).
7. D. B. Estreich, "A Monolithic Wide-Band GaAs IC Amplifier," *IEEE J. Solid-State Circuits*, **SC-17** (6), 1166–1173 (Dec. 1982).
8. K. B. Niclas, "Active Matching with Common-Gate MESFET's," *IEEE Trans. Microwave Theory Tech.*, **MTT-33** (6), 492–499 (June 1985).
9. H. Wallman, A. B. MacNee, and C. P. Gadsden, "A Low-Noise Amplifier," *Proc. IRE*, **36**, 700–708 (June 1948).
10. R. S. Pengelly, *Microwave Field-Effect Transistors—Theory, Design and Applications*, 2nd ed., Wiley, New York, 1986, pp. 51–64.
11. J. Orr, "A Stable 2-26.5 GHz Two-Stage Dual-Gate Distributed MMIC Amplifier," *IEEE Microwave Millimeter-wave Monolithic Circuits Symposium Digest of Papers*, June 1986, pp. 19–22.
12. E. M. Chase and W. Kennan, "A Power Distributed Amplifier Using Constant-R Networks," *IEEE Microwave Millimeter-wave Monolithic Circuits Symposium Digest of Papers*, June 1986, pp. 13–17.
13. R. D. Middlebrook, *Differential Amplifiers*, Wiley, New York, 1963.
14. E. J. Angelo, Jr., *Electronic Circuits*, 2nd ed., McGraw-Hill, New York, 1964.
15. S. A. Jamison, A. Podell, M. Helix, P. Ng. C. Chao, G. E. Webber, and R. Loken, "Inductively Coupled Push-Pull Amplifiers for Low Cost Monolithic Microwave ICs," *IEEE Gallium Arsenide Integrated Circuit Symposium Technical Digest*, October 1982, pp. 91–93.
16. V. Sokolov, R. E. Williams, and D. W. Shaw, "X-Band Monolithic GaAs Push-Pull Amplifiers," *1979 IEEE International Solid-State Circuits Conference Digest of Technical Papers*, February 1979, pp. 118–119.
17. A. Podell, "MMIC-Based Subsystem Case Study: The TVRO Chip," in *Monolithic Microwave Integrated Circuits: Technology & Design*, R. Goyal, ed., Artech House, Norwood, MA, 1989, Chapter 7.
18. C. R. Battjes, "Monolithic Wideband Amplifier," U.S. Patent 4,236,119 (Nov. 25, 1980).
19. B. W. H. Lai, "Amplification Beyond f_T Using Non-Inductive Circuit Techniques," Master of Science thesis, University of California, Berkeley, CA, May 1983.
20. C. R. Battjes, "Amplifier Circuit," U.S. Patent 3,633,120 (Jan. 4, 1972).
21. J. M. Schellenberg and T. R. Apel, "Monolithic Microwave Integrated Circuit Design," in *GaAs Integrated Circuits*, J. Mun, ed., Macmillan, New York, 1988, pp. 223–229.
22. R. A. Pucel, "Design Considerations for Monolithic Microwave Circuits," *IEEE Trans. Microwave Theory Tech.*, **MTT-29** (6), 513–534 (June 1981).
23. I. E. Ho and R. L. Van Tuyl, "Inductorless Monolithic Microwave Amplifiers with Directly Cascaded Cells," *1990 IEEE MTT-S International Microwave Symposium Digest*, May 1990, Vol. 1, pp. 515–518.
24. S. Hara, T. Tokumitsu, T. Tanaka, and M. Aikawa, "Broad-Band Monolithic Active Inductor and Its Application to Miniaturized Wide-Band Amplifiers," *IEEE Trans. Microwave Theory Tech.*, **MTT-36** (12), 1920–1924 (Dec. 1988).

25. N. Scheinberg, "Designing High Speed Operational Amplifiers with GaAs MESFETs," *1987 IEEE International Symposium on Circuits and Systems*, Philadelphia, May 4–7, 1987, Vol. 1, pp. 193–198.
26. C. Toumazou and D. G. Haigh, "High Frequency Gallium Arsenide Current Mirror," *Electron. Lett.*, **26** (21), 1802–1804 (Oct. 1990).
27. R. S. Muller and T. I. Kamins, *Device Electronics for Integrated Circuits*, 2nd ed., Wiley, New York, 1986, Chapter 3, pp. 125–166.
28. D. B. Estreich, "A Simulation Model for Schottky Diodes in GaAs Integrated Circuits," *IEEE Trans. Computer-Aided Design of Integrated Circuits & Systems*, **CAD-2** (2), 106–111 (April 1983).
29. D. P. Hornbuckle and R. L. Van Tuyl, "Monolithic GaAs Direct-Coupled Amplifiers," *IEEE Trans. Electron Devices*, **ED-28** (2), 175–182 (Feb. 1981).
30. K. Poulton, J. J. Corcoran, and T. Hornak, "A 1-GHz 6-bit ADC System," *IEEE J. Solid-State Circuits*, **SC-22** (6), 962–969 (Dec. 1987).
31. K. R. Nary and R. L. Van Tuyl, "An MMIC Amplifier for Automatic Level Control Applications," *IEEE 1990 Microwave & Millimeter-Wave Monolithic Circuits Symposium Digest of Papers*, Dallas, TX, May 7–8, 1990, pp. 73–76.

CHAPTER SIX

Wideband Amplifiers

DONALD ESTREICH
Microwave Technology Division
Hewlett-Packard Company
1412 Fountaingrove Parkway
Santa Rosa, CA 95403

6.1 INTRODUCTION

Wideband amplifiers, both lowpass and bandpass, are an important class of circuits for electronic systems. Several representative GaAs IC amplifiers are described herein illustrating the application of many of the FET building blocks discussed in Chapter 5. Section 6.2 begins the discussion with general principles and limitations governing wideband amplifier design. The topics include gain–bandwidth product, impedance-matching limitations, noise figure, and distortion in FETs. Also included in Section 6.2 is a brief survey of broadbanding techniques, many of which are applied in the following sections of this chapter.

The amplifiers discussed in Sections 6.3 and 6.4 are mostly of the Van Tuyl–Hornbuckle type [1–2]: voltage-gain amplification blocks working in high-impedance environments. The Van Tuyl–Hornbuckle-type amplifiers have greatly influenced the design of many wideband amplifiers since their introduction in and around 1980. Examples of more recent designs advancing the state of the art in the evolutionary chain are described in Sections 6.3.3 and 6.7 (transimpedance amplifiers). Two common types of amplifiers are *not* discussed in this chapter: the reflective or reactively

High-Frequency Analog Integrated-Circuit Design, Edited by Ravender Goyal
ISBN 0-471-53043-3

matched amplifier [3–5] and the distributed or traveling-wave amplifier [6–8]. Both types of amplifiers have been discussed extensively throughout the microwave amplifier literature and are not included in this chapter for that reason. Finally, Section 6.5 addresses the topic of gain-control amplifiers and Section 6.6 briefly describes the phase-splitting amplifier.

6.2 DESIGN CONSIDERATIONS

Important constraints and limitations pertaining to wideband amplifier design are briefly reviewed in this section. Our discussion begins with the gain–bandwidth product of the FET. The discussion next turns to impedance-matching constraints and broadbanding techniques. FET limitations due to noise, junction breakdown, and distortion are discussed before proceeding to amplifier design.

6.2.1. Amplifier Gain–Bandwidth Product

The active device in an amplifier is dominant in establishing the amplifier's response. A widely used figure of merit for high-frequency transistors is the *current gain–bandwidth product* or *transition frequency*. It is designated by the symbol ω_T ($=2\pi f_T$) and has units of frequency (radians per second) because current gain is a unitless ratio. Consider Figure 6.1*a*, where a

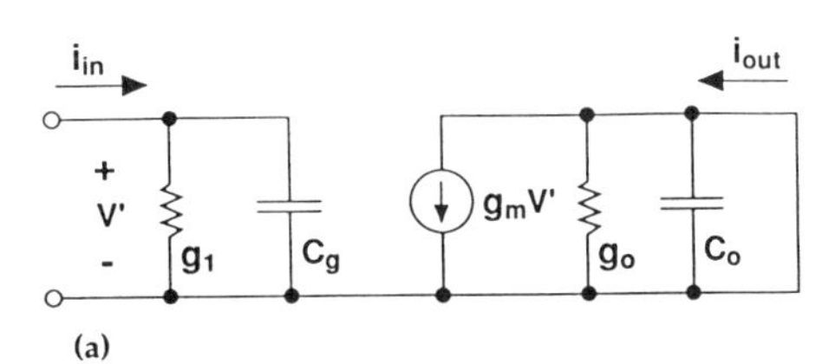

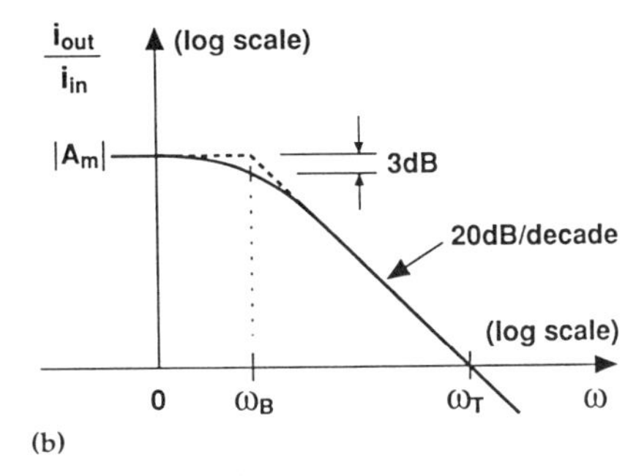

FIGURE 6.1 (*a*) Small-signal unilateral equivalent circuit of an active device such as an FET and (*b*) Bode plot of its frequency response.

small-signal unilateral equivalent circuit for an active device (e.g., GaAs FET) is shown with its output short-circuited. For this equivalent circuit[1] the current gain is

$$\frac{i_{\text{out}}}{i_{\text{in}}} = \frac{g_{\text{m}}}{g_1 + sC_{\text{g}}} \tag{6.1}$$

where $s = j\omega$ as usual. The *short-circuit forward current gain* is one of the *h parameters* and is often designated by the symbol h_{fs} (common-source connection for the three-terminal FET). At frequencies below ω_{B}, the current gain is a constant $|A_{\text{m}}|$. Above ω_{B} the current gain monotonically decreases at 20 dB/decade (=6 dB/octave). This appears as a straight line on a Bode plot (logarithmic magnitude and frequency) as drawn in Figure 6.1*b*. Extrapolating a straight line from $\omega = \omega_{\text{B}}$, where ω_{B} is the break frequency corresponding to $g_1 = |\omega_{\text{B}} C_{\text{g}}|$, yields the *gain–bandwidth product* (GBP)

$$\text{GBP} = \omega_{\text{B}} \cdot |A_{\text{m}}| = \frac{g_{\text{m}}}{C_{\text{g}}} \tag{6.2}$$

The frequency at which the current gain becomes unity, namely ω_{T}, is equal to the ratio

$$\omega_{\text{T}} = \frac{g_{\text{m}}}{C_{\text{g}}} \tag{6.3}$$

Hence, the intrinsic GBP of the active device is given by the intrinsic transconductance divided by its input capacitance. For a GaAs FET C_{g} is given by the sum of the gate-to-source capacitance and gate-to-drain capacitance ($C_{\text{g}} = C_{\text{gs}} + C_{\text{gd}}$) when the output is short-circuited.

Other active-device gain parameters exhibit similar behavior. Consider Figure 6.2 showing several gain parameters for a 1-μm × 500-μm ion-implanted GaAs MESFET. Each gain parameter [the common-source forward current gain, h_{fs}; forward voltage gain into a 50-Ω termination, $|S_{21}|^2$; maximum unilateral transducer power gain, $G_{\text{TU,max}}$; unilateral power gain, U-function; maximum available power gain (MAG)] decreases with increasing frequency (at −6 dB/octave except for MSG, which is −3 dB/octave [9]).

More generally, suppose the circuit in Figure 6.3 represents two cascaded active devices where a resistive interstage network is used between input

[1] Most active devices have a feedback capacitor in their equivalent circuit, which is often transformed using Miller's theorem. The feedback capacitor is then accounted for with shunt capacitors at input and output, with the input shunt capacitance usually the dominant element if the device has appreciable gain.

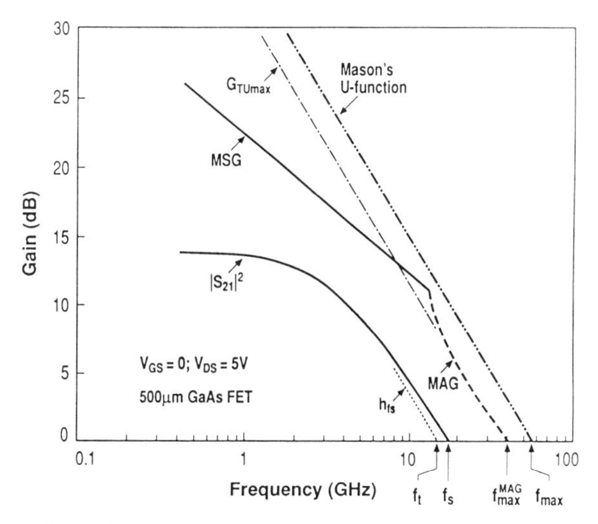

FIGURE 6.2 Several commonly used gain parameters versus frequency associated with active devices. The data is for a 1-μm × 500-μm ion-implanted GaAs FET. Note that all fall in magnitude with increasing frequency.

and output. For this case of *resistive broadening* the gain–bandwidth product is given by

$$\text{GBP} = \frac{g_m}{C_t} \tag{6.4}$$

where the voltage gain is given by

$$A_v(s) = \frac{-g_m R(1/RC_t)}{s + 1/RC_t} \tag{6.5}$$

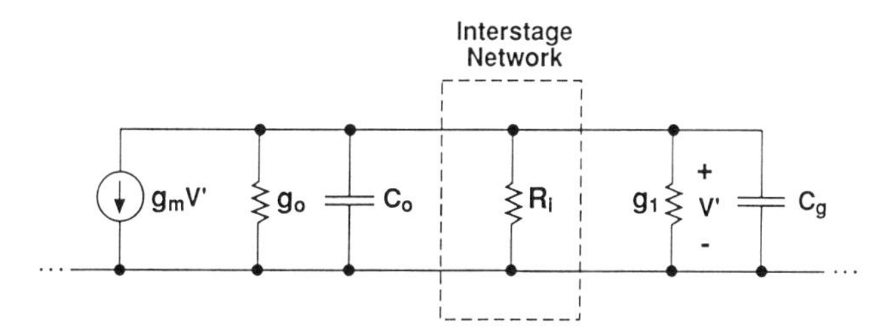

FIGURE 6.3 Resistive interstage network (shunt resistance R_i) shown for two identical cascaded active devices (equivalent circuit corresponds to that shown in Fig. 6.1). Sometimes this is called *resistive matching*.

the parallel combination of resistances is

$$R = \left(g_1 + g_0 + \frac{1}{R_i} \right)^{-1} \tag{6.6}$$

$$C_t = C_1 + C_0 \tag{6.7}$$

In Equation (6.6) R_i is the interstage load resistance; often R_i is much smaller than r_0 $(=1/g_0)$ or $1/g_1$, in which case, $R \cong R_i$. Reducing the magnitude of the interstage resistor R_i increases the −3-dB bandwidth, but lowers the voltage gain. Of course, the GBP remains unchanged. Equation (6.4) is true only for circuits consisting of a simple combination of R and C values.

When n identical single-pole gain stages are cascaded, the overall voltage gain is given by

$$A_V(s) = \frac{A_0^n (s_0)^n}{(s + s_0)^n} \tag{6.8}$$

where $A_0 = -g_m R$ and the pole $s_0 = (1/RC_t)$. The −3 dB bandwidth is given by

$$\omega_{3db} = s_0 \sqrt{2^{1/n} - 1} \cong \frac{s_0}{1.2\sqrt{n}} \tag{6.9}$$

The term multiplying s_0 in Equation (6.9) is the *shrinkage factor* for n identical cascaded stages. For a multistage amplifier the appropriate figure of merit is the *mean gain–bandwidth product* [10] as defined by

$$\text{MGBP} = \frac{g_m}{C_t} \sqrt{2^{1/n} - 1} \tag{6.10}$$

An often quoted figure of merit is the *gain–bandwidth factor*, which is equal to the term multiplying g_m/C_t. For a given bandwidth there is a maximum gain that cannot be exceeded. Likewise, there is a maximum attainable bandwidth for a specified gain. This occurs when each gain stage contributes a gain of 1.65 $(=\sqrt{e} \rightarrow 4.34\,\text{dB})$ [10]. The maximum bandwidth is determined from

$$\omega_{max} = \frac{(g_m/C_t)}{6.52\mathscr{A}} \tag{6.11}$$

where $\mathscr{A}$ is equal to square root of the natural logarithm of the overall or total amplifier gain. A gain of 1.65 is low; voltage gains larger than two (>6 dB) are generally used in practical amplifiers.

6.2.2 Impedance-Matching Limitation

Impedance matching is the use of a network to approximately transform a load impedance to a desired impedance level over some specific band of frequencies. *Matching networks* are filters that absorb reactance at one or both ports. The first to quantify the fundamental limitation of impedance matching networks was Bode [11]. For a simple parallel RC network the Bode gain–bandwidth restriction is

$$\int_0^\infty \ln\left|\frac{1}{\Gamma}\right| d\omega \leq \frac{\pi}{RC} \tag{6.12}$$

where Γ is the input reflection coefficient of the matching network. The integrand corresponds to the *return loss* (i.e., $RL = -20 \cdot \log_{10}|\Gamma|$) of the matching network. Hence, the wider the bandwidth, the less power flow due to a degraded impedance match. It is not possible to achieve a perfect match across a nonzero band of frequencies.[2] For the optimum case (i.e., a brickwall passband) of a constant $|\Gamma| = |\Gamma_m| < 1$ over frequency band ω_1–ω_2, but $|\Gamma| = 1$ outside this frequency band, the lowest value possible for $|\Gamma|$ corresponds to

$$\int_{\omega_1}^{\omega_2} \ln\left|\frac{1}{\Gamma}\right| d\omega = (\omega_2 - \omega_1) \ln\left|\frac{1}{\Gamma}\right| \leq \frac{\pi}{RC} \tag{6.13}$$

Of course, it is physically impossible to actually have a brickwall function with a network consisting of a finite number of components.

Whereas Bode treated only simple parallel R–C and series R–L networks, Fano [12] extended Bode's work to include a larger class of lossless matching networks (doubly terminated with resistances). Fano's integral relation for a resistor R presented with a shunt capacitance C through the lossless matching network is

$$\int_0^\infty \ln\left|\frac{1}{\Gamma}\right| d\omega \leq \frac{\pi}{RC} - \pi \sum_i s_i \tag{6.14}$$

where the s_i are the poles of $1/\Gamma$ in the right-half complex frequency plane. The available gain–bandwidth is reduced by the presence of the complex poles.

The impedance-matching limit expressed above can be visualized with the two reflection coefficient–frequency plots shown in Figure 6.4. Note that $|\Gamma_{m1}| < |\Gamma_{m2}|$, the response of Figure 6.4*a* implies a better match in the passband than the response of Figure 6.4*b*. The dashed line represents the brickwall passband approximation to the filter responses shown.

[2] Actually a matching network can achieve a perfect match at only a single frequency, and matching degradation follows at all other frequencies.

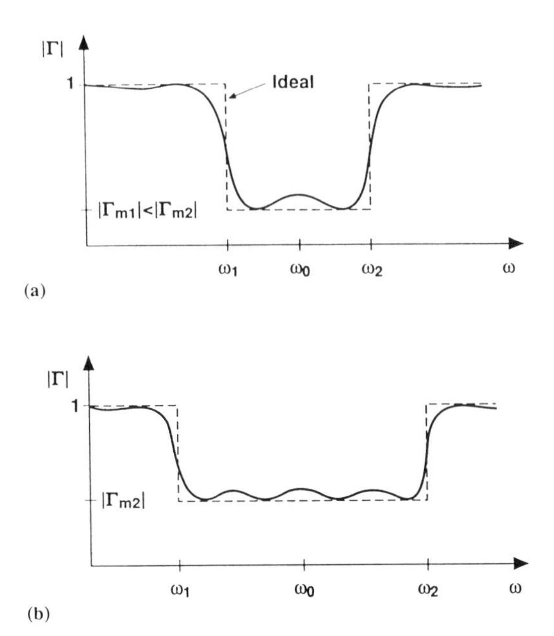

FIGURE 6.4 Reflection coefficient versus frequency for two matching networks showing the tradeoff between match and bandwidth.

Another viewpoint is in terms of load Q. Bode's relation is equivalent to the integral of the return loss over frequency being bounded by a constant, where this constant is inversely proportional to the load Q. When the load Q is decreased, a better match is achieved over a specified bandwidth, or the same match is realized over a wider bandwidth. Smaller load Q corresponds to less reactance (or susceptance) being absorbed by the matching network relative to the resistance (or conductance). For a shunt capacitance the best (i.e., smallest) reflection coefficient achievable is

$$\Gamma_m = \exp\left(-\pi \frac{Q_b}{Q_{load}}\right) \tag{6.15}$$

where the load Q[3] is $Q_{load} = R/X_c = \omega_0 RC$, and bandwidth Q is defined as $Q_b = \omega_0/(\omega_2 - \omega_1)$. Parameter Q_b is the reciprocal of the fractional bandwidth. When applying Equation (6.15) to the bandpass case, ω_0 is the geometric mean band-center frequency, $\omega_0 = \sqrt{\omega_1 \omega_2}$. For the lowpass case, $Q_b = 1$ because $\omega_1 = 0$ and the upper band edge $\omega_2 = \omega_0$.

[3] For parallel components load Q is of the form R/X, but X/R for series components. The reactance X is evaluated at the bandpass geometric mean frequency or lowpass upper-edge frequency.

6.2.3 Broadbanding Techniques

Most broadband design techniques fall into two categories: designing interstage networks (including the input and output networks) and the application of feedback (or both in combination). From a network viewpoint broadband design involves (1) removing or canceling the pole, or poles, nearest the $j\omega$ axis and (2) changing the positions of the remaining poles to realize the desired transfer characteristic [e.g., filter characteristics such as Butterworth or MFM (maximally flat magnitude), Chebyshev or equal-ripple, Thomson or MFD (maximally flat delay)] [13]. Both 1 and 2 (above) can be achieved by either the inclusion of proper interstage networks or by using feedback. Interstage networks can be either *lossless* (i.e., consist only of L and C elements) or *lossy* (i.e., include R elements with the L and C elements).

A practical limitation in broadbanding is the impedance arising from the input capacitance of the active device. The input impedance of an active device over much of its useful frequency range varies as $1/|\omega C_t|$, where ω is the angular frequency and C_t denotes the total input capacitance. Assuming an impedance to be constant in magnitude over a frequency band of $\mathcal{B}$ radians per second, the maximum impedance level obtainable with such a device is of the order of $2/\mathcal{B}C_t$.

There are many techniques for broadbanding transistor amplifiers. Some of the more commonly used or the most effective techniques for broadbanding are discussed below.

6.2.3.1 *Resistive Broadbanding* Figure 6.3 is an example of a resistive interstage network. While it is the simplest type of broadbanding, it is also the most inefficient (i.e., lower gain, lower output power, and degraded noise figure) and is seldom used in this simple form.

6.2.3.2 *Lossy Match* The *lossy match* [14] approach is superior to pure resistive broadbanding. Resistors are used in the matching networks to provide lossy gain compensation. To implement a lumped-element lossy match a series R–L branch is shunt connected to ground. The implementation of the lossy match technique at the input of the active device is shown in Figure 6.5*a*. At frequencies well below the upper edge of the band, the R–L branch behaves as a pure shunt resistance R. However, for frequencies approaching the upper band edge, the effective branch impedance increases ($|Z| \gg R$ at upper-band-edge frequency) because of the inductor's series impedance. If the lossy match network is properly designed the lossy match technique provides wideband impedance matching to 50-Ω systems.

At microwave frequencies, where distributed elements are practical, the resistor is series connected to an ac short-circuited transmission line (see Fig. 6.5*b*). The length of this transmission line is chosen to be one-fourth of a wavelength at the upper-band-edge frequency ω_2; therefore, the transmis-

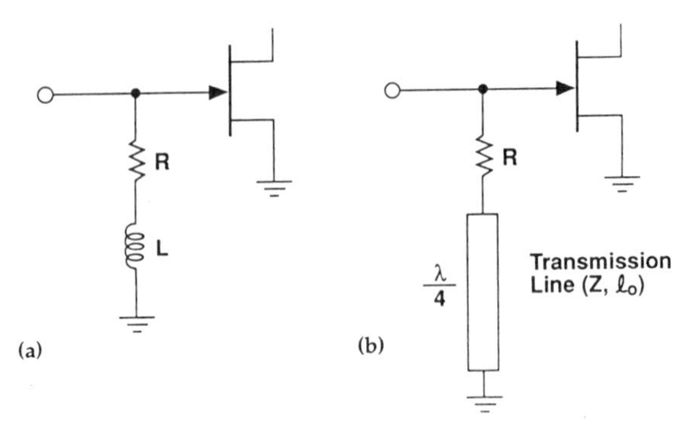

FIGURE 6.5 Lossy match configuration shown for (*a*) lumped-element realization and (*b*) distributed-element realization.

sion line rotates to an open circuit at ω_2 and effectively removes the shunting resistor R.

Lossy matching is especially easy to implement and gives a good impedance match over a wide frequency band. The disadvantages are (1) lower voltage gain than can be achieved with reactively matched amplifiers (although much wider bandwidths are possible using lossy match), (2) lower output power when a lossy match network is placed at an amplifier's output, and (3) the loss in a input lossy match network degrades the noise performance.

6.2.3.3 Shunt Peaking *Shunt peaking* [15, 16] or *shunt compensation* is briefly discussed in Section 5.4.3 of Chapter 5. It is the first step in circuit refinement beyond the basic resistance-coupled stage and is similar to the lossy match technique. Shunt peaking in its simplest form requires a single inductor in series with the interstage (or load) resistor (see Fig. 5.20*a*). The shunt-peak network forms a one-port interstage network. Inductive peaking theory [16] holds that the best possible bandwidth improvement using a one-port interstage network is a gain–bandwidth factor of 2.00 over the pure *RC* case. This compares with a gain–bandwidth factor of 1.76 for the shunt-peaked circuit of Figure 5.20*a* corresponding to the Butterworth (or MFM) filter characteristic that gives the highest gain–bandwidth factor in a shunt-peaked stage.

If more complex two-port interstage networks are used, an even higher gain-bandwidth factor can be achieved (values greater than 4 are theoretically possible). Reference [15] presents several examples of useful two-port interstage networks. Unfortunately, many of these networks become too complex for incorporation into GaAs monolithic microwave integrated circuits.

6.2.3.4 Series Peaking Closely related to shunt peaking is *series-peaking* [16], where an inductor splits the output and input capacitors of two

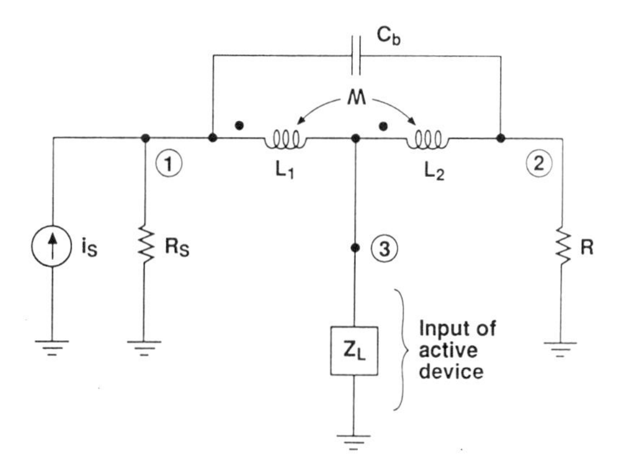

FIGURE 6.6 Bridged T-coil with load impedance Z_L. The input impedance at node 1 is equal to R with the proper choice of element values (see text).

cascaded stages. It is the simplest two-port interstage network possible. Series peaking[4] uses delay to increase the bandwidth (and speed up the risetime) by time separating the charging of capacitors. While series peaking offers an improvement in the gain–bandwidth factor as does shunt peaking, the shunt-peaking case gives better gain–bandwidth factors for comparable parameter values.

6.2.3.5 T-Coil Peaking Figure 6.6 shows the *bridged T-coil* network which is a controlled coupling transformer with the common terminal connected to the load. T-coil peaking [17] is capable of giving greater bandwidth extension than either series peaking or shunt peaking. The load impedance Z_L is the input impedance of the active device, and resistor R is typically 50 Ω. By proper choice of L_1, L_2, and the coupling M, the input impedance of the T-coil is equal to R and independent of frequency. Compared to the shunt-peaked case, the bridged T-coil produces nearly twice the bandwidth extension. A bridged T-coil can be realized (with limitations in monolithic form) using interwound spiral inductors on GaAs substrates.

6.2.3.6 Series-Connected Parallel R–C Network A parallel R–C series connected network to the active device's input has an impedance that decays at a rate of −6 dB/octave. Proper choice of the R and C values leads to pole-zero cancellation resulting in greater bandwidth.

6.2.3.7 Feedback Feedback is a very powerful technique for extending bandwidths among other properties (e.g., setting input/output impedances

[4] In fact, inductive peaking in general uses this principle to increase bandwidth. The limit is embodied by the lossless, distributed transmission line that in principle has infinite bandwidth.

and reducing distortion). Section 6.2.4 below is devoted to a discussion of feedback.

6.2.3.8 Other Broadbanding Methods There are of course other component inclusions and combinations that extend the bandwidth of amplifiers. These techniques can be combined to give still further bandwidth extension. For example, combining the bridged T-coil with series-peaking [17] leads to a bandwidth greater than that achievable with a T-coil alone. The next class of broadbanding techniques involves multiple-path (e.g., feedback) circuits [18–20] and is the subject of the next section.

6.2.4 Multiple-Path Circuits

Feedback is a powerful circuit technique because it produces several important circuit benefits—it can increase bandwidth, stabilize gain, establish and control input/output impedances, reduce parametric sensitivities, and reduce distortion. It involves returning a portion of the output signal to the input for further amplification. Consider the simplest of feedback circuits: an amplifier of open-loop gain $A(s)$ coupled to a single feedback network with transfer function $F(s)$, summing a feedback signal $F(s)v_0$ with the input signal v_i at the input of the amplifier. This is illustrated in Figure 6.7 in both block diagram and signal flow graph form [20]. The canonical feedback equation for the closed-loop gain $\mathscr{A}(s)$ is derived from the node equations

$$e = v_i - F(s)v_0 \tag{6.16a}$$

$$v_0 = A(s)e \tag{6.16b}$$

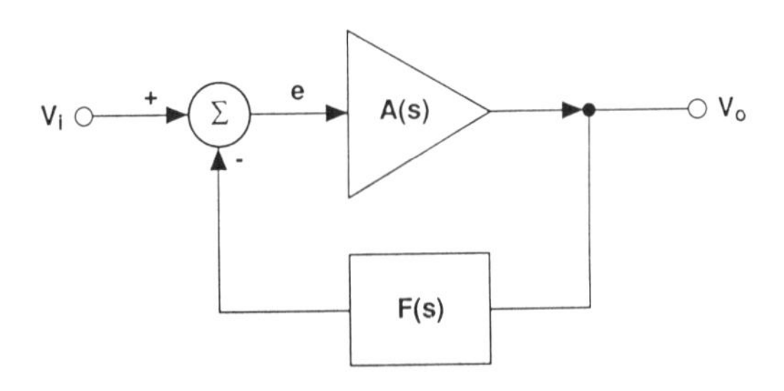

Signal flow graph:

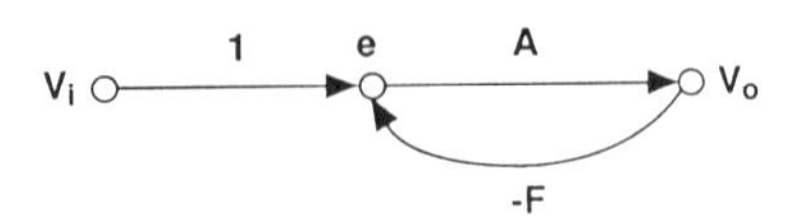

FIGURE 6.7 (*a*) Canonical feedback block diagram; (*b*) signal flow graph for the block diagram in part (*a*).

where e is the *error signal* fed to the input of the amplifier. Combining these two equations gives

$$\mathscr{A}(s) = \frac{A(s)}{1 + A(s)F(s)} \tag{6.17}$$

The term $A(s)F(s)$ is the *loop gain*. When the denominator of Equation (6.17) is greater than unity, the feedback is said to be *negative feedback*, and $|\mathscr{A}(s)| < |A(s)|$. *Positive feedback* corresponds to the denominator -being less than unity and can lead to stability problems.

Negative feedback increases the bandwidth at the expense of gain. Suppose the transfer function of the amplifier in Figure 6.7 is

$$A(s) = \frac{A_0}{1 + s/s_0} \tag{6.18}$$

and the feedback function is a constant [i.e., $F(s) = F_0$]. Substitution into Equation (6.17) gives

$$\mathscr{A}(s) = \frac{A_0/(1 + A_0F_0)}{1 + [s/(1 + A_0F_0)s_0]} \tag{6.19}$$

Inspection of Equation (6.19) reveals that the gain A_0 is reduced by the factor $[1/(1 + A_0F_0)]$, whereas the pole frequency is increased from $|s_0|$ to $(1 + A_0F_0) \cdot |s_0|$. The gain–bandwidth product is unchanged by the feedback.

There are four basic feedback configurations,[5] of which the shunt–shunt and series–series configurations are shown in Figure 6.8. Table 6.1 presents design equations for gain, input immittance, and output immittance for both configurations. An example of the shunt–shunt case is a common-source FET with a shunt feedback resistor R_f connected from the drain node back to the gate node [18,19]. If $A_v = -g_m r_{ds}$, $Y_f = 1/R_f$, $Y_0 = 1/r_{ds}$, $Y_{in} \cong 0$ because of the high input impedance of the FET, and $Y_l \cong 0$ corresponding to a large load impedance (approximated by operating into an open circuit), the closed-loop voltage gain becomes

$$\mathscr{A}_v = \frac{-g_m r_{ds}}{1 + (g_m r_{ds}/Y_s R_f)} \tag{6.20}$$

The signal inversion associated with the common-source FET results in negative feedback so long as excess phase shifts in the circuit are small [19]. The input impedance is determined by the amplifier open-loop gain and the

[5] The four categories are shunt–shunt, series–series, shunt–series, and series–shunt. However, not every feedback connection can be neatly classified into one of these four categories.

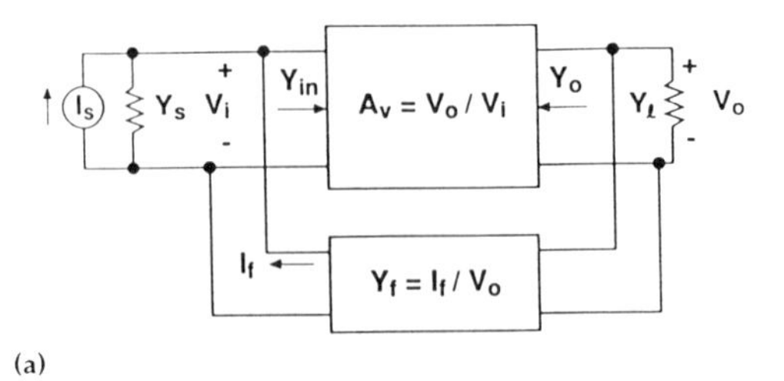

(a)

(b)

FIGURE 6.8 (*a*) Shunt–shunt feedback connection using *y*-parameter notation; (*b*) series–series feedback connection using *z*-parameter notation.

TABLE 6.1 Gain and Input/Output Immittances for Shunt–Shunt and Series–Series Configurations[a]

Shunt–Shunt	Series–Series
Voltage gain	
$\mathscr{A}_v = \dfrac{[Y_s A_v/(Y_s + Y_{in})]}{1 - \{(Y_f/Y_s)[Y_s A_v/(Y_s + Y_{in})]\}}$	$\mathscr{A}_v = \dfrac{-[Z_1 A_i/(Z_s + Z_{in})]}{1 - \{(Z_f/Z_s)[Z_1 A_i/(Z_s + Z_{in})]\}}$
Input immittance	
$Y_{in}^f = Y_{in}\left(1 - \dfrac{Y_f}{Y_{in}} A_v\right)$	$Z_{in}^f = Z_{in}\left(1 - \dfrac{Z_f}{Z_{in}} A_i\right)$
Output immittance	
$Y_0^f = Y_0\left(1 - \dfrac{1 + Y_1/Y_0}{1 + Y_s/Y_{in}} \cdot \dfrac{Y_f}{Y_{in}} A_v\right)$	$Z_0^f = Z_0\left(1 - \dfrac{1 + Z_1/Z_0}{1 + Z_s/Z_{in}} \cdot \dfrac{Z_f}{Z_{in}} A_i\right)$

[a] Symbols used in these equations are defined in Figure 6.8.

magnitude of R_f; hence

$$Z_{in} \cong \frac{R_f}{g_m r_{ds}} \tag{6.21}$$

Just as the input impedance is reduced by the feedback, the output impedance is also decreased:

$$Z_0 \cong \frac{r_{ds}}{1 + (g_m r_{ds} / Y_s R_f)} \tag{6.22}$$

An example of a series–series configuration would be a resistor in series with the source terminal of a common-source FET. This series resistor in the source branch results in negative feedback. In the series–series case the input and output impedances are increased as a result of the feedback. For a discussion of the shunt–series and series–shunt configurations and their respective examples, References [18] and [20] can be consulted.

A widely used technique to maximize the bandwidth of an FET feedback stage adds an inductor (L_D) in series with the drain node and shunt feedback formed by a series R–L branch connected around both L_D and the FET. The feedback elements are selected so that the resistance R_f sets the stage's gain at lower frequencies and the reactance of the inductance $|\omega L_f|$ peaks up at the upper band edge to reduce (ideally, eliminate) the negative feedback. The drain inductor L_D compensates for the FET's drain-to-source capacitance (i.e., output capacitance) at the upper band edge.

Sometimes more than one feedback path is used to take full advantage of the benefits of feedback. An example of this is the gain stage discussed in Section 6.3.3 (see Fig. 6.24). In addition to feedback, *feedforward* [20] has important applications in circuits (Fig. 6.24 also provides an example of the use of feedforward).

6.2.5 Wideband Noise Representation

At a specified frequency the noise of a two-port network is fully described by four noise parameters. For example, one commonly used representation for the noise generated by an active device is with partially correlated *equivalent input noise voltage* and *current generators* [21] as shown in Figure 6.9*a*. The complex correlation coefficient γ $(=\gamma_{re} + j\gamma_{im})$, along with the magnitudes of the series noise voltage generator and shunt noise current generator, are the four noise parameters required. For an active device the noise parameters are also a function of the bias state. Of course, there are several other equivalent representations used in the literature [21] for two-port noise (e.g., partially correlated input and output noise current generators).

The representation in Figure 6.9*a* can often be simplified for noise

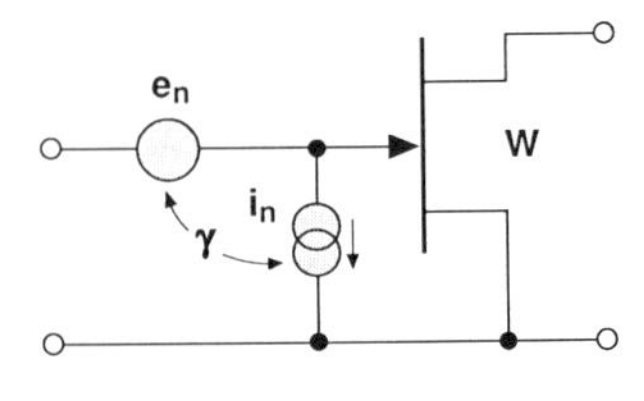

(a)

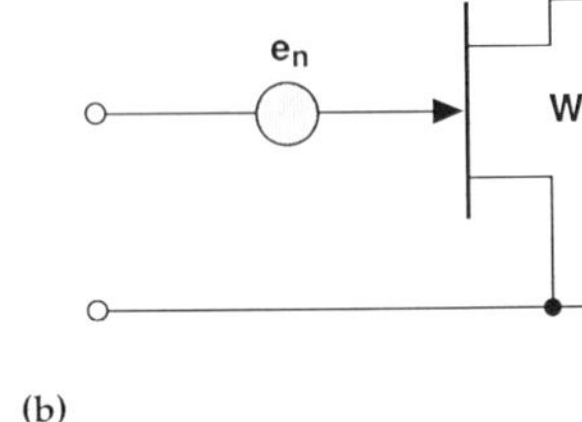

(b)

FIGURE 6.9 Representation of noise generation in FET: (*a*) representation by partially correlated equivalent input noise voltage and current generators; (*b*) representation simplified to a single equivalent input noise voltage generator.

calculations in wideband amplifiers. For instance, when high-input-impedance FETs are driven from real impedance sources of low value (e.g., 50 Ω) only the equivalent input noise voltage source e_n need be included in the analysis. The noise equivalent circuit then reduces to Figure 6.9*b*. Noise generator e_n is *white noise*[6] above the $1/f$ noise corner frequency. A rule of thumb for the mean-square noise voltage is

$$\langle e_n^2 \rangle = \frac{\mathcal{K}}{2W} \qquad [\text{nV}^2/\text{Hz}] \tag{6.23}$$

where W is the width of the FET in millimeters and $\mathcal{K}$ is in general a number that is IC-process-dependent and bias-state-dependent. Equation (6.23) clearly emphasizes the inverse dependence of $\langle e_n^2 \rangle$ on FET width W. This dependence follows from the available output noise power from an FET being proportional to the size of the FET. For a modern 1-μm GaAs IC foundry process a good value for $\mathcal{K}$ is $1\ \text{nV}^2 \cdot \text{mm/Hz}$ for purposes of performing approximate noise calculations. This value assumes a bias state where $V_{DS} > V_{knee}$, so the FET is solidly in saturation, and the drain current $I_D > 0.25 I_{DSS}$, so g_m is sufficiently large. Regarding high-performance submicrometer GaAs FETs and AlGaAs HEMTs (or MODFETs) [22], $\mathcal{K}$ will generally by smaller than $1\ \text{nV}^2 \cdot \text{mm/Hz}$ for these low-noise devices. As a convenient reference value, for a 500-μm GaAs FET ($W = 0.5$ mm), fabricated in a workhorse 1-μm IC technology, $\langle e_n^2 \rangle = 1\ \text{nV}^2/\text{Hz}$.

Below the $1/f$ noise corner frequency, denoted by f_c, the noise power

[6] White noise has a power spectral density that is constant with frequency.

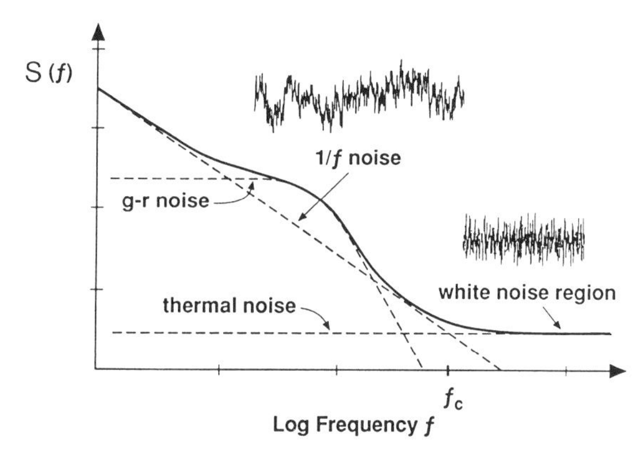

FIGURE 6.10 Noise power spectral density versus frequency for a typical active device showing the flicker or $1/f$ region superimposed with generation–recombination (g–r) noise.

increases as the frequency decreases as shown in Figure 6.10, where $\langle e_n^2 \rangle \approx 1/f^{\alpha}$ with α close to unity. It is not uncommon for generation–recombination noise to add bumps to the $1/f$ noise spectrum. Typical values for f_c range from 10 MHz to in excess of 100 MHz. An example of the low-frequency behavior of e_n for a 1-μm × 500-μm GaAs MESFET appears in Figure 6.11 [23]. For this FET the corner frequency f_c is approximately 40 MHz and $\mathcal{K} \cong 1.3\ \text{nV}^2 \cdot \text{mm/Hz}$ for the white-noise level.

The size of the FET is important in establishing the noise figure of gain

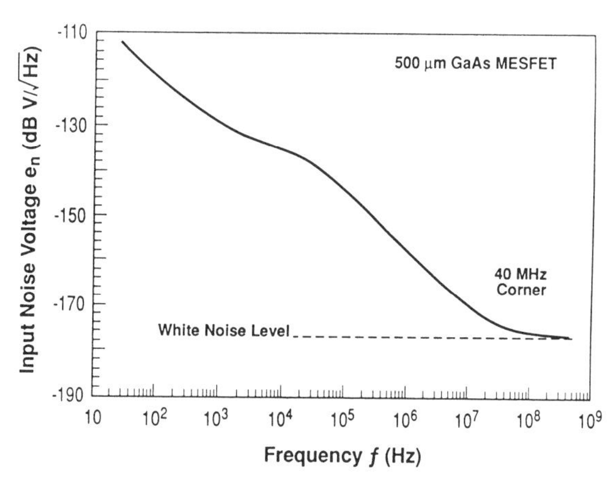

FIGURE 6.11 Measured input noise voltage spectrum for a 1-μm × 500-μm GaAs MESFET at 3.5 V on the drain and 50-mA drain current. Note the $1/f$ corner frequency f_c at approximately 40 MHz. (Courtesy of Brian Hughes, Hewlett-Packard Company; reproduced by permission.)

stages such as discussed in Chapter 5 and elsewhere in this book. To demonstrate this dependency consider the plot of noise figure versus FET gate width for five different gain cells in Figure 6.12. In constructing this figure the following assumptions hold: (1) the generator resistance R_s is $50\,\Omega$ (2) the broadband input impedance is $50\,\Omega$ for each gain cell, (3) $\mathcal{K} = 1\ \mathrm{nV}^2 \cdot \mathrm{mm/Hz}$ for all five gain cells, (4) each active load is one-half the width of the corresponding amplifying FET, and (5) passive resistors are chosen to equal $r_{ds}/2$ where r_{ds} is the output resistance of the amplifying FET. The two common-source FET gain cells without feedback use shunt resistors (i.e., *resistive match*) to establish the 50-Ω input resistance.

Several conclusions follow from Figure 6.12. For the common-source FET gain cells a larger FET width leads to a lower noise figure. This follows from the amplifier's output noise power from the thermal noise of R_s by itself being proportional to W, whereas the output noise power from noise

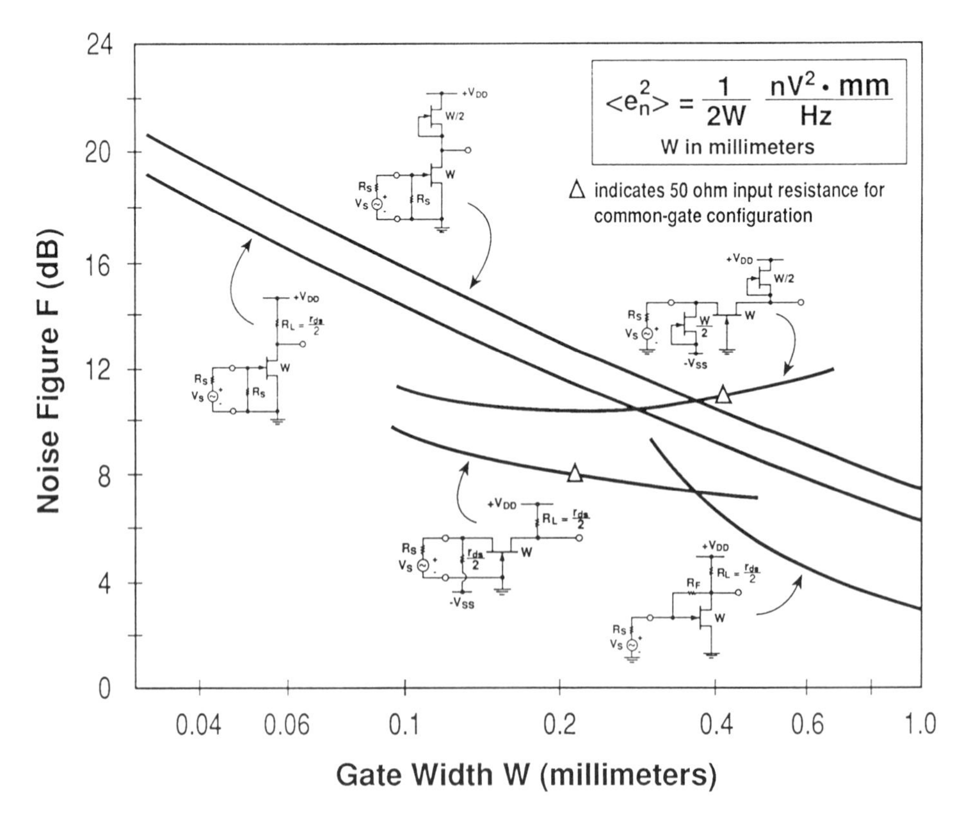

FIGURE 6.12 Midband noise figure versus gate width for five FET gain cells. See text for assumptions used in calculating the noise figures.

generation within the FET being independent of W (or $e_n \propto 1/\sqrt{W}$)[7]. Hence, the noise figure $\mathcal{F}$ goes as

$$\mathcal{F} = 1 + \frac{1}{W}\left[\frac{\langle e_n^2 \rangle}{\langle e_{ns}^2 \rangle}\right] \tag{6.24}$$

where $\langle e_{ns}^2 \rangle$ is the mean-square thermal noise voltage from R_s. Note in Equation (6.24) that the voltage gain cancels and does not appear in the equation. The two common-gate FET cells satisfy the 50-Ω input resistance requirement by choosing the proper FET width (see Section 5.3.1.2 of Chapter 5). This width is about 220 μm for the passive-load case and 400 μm for the active-load case for the FET parameters ($g_m r_{ds} = 10$ and $g_m = 100\,\text{mS/mm}$) assumed in constructing Figure 6.12. Obviously, active loads are more noisy that passive loads (but other engineering tradeoffs must be made as detailed in Chapter 5). For example, comparing the two common-gate cells, 60% of the noise power at the output of the cell with active load is generated by the load FET, whereas only 5% is generated by the passive load in the resistor loaded cell. Resistive matching (i.e., shunt 50-Ω resistor at the input) degrades the noise figure by 3–6 dB, with the worst degradation occurring in gain cells using small-width FETs (with the shunt 50-Ω resistor $\mathcal{F}$ now becomes $[2 + 4(\langle e_n^2 \rangle / \langle e_{ns}^2 \rangle)/W]$). The lowest noise figure shown in Figure 6.12 corresponds to the common-source FET cell with feedback and having a passive resistor load. The feedback resistor R_f sets the broadband 50-Ω input resistance. This stage has a lower noise figure because it does not have the shunting input 50-Ω resistor that degrades the noise figure (even though R_f does itself contribute a small amount of noise power to the amplifier's output).

6.2.6 FET Distortion and Breakdown

Distortion in single FETs results from three primary FET nonlinearities [24]. These are (1) the transconductance g_m nonlinearity (see Fig. 6.13), (2) the gate-to-source capacitance C_{gs} nonlinearity (Fig. 6.14), and (3) the

[7] One way to think about this is to imagine a single FET being divided into m identical FETs, each of width W, connected in parallel. Assume that each small FET slice has transconductance g_m and output resistance r_{ds}. The noise voltage generators associated with these FET slices are uncorrelated, so the mean-square noise currents add at the output. The output noise power then is the sum of m mean-square noise currents multiplied by the equivalent output resistance $[m(g_m^2 \langle e_n^2 \rangle) r_{ds}/m]$. The factor of m cancels, giving a constant output noise power independent of W. However, the output noise power due to resistor R_s is proportional to W as expected because all m FET slices see the same noise signal (i.e., fully correlated); hence, the output noise power is $[m^2(g_m^2 \langle e_{ns}^2 \rangle) r_{ds}/m]$.

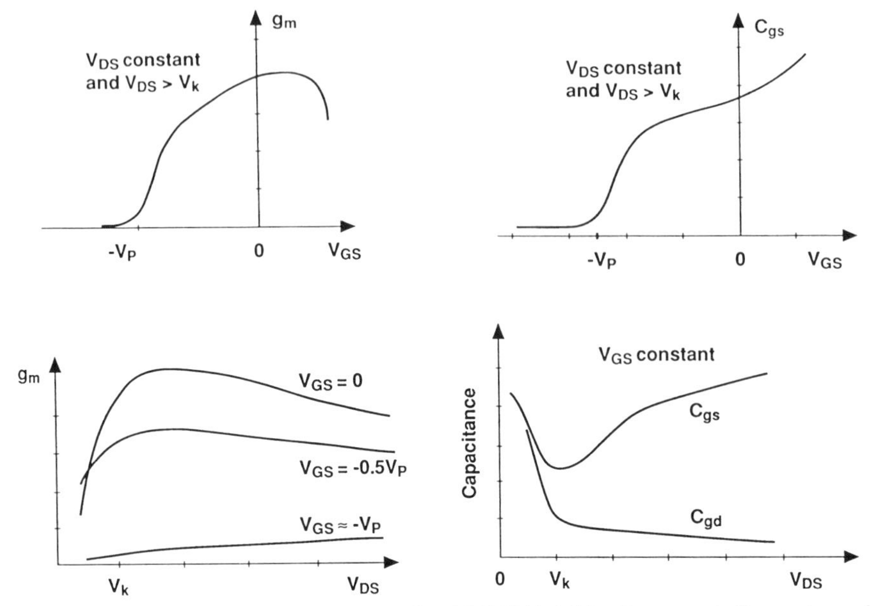

FIGURE 6.13 Representative gate and drain voltage dependence of a GaAs FET's transconductance g_m. Voltage V_K is the knee voltage corresponding to the threshold of saturation.

FIGURE 6.14 Representative gate and drain voltage dependence of a GaAs FET's gate-to-source capacitance C_{gs}. Also shown is the drain voltage dependence of the gate-to-drain capacitance C_{gd}. Voltage V_K is the knee voltage corresponding to the threshold of saturation.

output conductance g_{ds} $(=1/r_{ds})$ nonlinearity (Fig. 6.15). At low frequencies the g_m nonlinearity tends to dominate, especially for low-gain amplifiers. As the voltage gain increases, the g_{ds} nonlinearity becomes more important. The reason is that in high-voltage-gain amplifiers the drain-to-source voltage swing is much larger than the gate-to-source voltage swing. From Figures 6.13 and 6.15 it is clear that harmonic distortion in FET-based circuits is a strong function of bias state. In fact, it is possible to actually have distortion

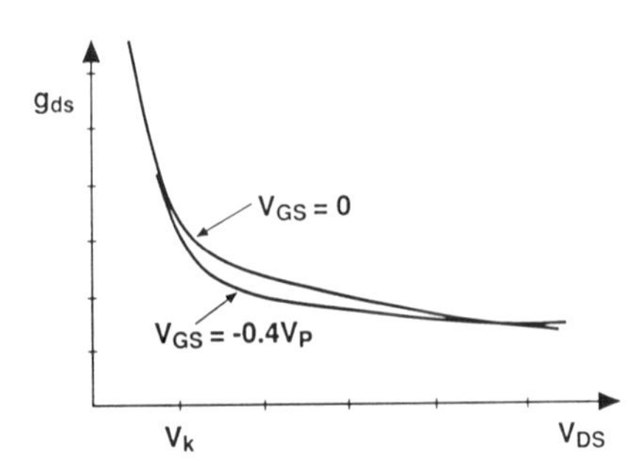

FIGURE 6.15 Representative gate and drain voltage dependence of a GaAs FET's output (drain-to-source) conductance g_{ds}. Voltage V_K is the knee voltage corresponding to the threshold of saturation.

from the g_m nonlinearity mostly cancel the distortion from the g_{ds} nonlinearity if the bias point and load to the FET are carefully selected. This second harmonic reduction occurs because both distortion generating nonlinearities flatten opposite sides of a sinusoidal signal.

At high frequencies the capacitance nonlinearity becomes important. A nonlinear voltage waveform is generated at the input of FET leading to distortion being carried through to the output. Distortion from nonlinear C_{gs} is more prominent in high impedance circuits and is of opposite sign to the g_m nonlinearity [24]. Further discussion of distortion in FET circuits, including slew-rate limiting, appears in Section 6.3.2.3 of this chapter.

In addition to these parametric nonlinearities, current and voltage limits associated with a FET generated distortion under large-signal conditions. Consider Figure 6.16, showing the *I–V* characteristics of the FET (see also Section 5.2.1), including the maximum current $I_{D,max}$ available (corresponding to a forward gate voltage of V_F) and the drain-to-source breakdown V_{BR}. Many GaAs FETs have a soft breakdown characteristic, and so the value of V_{BR} is open to interpretation. Avalanche breakdown is generally the dominant mechanism for breakdown; the avalanche buildup time tends to increase the breakdown voltage as the frequency increases into the microwave region. The load line drawn as a solid line in Figure 6.16 represents the *maximum power load line.* At low to midrange frequencies the maximum output power from an FET is determined by $I_{D,max}$ and the maximum voltage swing possible, which is $(V_{BR} - V_K)$, where V_K $(=V_{knee})$ is the *knee voltage* corresponding to the drain voltage beyond which the FET is operating in saturation. The maximum RF output power is estimated from $I_{D,max}[(V_{BR} - V_K)/8]$. Driving the input so that the output attempts to swing beyond these limits results in clipping and added distortion.

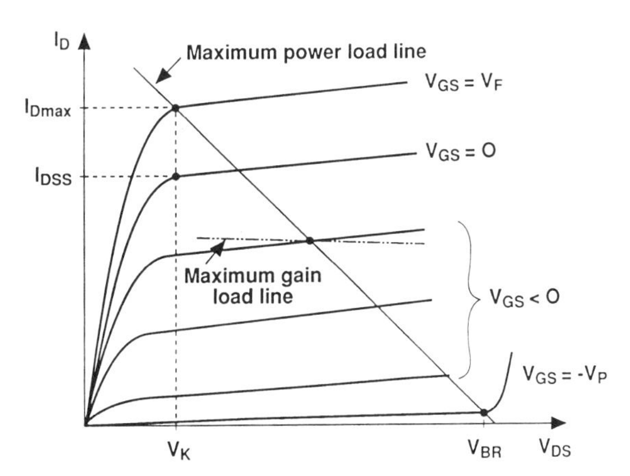

FIGURE 6.16 FET *I–V* characteristic showing maximum power load line with its limits determined by the maximum current $I_{D,max}$, and maximum voltage set by the FET's drain breakdown voltage V_{BR}.

Further discussion on distortion appears in Section 6.3.2.3 featuring distortion in the Van Tuyl–Hornbuckle FET voltage amplifiers.

6.3 DIRECT-COUPLED AMPLIFIERS

Among the earliest (ca. 1977–1978) GaAs IC amplifiers reported [25–27] were a class of direct-coupled (dc) MESFET amplifiers emphasizing small chip size for lower cost. These amplifiers were a deliberate departure from the design philosophy governing hybrid microwave integrated-circuit (HMIC) technology [28].

6.3.1 Historical Perspective

The first truly monolithic analog amplifier was an X-band (8–12 GHz) amplifier [29] reported by Plessey in 1976. This amplifier, and many analogous MMIC designs following [30,31], were designed using passive tuning elements in the microwave bands. Reactively matched amplifiers generally must allocate a substantial area to accommodate physically large matching elements. Many GaAs amplifiers designed in the late 1970s and early 1980s were translations of HMIC designs. This approach demonstrated the feasibility of IC technology, but was not especially cost-effective because it did not play directly to the strength of the newer IC technology.

Van Tuyl [25] first reported on GaAs IC voltage gain blocks that did not use passive tuning elements, but rather operated directly within the high-impedance environment characteristic of the GaAs MESFET. In fact, they were a spinoff of a buffered FET logic NOR gate [32] biased in the transition (linear) region. This approach emphasized lower cost by area reduction and promoted higher integration levels [33]. Section 6.3.2 is devoted to the design and analysis of the Van Tuyl–Hornbuckle gain blocks. These gain blocks are described in some detail because they have been used by many designers as the basic gain blocks for direct-coupled, wideband amplifiers. As an example Section 6.3.3 describes one class of design modifications using the cascode configuration in the design of direct-coupled gain blocks.

6.3.2 Van Tuyl–Hornbuckle Gain Blocks

The basic amplifying configuration in these voltage-gain blocks is the common-source FET inverter (the CSF is discussed in Section 5.3 of Chapter 5) combined with an active-load FET (see Section 5.4). Figure 6.17 schematically illustrates this configuration. The active load has several advantages over a passive resistor. It provides superior large-signal performance when selected for identical small-signal gains. It also consumes less area and allows for a reduction in power supply voltage, leading to

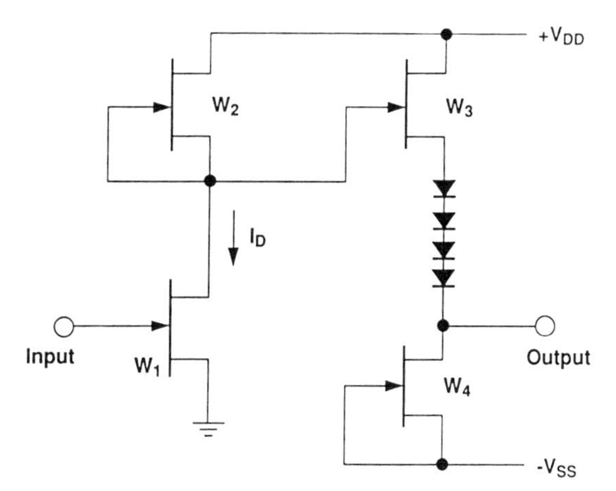

FIGURE 6.17 Schematic circuit of the SA-1 gain block.

reduced dc power consumption. Scaling the width of the active load, relative to the inverter FET's width, allows the drain current I_D of the inverter (FET1) to be set to any value from slightly greater than I_{DSS} down to a small fraction of I_{DSS}. Choosing the drain current in the range of one-half to two-thirds of I_{DSS} approximately centers the bias point, thereby equalizing drain voltage excursions under large-signal conditions (see Fig. 5.19 in Chapter 5 for an illustration of biasing with an active load). The CSF with active load (henceforth referred to as CSAL) is an inverter with a high output impedance and a capacitive input impedance determined by the width of the CSF. It is therefore highly susceptible to capacitive loading resulting in bandwidth shrinkage.

To cascade CSAL stages for high voltage gain, and still drive loads with reasonable capacitance, it is necessary to use a buffer between the CSAL stages to minimize bandwidth shrinkage. A source follower (see Section 5.3) performs this function quite well because of its very low input capacitance and low output impedance. Low output impedance is necessary for driving capacitive or low-impedance loads. Figure 6.17 shows the CSAL gain stage (i.e., CSF FET1 and active-load FET2) with buffer–level-shift network (source follower FET3 and current source FET4). FET4 establishes the quiescent current in FET3 and the level-shifting diodes (refer to Sections 5.5 and 5.6 for discussion) provide the dc voltage shift required for stage-to-stage dc connection. In the original implementation of the SA-1 gain stage,[8] shown in Figure 6.17, the gate-to-source voltage supplied to FET1 was nominally -0.7 to $-1.0\,V$ for a GaAs IC process with a mean pinchoff voltage of -2.1 V [34]. The drain supply voltage V_{DD} typically ranges from 6 to 7 V, and V_{SS} is nominally -4 V.

[8] The SA-1 designation stood for "single-ended amplifier of type 1."

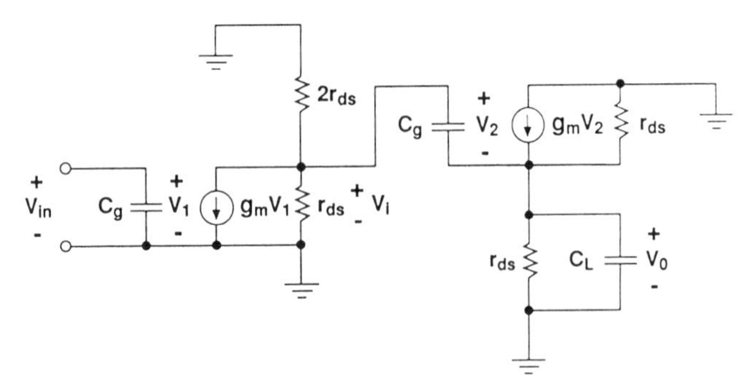

FIGURE 6.18 Small-signal equivalent circuit for the SA-1 gain block. Series resistance of the four-diode stack is omitted in the equivalent circuit. A pure capacitive load is assumed.

6.3.2.1 SA-1 Gain and Bandwidth The midband gain of the SA-1 gain block is estimated as follows. Considering first the CSAL inverter, its voltage gain is

$$G_{\mathrm{CSAL}} = \frac{g_{\mathrm{m}}}{g_{\mathrm{ds}} + \frac{1}{2} g_{\mathrm{ds}}} = \frac{2}{3}\left(\frac{g_{\mathrm{m}}}{g_{\mathrm{ds}}}\right) \tag{6.25}$$

assuming the width of FET2 to be one-half that of FET1. The ratio $g_{\mathrm{m}}/g_{\mathrm{ds}}$ is simply the FET's intrinsic gain and is determined by the IC process. For $g_{\mathrm{m}}/W = 150\,\mathrm{mS/mm}$ and $g_{\mathrm{ds}}/W = 10\,\mathrm{mS/mm}$, G_{CSAL} is 10.0 by calculation, or 20.0 dB when taking $20 \cdot \log_{10}(G_{\mathrm{CSAL}})$. Note that G_{CSAL} is independent of the absolute value of the CSF's gate width, but is a function of the ratio of the gate widths. The gain of the CSAL stage must be multiplied by the gain (actually a loss) G_{BLS} of the buffer–level-shift (BLS) stage

$$G_{\mathrm{BLS}} = \frac{g_{\mathrm{m}} R_{\mathrm{e}}}{1 + g_{\mathrm{m}} R_{\mathrm{e}}} \cong 0.88 \tag{6.26}$$

where $R_{\mathrm{e}} = (2g_{\mathrm{ds}})^{-1}$ is used in Equation (6.26). This is a loss of -1.1 dB. In addition, the parasitic series resistance of the diode stack results in additional loss not accounted for in Equation (6.26) (typically an additional -1.0 dB to as much as -2.5 dB depending on the diode design and layout). Assuming the total loss of the BLS block to be -2.5 dB, the total midband gain (i.e., the product $G_{\mathrm{CSAL}} \cdot G_{\mathrm{BLS}}$) of the SA-1 becomes 17.5 dB.[9]

Figure 6.18 shows a simplified SA-1 small-signal equivalent circuit used to

[9] The measured midband voltage gain of the SA-1 reported in (Hornbuckle and Van Tuyl [27]) was 13 dB. The calculated midband voltage gain in the text is higher (i.e., 17.5 dB vs. 13 dB) because we have assumed the intrinsic gain of the FET to be 15, rather than 10, which was true of the process used in (Hornbuckle and Van Tuyl [27]). We use the higher value because it is more representative of 1-μm GaAs IC processes available from foundries.

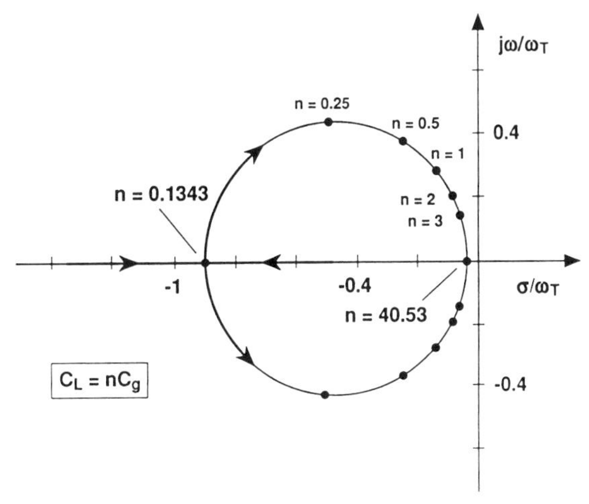

FIGURE 6.19 SA-1 pole locations plotted as a function of the capacitance ratio n.

estimate bandwidth. Devices FET1, FET3, and FET4 are all identical in width, whereas the active-load FET2 is one-half their width. The impedance presented to the SA-1's output is assumed to be primarily capacitive such as looking into the gate of an FET. The loading capacitance C_L is a key parameter in limiting bandwidth.[10] For convenience C_L is expressed as a multiple of the input gate capacitance C_g of FET1, that is, $C_L = nC_g$ by definition where n is ratio of the load capacitance to the input capacitance. Using Kirchhoff's current law, the two nodal equations are

$$g_m v_{in} = -\frac{3}{2} g_{ds} v_i - sC_g(v_i - v_0) \tag{6.27}$$

$$g_m(v_i - v_0) + sC_g(v_i - v_0) = (2g_{ds} + sC_L)v_0 \tag{6.28}$$

Noting that $v_2 = v_i - v_0$, solving Equation (6.28) for v_i, and substituting this expression for v_i into Equation (6.27) yields the solution for the small-signal voltage gain $G_{SA\text{-}1}$ $(=v_0/v_{in})$.

$$G_{SA\text{-}1} = \frac{-g_m(g_m + sC_g)}{nC_g^2 s^2 + \{(3n+7)/2\} g_{ds} C_g s + 3(g_m + 2g_{ds})g_{ds}/2} \tag{6.29}$$

The zeros of the denominator are the poles of the circuit. For $n \leq 0.1343$ and $n \geq 40.53$ the poles fall on the real axis (σ axis). However, when n is in the range, $0.1343 < n < 40.53$, the two poles are complex as shown in Figure 6.19. Figure 6.19 plots the locus of the two poles in the complex $(\sigma + j\omega)$

[10] Parasitic interconnection and layout capacitances have been ignored in the equivalent circuit; however, they are easily included by a designer.

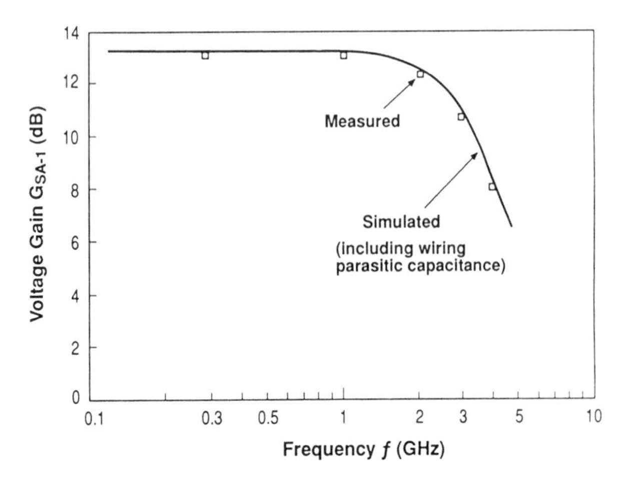

FIGURE 6.20 SA-1 small-signal voltage gain versus frequency ($n = 1.5$) comparing simulated with measured response (after Hornbuckle and Van Tuyl [27], © IEEE; reproduced by permission).

plane. The axes are normalized to ω_T, which is $15.9(2\pi) \cdot 10^9$ radians/s (where $\omega_T = g_m/C_g$ by definition). When $n = 0.1343$, both poles coincide on the real axis with position $s_{1,2} = -0.919 \cdot \omega_T$, and for $n = 40.53$, $s_{1,2} = -0.0530 \cdot \omega_T$.

The CSAL inverter by itself has a real-axis pole determined by the capacitive loading at the internal node, and in isolation, the BLS stage has a real-axis pole from its load capacitance.[11] However, the SA-1 has a complex pole pair because of strong coupling between the CSAL and BLS stages due to the gate capacitance of the source follower FET3. Typical values of n range from approximately unity (e.g., when the BLS stage drives a FET slightly smaller than FET1) to perhaps 4 or 5. Larger values of n result in too much bandwidth shrinkage to be practical. For nominal values of n encountered in applications the complex pole pair exists. The −3-dB frequency f_c^{bw} is estimated from

$$f_c^{b2} = (2\pi)^{-1}\sqrt{(\sigma')^2 + (\omega')^2} \tag{6.30}$$

for a complex pole pair at $s_{1,2} = (\sigma' \pm j\omega')$. Without the BLS stage the pole of the CSAL stage is located at $-(3g_{ds}/2nC_g)$. If $n = 1$, this pole is located at $\sigma/\omega_T \cong -0.1$. Inclusion of the BLS stage increases the bandwidth by a factor of >3 for the same capacitive loading.

The voltage gain–frequency response of the SA-1 appears in Figure 6.20. The simulated response [on SPICE (Simulation Program with IC Emphasis)]

[11] For the parameter values used in this section the real-axis pole of the CSAL is at −0.1 and for the BLS the real-axis pole falls at −1.13. This assumes n to be unity.

is compared with the measured response using ICs fabricated with the 1-μm GaAs IC process reported by Van Tuyl et al. [34]. In performing the simulation parasitic capacitance from the IC layout was included with the capacitance to ground at the SA-1's internal node being the strongest parasitic in reducing the bandwidth. The measured −3-dB bandwidth is approximately 3.2 GHz with a capacitive loading equivalent to $n \cong 1.5$. Without the inclusion of the parasitic wiring capacitances the simulated bandwidth is approximately 30% higher than the measured bandwidth.

6.3.2.2 SA-2 Gain and Bandwidth Feedback is often used to achieve greater bandwidths. Figure 6.21 shows the SA-2 gain block where a single FET is added to provide negative feedback. This FET is conveniently inserted because the level-shift network provides the proper dc gate voltage to FET5. The magnitude of the feedback is set by the ratio of the width of FET5 to that of FET1. Typical values for the W_5/W_1 ratio range from 0.15 to 0.30, with a value of 0.25 used in the measured results presented in this chapter. To estimate bandwidth improvement from feedback, consider a constant magnitude feedback factor F [i.e., $F \neq F(s)$]. The gain expression for the SA-1 is a ratio of polynomials in s of the form [e.g., Equation (6.29)]

$$G_{\text{SA-1}} = \frac{N(s)}{D(s)} \tag{6.31}$$

Using the standard feedback equation the voltage gain of the SA-2 becomes

$$G_{\text{SA-2}} = \frac{N(s)/D(s)}{1 + F \cdot N(s)/D(s)} = \frac{N(s)}{D(s) + F \cdot N(s)} \tag{6.32}$$

If the poles of $G_{\text{SA-1}}$ are [using the standard quadratic formula on the

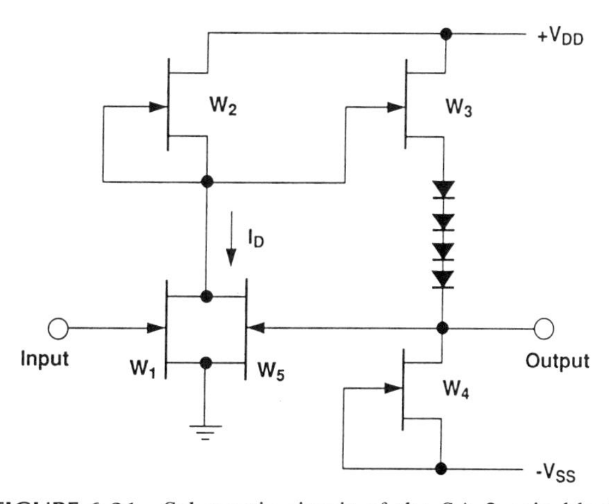

FIGURE 6.21 Schematic circuit of the SA-2 gain block.

denominator, $D(s) = As^2 + Bs + C$]

$$s_{1,2} = \frac{-B \pm \sqrt{B^2 - 4AC}}{2A} \tag{6.33}$$

then the poles for $GA_{\text{SA-2}}$ are found from the modified form

$$s_{1,2} = \frac{-(B + F\mathscr{B}) \pm \sqrt{(B + F\mathscr{B})^2 - 4A(C + F\mathscr{C})}}{2A} \tag{6.34}$$

where $N(s) = \mathscr{B}s + \mathscr{C}$, and F is a constant, as is the case for the SA-2. The circle defining the locus of poles for the SA-2 is enlarged by the addition of the feedback as shown in Figure 6.22. The first-order feedback increases the magnitude of the imaginary part of the pole's coordinates and correspondingly increases the bandwidth.

The midband gain of the SA-2 is easily estimated using the circuit schematic in Figure 6.23. By inclusion of FET5 the gain of the inverter decreases because the internal node admittance Y_i increases. The voltage gain of the SA-2's CSAL stage is

$$G_{\text{CSAL}}^{\text{SA-2}} = -\frac{g_{m1}}{Y_i} \tag{6.35}$$

where g_{m1} is the transconductance of FET1 and the internal node admittance Y_i is a sum of conductances

$$Y_i = g_{ds1} + g_{ds2} + g_{ds5} + G_{BLS} \cdot g_{m5} \tag{6.36}$$

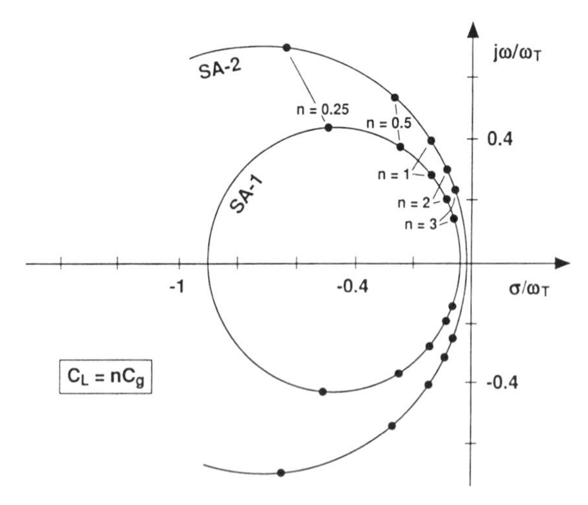

FIGURE 6.22 Pole locations of the SA-2 compared to the pole positions of the SA-1 (cf. Fig. 6.19).

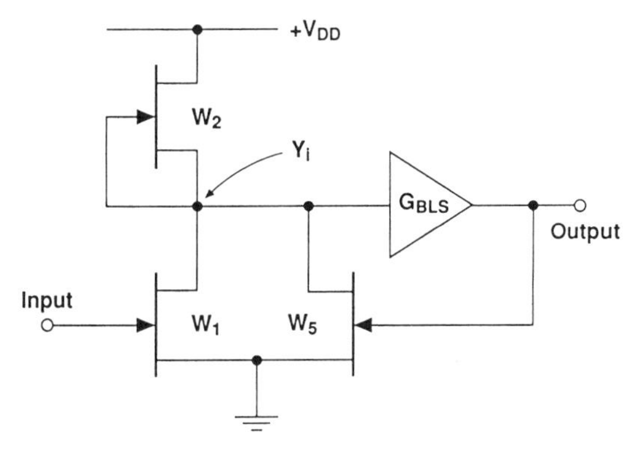

FIGURE 6.23 Simplified schematic circuit used for the gain calculation of the SA-2.

The numeric part ($_x$) of the subscript appended to each FET parameter corresponds to FETX (W_x). Consider the example where the sum of the widths of FET1 and FET5 in an SA-2 is equal to the width of FET1 in the SA-1, and the width ratio $(W_5/W_1)^{\text{SA-2}}$ is $\frac{1}{4}$. Again assuming G_{BLS} to be 0.75 (or -2.5 dB), Equation (6.35) becomes

$$G_{\text{CSAL}}^{\text{SA-2}} = \frac{-0.8g_{\text{m}}}{1.5g_{\text{ds}} + (0.75)(0.2g_{\text{m}})} = -3.2 \tag{6.37}$$

where as before $g_{\text{m}} = 150$ mS/mm and $g_{\text{ds}} = 10$ mS/mm, and the minus sign denotes inversion. This value for $|G_{\text{CSAL}}^{\text{SA-2}}|$ corresponds to 10.1 dB. For the overall midband gain of the SA-2 the loss of the BLS stage must be included, again is assumed to be -2.5 dB:

$$G_{\text{SA-2}} = 10.1\ \text{dB} - 2.5\ \text{dB} = 7.6\ \text{dB} \tag{6.38}$$

Figure 6.24 shows both simulated (with loading capacitance ratio $n = 1.5$) and measured voltage gain versus frequency for an SA-2 gain block fabricated using the IC process in Van Tuyl et al. [34]. The measured midband gain of the SA-2 is approximately 7.0 dB with a -3 dB bandwidth of nearly 5.2 GHz. In comparison to the SA-1, feedback has increased the -3 dB bandwidth from 3.2 to 5.2 GHz, but the tradeoff is a reduction in gain.

To achieve the best fit between simulated and measured gain response it is necessary to include parasitic wiring and layout capacitances. These SA-2 parasitic elements are shown in lumped form in Figure 6.25. The simulated results shown in Figure 6.24 used the FET sizes and parasitic capacitor values listed in Table 6.2. The peaking in the response centered around 3 GHz is caused by load capacitance and the delay time parameter τ_{d} used in the GaAs FET model [35,36]. At low frequencies feedback produces a natural 180° phase shift because FET5 is connected as an inverter. As the

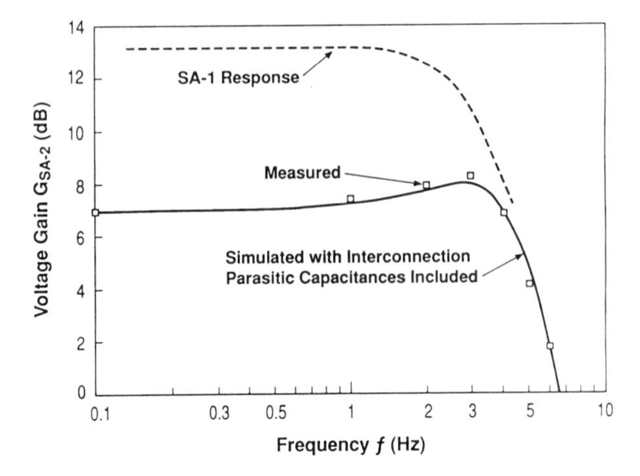

FIGURE 6.24 SA-2 small-signal voltage gain versus frequency ($n = 1.5$) comparing simulated with measured response.

frequency increases, load capacitance produces additional phase shift. Another contributor to the excess phase shift is the internal time delay τ_d associated with the FET3 and FET5. When sufficient excess phase shift is developed, the intended negative feedback advances toward positive feedback. The result is the peaking in the SA-2's response. For the simulated results appearing in Figures 6.20 and 6.24, τ_d is 5 ps (picoseconds). For especially large values of the loading capacitance ratio n, the resulting excess

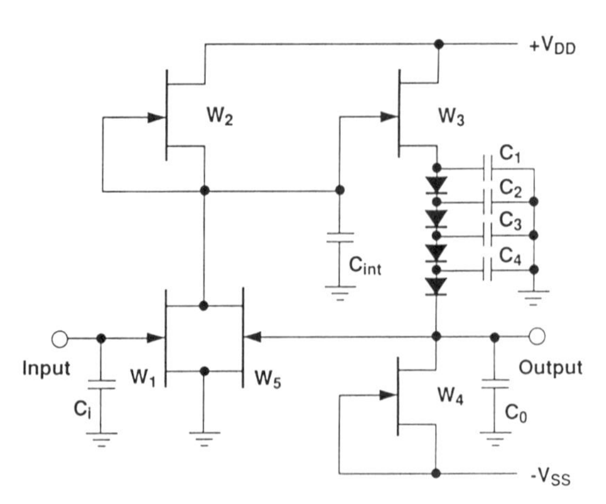

FIGURE 6.25 Parasitic capacitances included in the SA-2 simulated response shown in Figure 6.24. Table 6.2 lists component values.

TABLE 6.2 FET Widths and Parasitic Capacitances for SA-2 Simulation

FETX (W_x)	Capacitor	Value
FET1 (CSF)	—	32 μm
FET2 (AL)	—	20 μm
FET3 (SF)	—	40 μm
FET4 (CS)	—	40 μm
FET5 (feedback)	—	8 μm
—	C_i	37.5 fF
—	C_{int}	5.7 fF
—	C_1	8.6 fF
—	C_{2-4}	4.0 fF each
—	C_0 [a]	15 fF

[a] Output node also drives a 40-μm source follower provided for minimal loading at measurement.

phase shift may produce sufficient positive feedback to approach becoming regenerative (although this was not observed experimentally by Van Tuyl or Hornbuckle).

6.3.2.3 Distortion in the SA-1 and SA-2 The principal causes of distortion in discrete GaAs FETs are discussed in Section 6.2.6. The generation of distortion in multiple FET circuits is complicated because of multiple sources of distortion and interactions between nonlinearities. As a simple example, the use of an active load, rather than a linear passive load, increases third-order distortion because of the inherent nonlinearity in the active load's g_{ds}. This adds directly to the third-order distortion generated by the inverter FET.

The SA-1 voltage transfer curve shown in Figure 6.26 illustrates the strong nonlinearity attendant with large voltage deviations about the quiescent point Q ($V_i = 3.0$ V when $V_{in} = -1.2$ V dc). Voltage V_i represents the SA-1's internal node voltage (i.e., the drain of FET1) and V_{out} is the output voltage taken at the bottom of the diode string. Voltage V_{out} is level-shifted from V_i by the stack of four diodes—the voltage shift is 4.3 V as drawn in Figure 6.26. The slope at the quiescent point Q is the small-signal voltage gain of the SA-1 and is negative because the SA-1 is an inverting amplifier. The transfer curve shows the onset of output voltage limiting for input voltage swings approaching $(V_{in} \pm v_{in}) = (-1.2 \pm 0.35)$ V. When point Q is centered (i.e., $V_i = \frac{1}{2}V_{DD}$) on the transfer characteristic, driving the input hard leads to third-order distortion. However, if point Q is allowed to shift from the center, the result is the premature onset of asymmetric clipping that generates second-order harmonic distortion. Controlling the location of the static bias point is especially important when cascading direct-coupled gain stages.

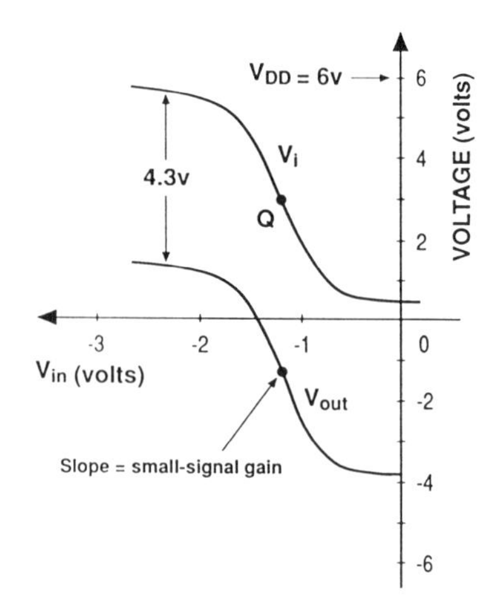

FIGURE 6.26 Transfer characteristic of the SA-1 with $V_{DD} = 6$ V and $V_{SS} = -4$ V.

Another source of distortion comes from the source follower buffer stage [37]. Here the source follower typically drives a large capacitive load that, if the load is the gate of another FET, is a nonlinear capacitance (refer back to Section 6.2.6 for an example of the C_{gs} nonlinearity). At higher frequencies, as the source follower is required to supply greater current, a distorted voltage waveform appears across the gate capacitance of the load FET. This voltage waveform's second harmonic is in phase with any second-order distortion due to the load FET's g_m nonlinearity. They add together resulting in greater second harmonic distortion.

The source follower can be driven at sufficiently high frequencies and signal levels that load current demand exceeds its ability to deliver. The "slew-rate limit" [27] is estimated from

$$\left[\frac{dV}{dt}\right]_{\max} = \frac{I_{D,\max}}{C_{gs}} \tag{6.39}$$

where $I_{D,\max}$ is the maximum operable drain current available from a FET (see Section 5.2.1) and C_{gs} is the gate capacitance of the load FET. Under these extreme, but not uncommon, drive conditions the source follower switches between $I_{D,\max}$ and complete cutoff. The resulting waveform is similar in appearance to a sawtooth-like waveform. A sawtooth-like waveform does not necessarily have poorer harmonic performance compared to a square waveform resulting from signal clipping (as happens at lower frequencies), but it does limit the saturated output power available. When this occurs the saturated output power decreases at -6 dB/octave with frequency as confirmed by measurements (see Fig. 6.38 for experimental verification).

Measured second harmonic and third harmonic distortion performances of both the SA-1 and SA-2 are shown in Figure 6.27. Harmonic levels

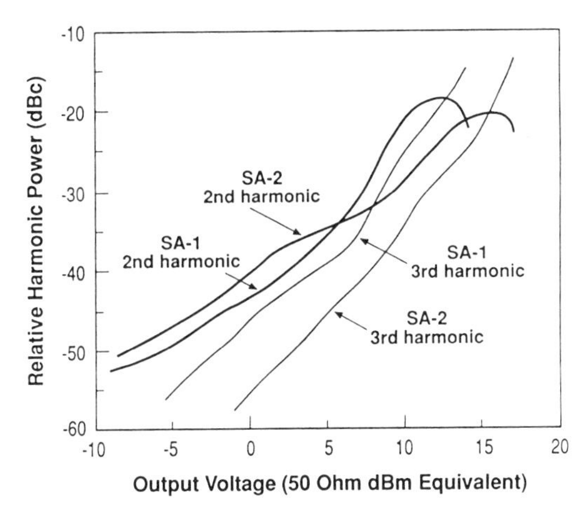

FIGURE 6.27 Comparison of the harmonic distortion observed in the SA-1 and SA-2 gain blocks. Harmonic powers are plotted as a function of the rms output voltage translated to a 50-Ω dBm equivalent power (after Hornbuckle and Van Tuyl [27], © IEEE; reproduced by permission).

relative to the output signal amplitude at the fundamental frequency are plotted as a function of the output voltage (50-Ω dBm equivalent because the SA-1 and SA-2 are voltage amplifiers). The SA-2's use of feedback is expected to produce lower distortion.[12] This is not the case for the second harmonic distortion as measured at lower output amplitudes—the SA-1 (open loop) has better second harmonic below 5 $\mathrm{dBm}_{(\mathrm{equiv})}$. According to Hornbuckle and Van Tuyl [27], the reason is a cancellation between the g_m and r_{ds} nonlinearities. This is verified by the strong dependence of the r_{ds} nonlinearity on V_{DD} and the experimental finding [27] that maximum cancellation occurred when V_{DD} was increased to 6.25 V (from 6.0 V nominal). In the SA-2 the feedback lowers the impedance at the internal node (see Fig. 6.23) so that this interaction is reduced. Also, the g_m nonlinearity dominates at lower drive levels.

At higher signal amplitudes the SA-2 clearly has superior harmonic performance. This follows from feedback softening the effect of the nonlinear elements on the signal distortion. As the output signal level increases there is a point beyond which the third harmonic exceeds the second harmonic level. In fact, for high enough output signal amplitude the second harmonic actually decreases with increasing output level. Strong saturating nonlinearities produce a more symmetric waveform so that the second harmonic decreases at the expense of third harmonic generation.

[12] However, the distortion improvement in the SA-2 from feedback is not as good as one would expect from resistive feedback. Feedback through a linear network, such as a resistor, is linear. The FET in the feedback loop is inherently nonlinear to begin with.

This follows from symmetric waveforms being richer in odd-order harmonics, whereas asymmetric waveforms are dominated by the even-order harmonics.

6.3.2.4 SA-1 and SA-2 Noise Behavior Figure 6.28 shows the CSAL inverter circuit schematic with noise generators included. For calculating the noise figure of many wideband amplifiers (including those discussed herein) a single noise voltage generator in series with the gate of the FET is adequate (refer to Section 6.2.5 for a brief discussion). In the SA-1 and SA-2 gain blocks the noise from the common-source inverter dominates the noise figure. A source resistance R_s is included with its thermal noise voltage generator e_{ns}. For $R_s = 50\,\Omega$, $\langle e_{ns}^2 \rangle = 0.828\,\text{nV}^2/\text{Hz}$ at $T = 290$ K (standard noise temperature). The mean-square noise voltage of an FET is inversely proportional to the gate width W, and for the analysis below the following rule of thumb is used (see Section 6.2.5):

$$\langle e_n^2 \rangle \sim \frac{1}{2W}\,[\text{nV}^2/\text{Hz}] \tag{6.40}$$

where width W is in units of millimeters and the brackets (i.e., $\langle \cdots \rangle$) denote the mean-value operation. As a reference value, a 500-μm FET corresponds to $\langle e_n^2 \rangle = 1\,\text{nV}^2/\text{Hz}$. Therefore, the mean-square noise voltage of the active-load FET2 is β times larger because the width of FET2 is (1/β) times the width of FET1.

Summing the noise powers at the output of the CSAL inverter, we obtain

$$\langle e_{n0}^2 \rangle = \left[\left(\frac{\beta g_m r_{ds}}{\beta + 1} \right)^2 + 2 \left(\frac{g_m r_{ds}}{\beta + 1} \right)^2 \right] \langle e_n^2 \rangle + \left[\frac{\beta g_m r_{ds}}{\beta + 1} \right]^2 \langle e_{ns}^2 \rangle \tag{6.41}$$

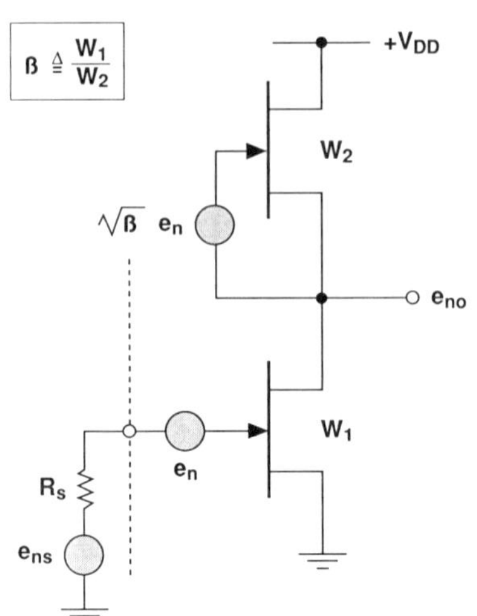

FIGURE 6.28 CSAL inverter with noise generators included. FET2 is $1/\beta$ smaller in width than FET1.

The last term in Equation (6.41) is the output noise power due to source resistance R_s. Assuming a β of 2 (i.e., $W_1 = 2W_2$), FET intrinsic gain $g_m r_{ds}$ of 15, and the widths of FET1 and FET2 to be 40 and 20 μm, respectively, the midband noise figure is 13.7 dB.[13] If the noise contribution of the BLS network is added to complete the SA-1 topology, and the BLS stage's loss included in the expression for $\langle e_{ns}^2 \rangle$, the noise figure is only slightly increased to 14.0 dB. This noise figure is large because small FET widths are used in the inverter [27]. For example, if the size of FET1 were increased to 200 μm, and FET2 set at 100 μm while retaining the other parameter values, the noise figure would reduce to 8.3 dB. Measured noise figures in Reference [27] were higher because the 1-μm GaAs IC process used to fabricate the ICs had 50% higher FET $\langle e_n^2 \rangle$ values than predicted from Equation (6.40). However, Equation (6.40) gives a better estimate for the GaAs IC processes used today.

For the SA-2 the added noise from the feedback FET5 must be included and the lower overall gain accounted for in the noise figure calculation. Assuming the inverter FET1 to be 32 μm, the feedback FET5 to be 8 μm, and the active-load FET2 to remain at 20 μm, the computed SA-2 midband noise figure is 15.6 dB (or 1.6 dB greater than the SA-1 gain block). The BLS network contributes approximately 18% of the total output noise power in the SA-2, but only about 5% of the total noise output power in the SA-1. Although FET5 is the most noisy FET because it is the smallest device, it contributes only 6% of the total output noise power because of its small loop gain. The SA-1 and SA-2 gain blocks have nearly identical $1/f$ noise behaviors with a corner frequency of the order of 10–100 MHz depending on the IC process (see Fig. 6.42 in Section 6.4.1 for an example of the $1/f$ noise characteristic).

6.3.2.5 SA-1 and SA-2 Performance The measured performances of both the SA-1 and SA-2 gain blocks are summarized in Table 6.3. Data in Table 6.3 are taken from the original papers by Hornbuckle and Van Tuyl [26,27]. Some general observations are (1) the voltage gain is low primarily because the transconductance g_m of the GaAs FET is inherently low, (2) the noise figure is high for this arrangement and GaAs suffers from very high $1/f$ or flicker noise, and (3) the dc power dissipation is high because depletion-mode FETs are exclusively used and the BLS network dissipates the majority of the dc power (in fact, it dissipates approximately 75% of the dc power given $V_{DD} = 6$ V and $V_{SS} = -4$ V). Section 6.4.1 describes multi-stage amplifiers using combinations of the SA-1 and SA-2.

With regard to low transconductance in GaAs FETs comment above, one reason why Van Tuyl and Hornbuckle chose the buffered voltage amplifier design was precisely because of this low transconductance. However, state-

[13] This value does not correspond to the CSAL gain cell noise figure in Figure 6.12 because of the 50 Ω shunting resistor included for input matching. The added resistor degrades the gain and noise figure.

TABLE 6.3 Comparison of SA-1 and SA-2 Measured Performance [26,27][a]

Parameters	SA-1	SA-2
Midband voltage gain	13 dB	7.2 dB
Bandwidth (−3 dB)	3.2 GHz	5.2 GHz
Midband noise figure	19.3 dB	—
$1/f$ noise corner frequency	10–50 MHz	10–50 MHz
Output power at −1 dB compression (note 2)	8 dBm	13 dBm
Second harmonic distortion	Best at low signal levels	Best at high signal levels
Third harmonic distortion	—	Better by 6–10 dB at all signal levels
Active area (notes 1, 3)	$190 \times 65\ \mu m^2$	$190 \times 65\ \mu m^2$
DC power dissipation (note 4)	104 mW	104 mW

[a] *Notes:*

1. SA-1 FET widths: FET1 is 40 μm; FET2 is 20 μm, FET3 and FET4 are both 40 μm; for SA-2, FET1 is 32 μm, and FET5 is 8 μm.
2. Power quoted as 50-Ω dBm equivalent [27].
3. Without bonding pads included.
4. Bias supply voltages: $V_{DD} = +6$ V and $V_{SS} = -4$ V.
5. FET parameters: $I_{DSS} = 200$ mA/mm, $g_m = 135$ mS/mm, $C_{gs} = 1$ pF/mm, $g_{ds} = 10$ mS/mm, and $V_p = -2.1$ V [34].

of-the-art HEMTs (or MODFETs) today have transconductances per unit width that are several times larger than a 1-μm GaAs FET technology. This allows the freedom to design with other amplifier topologies with much better performance.

6.3.3 Direct-Coupled Cascode Gain Blocks

Many direct-coupled GaAs IC amplifiers use a topology either identical to or similar to the Van Tuyl–Hornbuckle SA-1 and SA-2 gain blocks. However, numerous modifications intended to improve performance have been reported in the literature. Two direct-coupled cascode gain blocks are described below which advance the design art.

6.3.3.1 Colleran–Abidi Cascode Gain Block The Colleran–Abidi gain block [37] uses a cascode arrangement to improve the gain–bandwidth product. With the cascode configuration there are effectively two stages of gain, but only a single inversion which simplifies the application of feedback around the gain block. Figure 6.29*a* shows the simple cascode configurations discussed in Chapter 5 (Section 5.3.2.1). Colleran and Abidi [37] used the modified cascode arrangement shown in Figure 6.29*b*, where FET2 is smaller than FET1 by design. FET1 sinks the current of both active loads (i.e., FET3 and FET4). FET4 principally supports the dc current in FET1

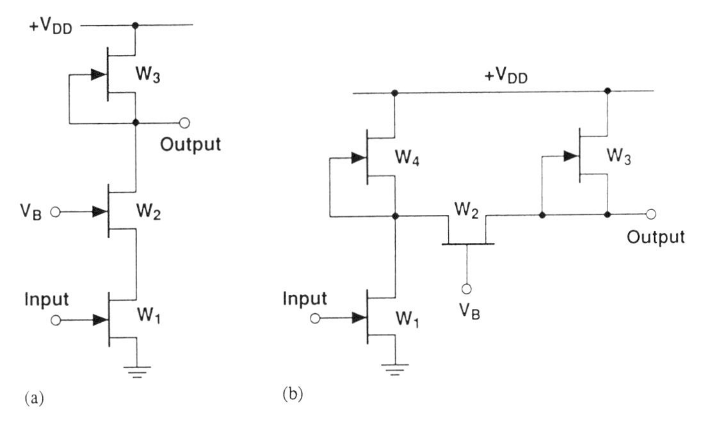

FIGURE 6.29 (*a*) Cascode gain cell with a single active-load transistor (as presented in Section 5.3.2.1 of Chapter 5) compared with (*b*) the Colleran–Abidi cascode gain cell.

while the greater part of the ac current flows in FET2. The voltage gain of the modified cascode cell is

$$G_{\text{cascode}} = -\frac{g_{\text{m1}}}{g_{\text{ds3}}} \cdot \frac{g_{\text{in}}}{g_{\text{in}} + g_{\text{ds1}} + g_{\text{ds4}}} \tag{6.42}$$

where g_{in} is the conductance looking into the source terminal of the common gate FET2

$$g_{\text{in}} = \frac{g_{\text{m2}} + g_{\text{ds2}}}{1 + (g_{\text{ds2}}/g_{\text{ds3}})} \tag{6.43}$$

By letting FET3 be small (i.e., conductance g_{ds3} approaches zero) the gain is maximized,

$$G_{\text{cascode}} \cong \frac{-g_{\text{m1}} g_{\text{m2}}}{(g_{\text{ds1}} + g_{\text{ds4}}) g_{\text{ds2}}} \tag{6.44}$$

A practical realization of this approximation is to choose the width of FET3 to be about one-tenth the combined widths of FET1 and FET4. As an example from Colleran and Abidi [37], FET1 and FET4 are 500 and 150 μm, respectively, while FET3 is 70 μm and FET2 is 200 μm. The cascode cell has better frequency response compared to the single-CSF stage because Miller multiplication of the drain-to-gate capacitance (FET1) is decreased. In the cascode configuration of Figure 6.29*b*, three poles are established by the time constants associated with the gate node of FET1, the drain node of FET1, and the drain node of FET2. These poles may be moved about by changing the FET widths, thereby eliminating any single dominant pole limiting the bandwidth [37]. A disadvantage to the modified

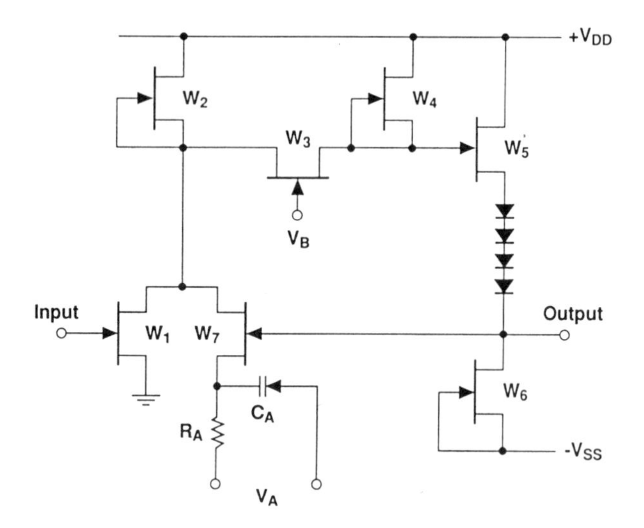

FIGURE 6.30 Complete cascode gain block with buffered level-shift network and feedback path (after Colleran and Abidi [37], © IEEE; reproduced by permission). The current source using FET6 is actually a cascoded current source in Reference [37], and bias details are omitted here.

cascode cell is the added phase delay because of the additional FET in the closed-loop feedback path.

The basic Colleran–Abidi gain block with feedback is shown in Figure 6.30. A conventional BLS network follows the cascode stage. Feedback FET7 is similar to that used in the SA-2, except a parallel R–C degeneration network is placed in series with FET7's source. Capacitor C_A is a voltage-variable capacitance formed by the junction of a Schottky barrier diode. This allows for some voltage-controlled variable peaking at the upper band edge. At high frequencies the excess phase shift[14] around the feedback loop becomes so large that substantial peaking may result (of course, this can lead to instability). The excess phase shift is reduced or counteracted by introducing a 90° "phase lead" from the presence of the capacitor C_A [i.e., FET7's drain current $i_D(W_7) \cong j\omega v_g(W_7)$, where v_g is the voltage at the gate terminal]. An alternate viewpoint is that the R–C network introduces a zero in the feedback function that pulls the complex closed-loop poles away from the $j\omega$ axis. A reduction in the amount of peaking in the response follows.

The improvement in gain–bandwidth product with the cascode cell was demonstrated by Colleran and Abidi by fabricating two separate amplifiers for comparison. A two-stage amplifier, an SA-1-type gain block driving an SA-2-type gain block, exhibited a 15.3-dB gain and a −3-dB bandwidth of 2.5 GHz. The other two-stage amplifier used cascode gain blocks similar to

[14] For pure negative feedback the phase of the feedback FET's drain current is opposite (i.e., 180°) to the phase of the inverter's drain current.

those shown in Figure 6.30 and gave 26 dB of voltage gain and a −3-dB bandwidth of 3.2 GHz. Further details appear in Reference [37].

6.3.3.2 Cascode Feedforward Gain Block In the cascode cell the direct connection between the output node of the inverter to the input node of the common-gate FET sacrifices voltage gain. The lost voltage gain results from a low impedance being presented to the inverter's output. Mauri et al. [38,39] added an intermediate buffer stage between the inverter and the common-gate FET. Figure 6.31 shows the gain block used in the work of Mauri et al. The intermediate buffer is formed by the combination of source follower FET3 and current source FET4. An appropriate number of diodes (schematically represented by diode D_1) are included for dc level shifting. Level shifting has the benefit of allowing for a smaller supply V_{DD} (e.g., $V_{DD} = +5$ V is a convenient choice). A disadvantage of the intermediate buffer is the introduction of even more phase shift with the application of feedback (via FET5 in Fig. 6.31). Mauri et al. fabricated gain blocks both with and without the intermediate buffer and demonstrated approximately a 3-dB improvement in voltage gain when the buffer is included.

To compensate some of the excess phase shift Mauri et al. used the R–C degeneration network of Colleran and Abidi (R_A in parallel with C_A) in the source lead of the feedback FET. The introduction of this network reduces the gain–bandwidth product by about 30% and also degrades the noise figure. With the intermediate buffer between the inverter and CGF stage

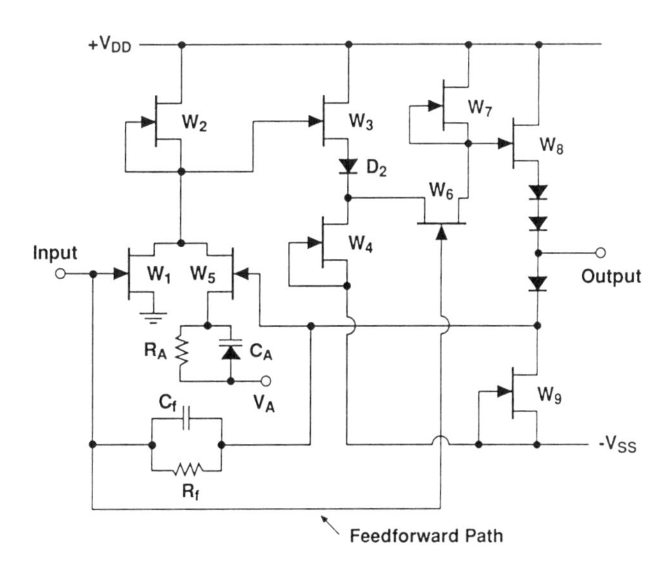

FIGURE 6.31 Cascode gain block with feedforward path and added buffer between inverter and common-gate stages (after Mauri et al. [38,39]).

instability is an even greater problem. To recover stability Mauri et al. introduced a *feedforward path* connected to the gate of the common-gate stage (viz., FET6 in Fig. 6.31). The feedforward path injects a signal that bypasses the inverter and intermediate buffer, and thereby avoids the phase shift from those two elements. The feedforward signal is vector combined with the inverter's amplified signal—it is not regenerative as long as it does not become too large a fraction of the total signal. Feedforward compensation resulted in a stable gain block [38,39].

An additional feedback network, the R_f–C_f parallel network in Figure 6.31, is connected back to the gate (input) of the inverter FET1. Colleran and Abidi had earlier introduced this network in their cascode gain block (although not shown in Fig. 6.30 for simplicity) for controlling the input impedance of their cascode gain block. With the inclusion of resistor R_f, the gain block takes the form of the familiar transimpedance amplifier (see Section 6.7 for a brief introduction to transimpedance amplifiers). Mauri et al. [38,39] included this R_f–C_f network for operation as a low-noise transimpedance amplifier.

6.4 MULTISTAGE AMPLIFIERS

Often a single stage of gain is not sufficient for a given application. The solution is to form multistage amplifiers using standard gain blocks. Four examples of direct-coupled, multistage amplifiers are described in Section 6.4.1. These amplifiers use SA-1 and SA-2 gain blocks in various combinations. In Section 6.4.2 an ac-coupled amplifier is described using a modified SA-2-type gain block.

6.4.1 Van Tuyl–Hornbuckle Multistage Amplifiers

In Section 6.3 the SA-1 and SA-2 gain blocks were described in some detail. Van Tuyl and Hornbuckle [40,41] used the SA-1 and SA-2 gain blocks in combinations to form higher-gain amplifiers. Four amplifiers from their work are briefly described and compared in this section. None of the four amplifiers are impedance-matched to 50 Ω at their input terminals; all have high input impedance (i.e., a small capacitive impedance equal to the input FET's C_{gs} plus parasitic wiring capacitance). Table 6.4 lists the sequence of stages and their respective FET widths for each of the four amplifiers. A primary attribute of these amplifiers is small physical size for low cost.

In cascading gain blocks one of the design variables is the size (i.e., width of input FET) of each succeeding stage relative to the stage driving it. This scaling of the gain block chain is shown schematically in Figure 6.32 for three cascaded SA-1 stages where each subsequent stage is n times larger than the stage driving it. Obviously the number of stages is dictated by the overall gain requirement. The output current required to drive a specified

TABLE 6.4 Summary of Direct-Coupled GaAs IC Amplifier Configurations

Amplifier	Stage 1	Stage 2	Stage 3	Output FET
A10	SA-1/30 μm (note 1)	SA-1/30 μm	SA-2/40 μm	(100 + 100) μm (note 2)
A60	SA-2/120 μm (L = 13 nH) (note 3)	—	—	500 μm
A61	SA-1/120 μm with resistive load (note 4) (L = 13 nH)	—	—	500 μm
A62	SA-2/80 μm	SA-2/120 μm (L = 13 nH)	—	500 μm

Notes:

1. FET width listed for SA-1 is FET1, and for SA-2 the combined width of FET1 and FET5 is given.
2. A10 amplifier has two 100-μm-wide output FETs that may be connected in parallel or used for dual outputs.
3. A60, A61, and A62 amplifiers each have 13-nH series peaking inductors.
4. A61 is an SA-1 except for passive resistor substituted for active load.
5. SA-1 and SA-2 bias voltages: $V_{DD} = 6$ V, $V_{SS} = -4$ V, and the output FET is biased at $V_{D,out} = 3$ V.
6. The A10, A60, and A61 were designed and fabricated in 1978 on the R2D2 mask set at Hewlett-Packard in Santa Rosa, CA. The A62 was designed and fabricated later in 1980 on ENA-1 mask set.

load determines the size of the last stage (e.g., width n^2W in Fig. 6.32). Several considerations influence the choice of n and involve selective engineering tradeoffs. As n increases, the overall amplifier bandwidth shrinks because the load capacitance C_L presented to each preceeding stage increases (e.g., loading capacitance ratio $n = C_L/C_{gs}$ as defined in Section 6.3 assuming wiring capacitance to be negligible). Smaller-width FETs have higher noise figures (as demonstrated in Fig. 6.12). Noise figure considerations, when important, dictate the input stage design. With the output stage

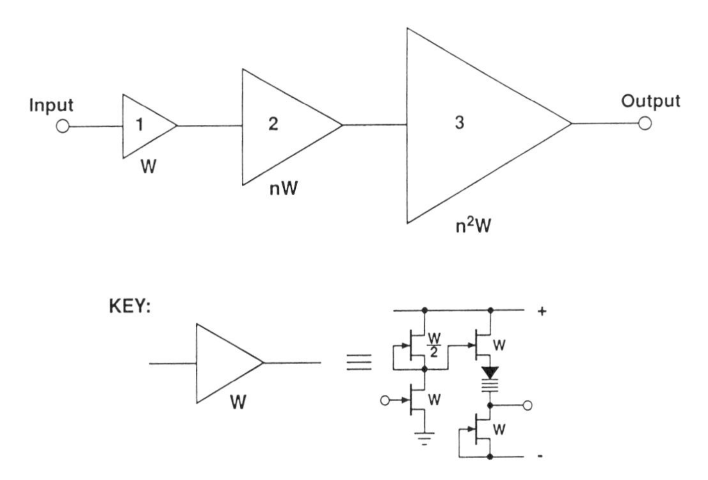

FIGURE 6.32 Cascaded gain stages with size scaling parameter n.

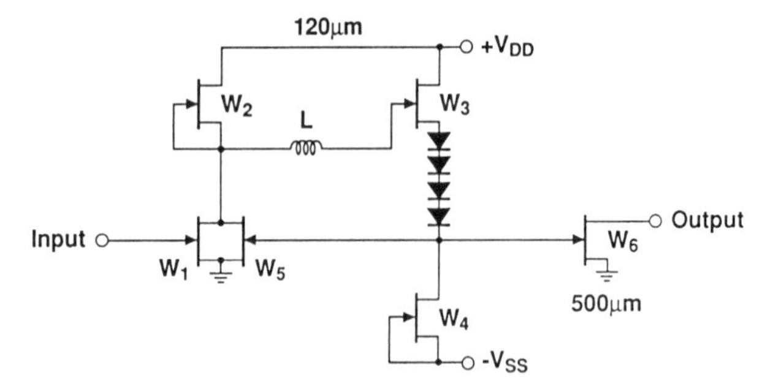

FIGURE 6.33 A60 direct-coupled amplifier schematic circuit (after Hornbuckle and Van Tuyl [25,26], © IEEE; reproduced by permission).

design established, larger n values lower the overall amplifier dc power dissipation and reduce the active area consumption because the earlier stages in the chain use smaller FETs (i.e., smaller drain currents).

The amplifiers described below use a single common-source FET (CSF) for the output stage. This is illustrated in Figure 6.33, where the A60 direct-coupled amplifier [40,41] consists of a single SA-2 stage driving a 500-μm-wide CSF output stage. The drain node of the 500-μm FET is the output terminal—no load element is provided on-chip. A separate bias connection, drain voltage $V_{D,out}$, is required along with an external impedance. While this can be an inconvenience in some instances, it does allow for a variety of load impedances and drain voltage settings depending on the application. For example, in Figure 6.34 a simple resistor load (R_L) is shown. In practice this resistor might be a thin-film resistor on an alumina or sapphire substrate, mounted as close as physically possible to the IC, and

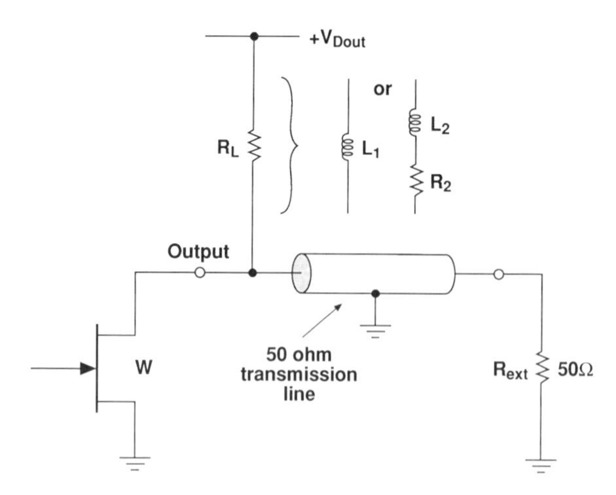

FIGURE 6.34 Common-source output FET with several load impedance options shown. When R_L and R_{ext} are both 50 Ω, the amplifier is said to be *reverse-terminated*.

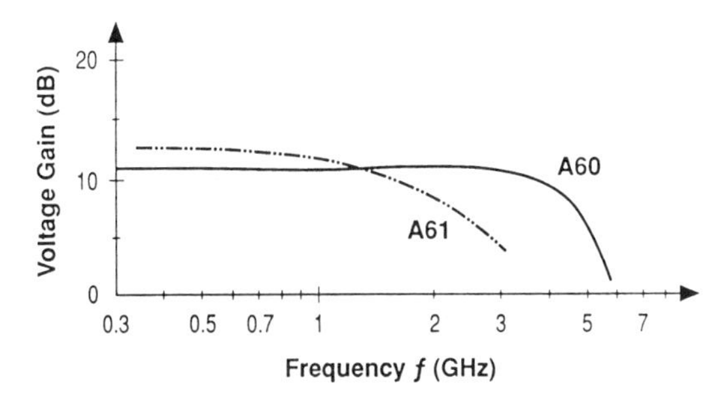

FIGURE 6.35 Comparing voltage gain versus frequency responses of A60 and A61 direct-coupled amplifiers.

connected by short wire bonds. A 50-Ω transmission line is shown connecting the amplifier's output node to a remote resistor R_{ext}, which is almost always 50 Ω. If R_{L} is chosen so that $R_{\text{L}} \| r_{\text{ds}}$ is equal to 50 Ω, the output is said to be *reverse-terminated*; that is, the amplifier's output resistance is matched to the impedance looking into the loaded transmission line (i.e., 50 Ω). The reverse-terminated case presents a 25-Ω impedance to the FET, so its voltage gain is $20 \cdot \log_{10}(25 \cdot g_{\text{m}})$ dB. This results in a 6-dB gain loss compared to the straightforward case where 50 Ω is directly connected to the drain node of the output FET. Other possible loads include a series R–L to extend the bandwidth via inductive peaking or a single inductor when the low frequency response is not important.

In the A60 direct-coupled amplifier a 13-nH spiral inductor is used to extend the bandwidth by inductive peaking (refer to the discussion in Section 5.4.3). The principle behind inductive peaking is to delay in time the charging of critical node capacitance, thereby improving the gain–bandwidth product of the stage. One consequence of inductive peaking is the introduction of additional phase shift, which in the limit can lead to positive feedback in an SA-2 stage. Figure 6.35 plots the voltage gain versus frequency of the A60 demonstrating a −3-dB bandwidth of approximately 4.5 GHz [40]. For the A60 the width of the inverter FET1 is 96 μm, 60 μm for the active-load FET2, and 24 μm for FET5. The source follower and level shift FETs (FET3 and FET4) are both 120 μm, and the output FET6 is 500 μm. Figure 6.36 is a photomicrograph showing the layout of the A60. The entire chip is 0.65 mm × 0.30 mm with the SA-2 stage positioned on the left side and the output FET to the right. Square bonding pads are 60 μm × 60 μm. The 13-nH spiral inductor measured 0.18 mm × 0.18 mm using 3-μm conductor traces and a 6-μm pitch (the inductor clearly dominates in area consumption).[15] Nominal dc power dissipation is about

[15] This is a good example of an expensive passive component as compared to inexpensive FETs (cost is proportional to the area consumed on an IC layout). As another example, active loads are less expensive than passive resistors because they achieve higher resistance values per unit area.

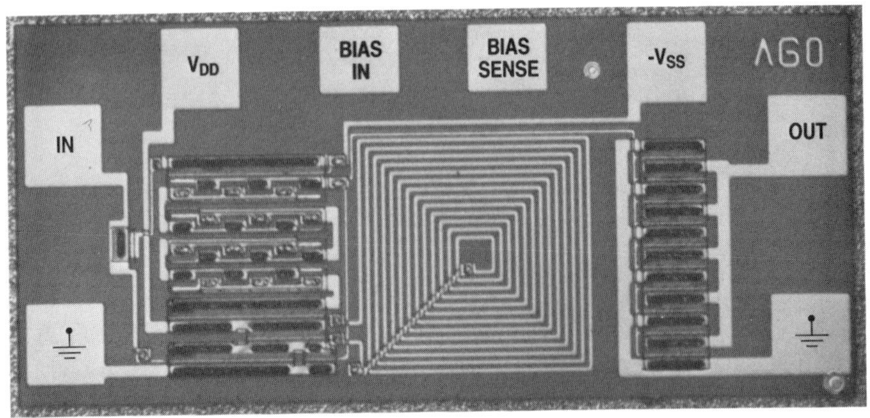

FIGURE 6.36 Photomicrograph of A60 direct-coupled amplifier from Hewlett-Packard R2D2 mask set (1978). (Reproduced with permission from Hewlett-Packard Company).

0.9 W under the conditions of $V_{DD} = 6$ V, $V_{SS} = -4$ V, and for the output FET, $V_{D,out} = 3$ V.

For another example consider the A61 direct-coupled amplifier consisting of a single SA-1 stage, but with the active load replaced with a 300-Ω passive resistor. The resistively loaded SA-1 gain block also includes a 13-nH peaking inductor. For comparative purposes the gain responses for both the A60 and the A61 are plotted in Figure 6.35. Although a passive resistor is more linear than an active load, the distortion performance of the A61 is not as good as the A60's distortion performance because the A60 uses feedback [40]. Furthermore, feedback in the A60 results in greater bandwidth. The noise figure for the A60 is 16.4 dB, while that for the A61 is about 1–2 dB lower. The passive resistor load contributes less noise than an active load.

An example of a two-stage amplifier is the A62 direct-coupled amplifier shown in Figure 6.37. It uses two cascaded SA-2 gain blocks followed by a single CSF output stage. The A62 has a series peaking inductor in the second SA-2 stage. The output FET is a 500-μm-wide device so chosen to

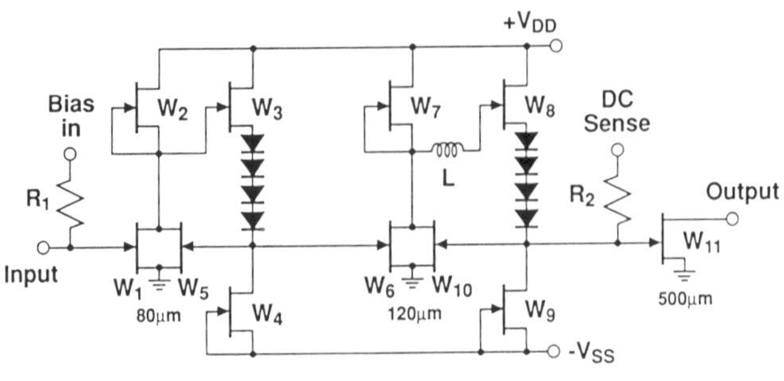

FIGURE 6.37 Two-stage A62 direct-coupled amplifier schematic circuit.

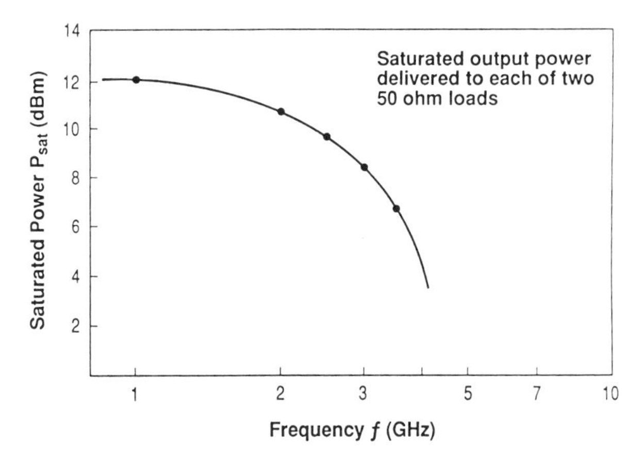

FIGURE 6.38 Saturated output power P_{sat} of A62 amplifier from 1 to 4 GHz.

provide higher output power capability. In a reverse-terminated connection the A62 delivers a monotonically decreasing saturated output power (P_{sat}) from 1 to 4 GHz as shown in Figure 6.38. For frequencies below 1 GHz the A62 delivers a saturated power greater than 12 dBm. The falloff in P_{sat} with frequency occurs because of slew-rate limiting. The midband small-signal voltage gain of the A62 is approximately 9 dB with a −3-dB bandwidth of 3.8 GHz. The A62's low voltage gain of 9 dB is with a 25-Ω load, whereas the voltage gain data in Figure 6.35 for the A60 and A61 are for a 50-Ω load. With a 50-Ω load the A62 would have approximately 15 dB of voltage gain. The dc power dissipation is close to 0.7 W with $V_{DD} = 6$ V, $V_{SS} = -4$ V, and for the output FET, $V_{D,out} = 3$ V at $I_{D,out} \cong 60$ mA.

A three-stage direct-coupled amplifier is shown in Figure 6.39. The A10 amplifier consists of two SA-1 gain blocks driving a single SA-2 gain block. No peaking inductors are used in the A10, and the output FET is a split into two identical 100-μm FETs (FET14 and FET15). This allows for either dual

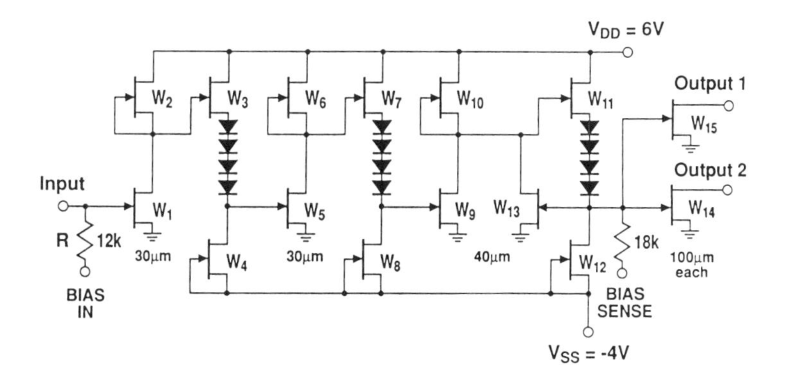

FIGURE 6.39 Three-stage A10 direct-coupled amplifier schematic circuit.

outputs or both FETs can be connected together to function as a single 200-μm output stage. Figure 6.40 is a photomicrograph of the A10 amplifier showing its small active area (the chip size is twice that of the A60 in Fig. 6.36). Voltage gains (vs. frequency) for both the A10 and A62 amplifiers are compared in Figure 6.41. The much smaller output FET (loaded with 25 Ω during the measurement) of the A10 accounts for its relatively low voltage gain. The bandwidth of the A10 is only 1.3 GHz, largely because the SA-1 gain block is used for the first two stages and wiring capacitance is more dominant because smaller FETs are used. Other similar two-stage and three-stage direct-coupled amplifier examples are presented in References [40] and [41].

A limitation of all these small FET direct-coupled amplifiers is high noise figure. This is illustrated in Figure 6.42 for the A10, where the noise figure is

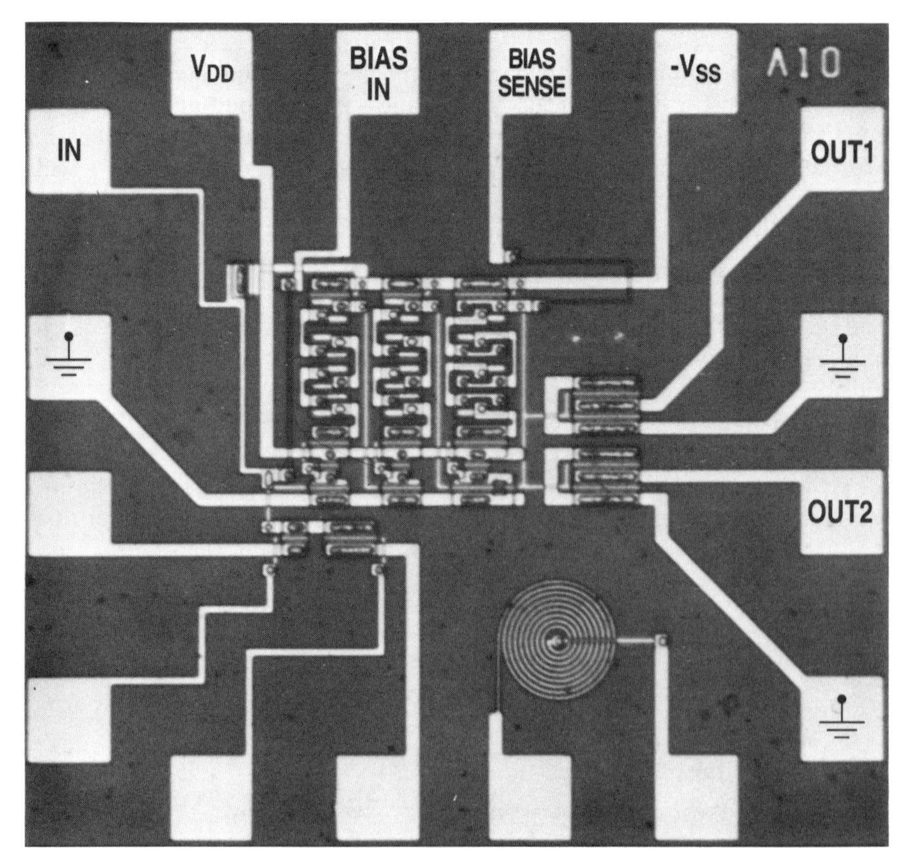

FIGURE 6.40 Photomicrograph of A10 direct-coupled amplifier chip. Notice the separate spiral inductor test pattern in the lower right and an optional attenuator–modulator in the lower left part of the tile. The chip size is $0.60 \times 0.65\,\text{mm}^2$. (Reproduced with permission from Hewlett-Packard Company).

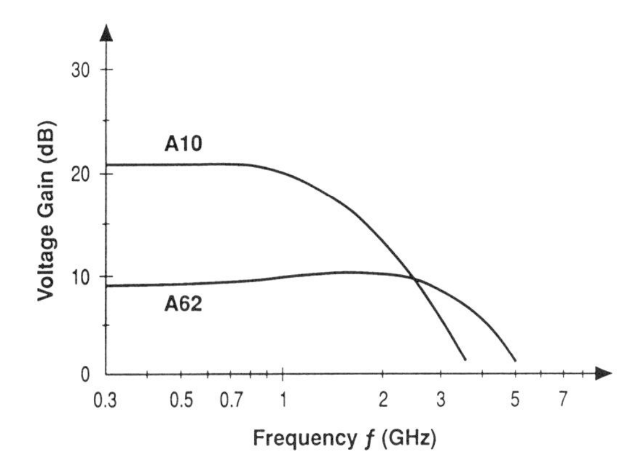

FIGURE 6.41 Plot of voltage gain versus frequency for both the A62 and A10 amplifiers.

plotted from 10 kHz to 1 GHz. Above the $1/f$ noise corner frequency the noise figure is of the order of 25 dB, an extraordinarily large value. This compared with about 18–20 dB for the A60 amplifier at 1 GHz. Measured noise figure data from two wafers are compared to a SPICE simulation in Figure 6.42. An equivalent input noise voltage[16] of $1.5\,\text{nV-mm}^{1/2}/\sqrt{\text{Hz}}$ was used in the SPICE simulation. This value was experimentally determined from FET test patterns using the fabrication process reported in Van Tuyl et al. [42]. The agreement between measured and simulated midband (e.g.,

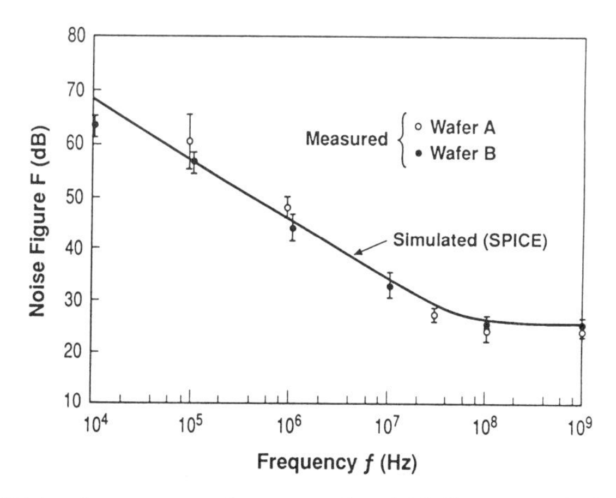

FIGURE 6.42 Noise figure versus frequency for A10 direct-coupled amplifier. Note the very large $1/f$ noise component.

[16] This equivalent input noise voltage is somewhat higher than what is achieved by most GaAs fabrication facilities nowadays. It was representative of the process under development in 1979. It is 50% higher than the nominal value used in Section 6.2.4.

1 GHz) noise figure is within about 1 dB. The $1/f$ noise corner frequency is approximately 40 MHz, which is representative of many GaAs IC processes.

Stability in the static bias point (i.e., offsets and drift) is another problem in direct-coupled, multistage amplifiers. An offset voltage at the output of one stage will be amplified by the next stage to produce an even larger offset or error voltage.[17] There are many possible causes of offsets—for example, nonrepeatibility in the diode stack level-shift voltage, mismatches in FET parameters (e.g., I_{DSS} or pinchoff voltage), changes in power supply voltages, or backgating. Changes in FET drain current over temperature causes dc output voltage drift in the SA-1 and SA-2 gain blocks. One approach to stabilizing bias points is to use an operational amplifier in a feedback path to stabilize the static bias states. Referring to the A62 schematic diagram in Figure 6.37, the "DC sense" resistor R_2 provides the monitor signal to the operational amplifier, while the error correction signal out of the operational amplifier is sent to the "bias in" resistor R_1. This scheme stabilizes the dc bias but prevents the amplifier from being used at frequencies below the loop bandwidth of the feedback loop. Another option is to include still another small FET in parallel with the inverter FET with the feedback connected to the gate of this FET to stabilize the bias point. Of course, if gain is not required all the way down to dc, an especially simple and effective solution is to ac-couple the stages (an example of this appears in Section 6.4.2).

Another approach is to add components within the basic SA-1 and SA-2 gain blocks to stabilize the dc bias state. An example is shown in Figure 6.43 for the SA-1 cell. Symmetry is used in the circuit (component group FET3–D_1–D_2 is identical to FET4–D_3–D_4) to render the dc output voltage insensitive to power supply variations or changes in FET pinchoff voltage [43]. The ratio of resistors R_1 and R_2 establishes ΔV_4, and $\Delta V_3 = \Delta V_4$ because $I_D(W_3) = I_D(W_4)$. Assuming matched components, the dc output voltage is maintained at 0 V for the topology shown in Figure 6.43. Of course, mismatches among the components in the BLS stage can still produce unwanted offsets.

These direct-coupled, multistage amplifiers have had considerable influence on wideband GaAs IC amplifier design in the 1980s. Table 6.5 summarizes the practical performance achievable with this class of multistage amplifiers with a standard 1-μm MESFET process. One of the original goals in the work of Van Tuyl and Hornbuckle was small chip size for low cost. The small size of an IC such as the A60 amplifier (Fig. 6.36) allows approximately 20,000 copies of this circuit on a 3-in. GaAs wafer. Assuming

[17] The SA-1 transfer curve in Figure 6.26 is useful in visualizing this phenomenon. For example, a change in the level-shift voltage produces a change in the output voltage even though bias point Q may remain stable. Likewise, a change in point Q will generally lead to changes in the output voltage. Unfortunately, shifts in point Q are seldom canceled by an equal and opposite change in the level-shift voltage.

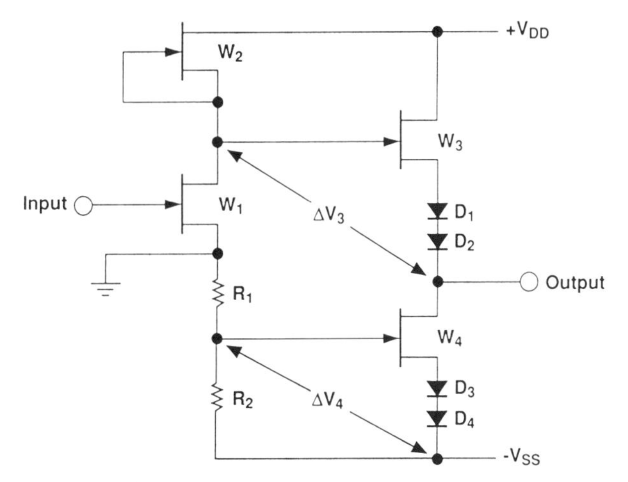

FIGURE 6.43 Symmetric SA-1-type gain block with controlled dc offset voltage (after Vella-Coleiro [43]).

1994 wafer costs, this suggests amplifiers at under one-third dollar per IC with wafer testing costs included.

6.4.2 A 10-MHz–3-GHz Wideband AC Amplifier

An ac-coupled, multistage GaAs IC amplifier [44,45] is described in this section that operates from ±5-V power supplies, is input matched to 50 Ω, has lower noise figure than the Van Tuyl–Hornbuckle multistage amplifiers, has a dc power dissipation under 1 W, and has 26 dB of voltage gain from 5 MHz to greater than 3 GHz. Figure 6.44 shows the block diagram of the amplifier. It is a four-stage amplifier with a common-gate input stage for lower noise and two intermediate stages that are "modified SA-2" stages. The output stage is a CSF identical to that described in Section 6.4.1. The complete amplifier schematic is shown in Figure 6.45, and Table 6.6 lists all FET widths used in the amplifier.

The "modified SA-2" gain block is described first. Figure 6.46 shows the schematic circuit of the "modified SA-2' with $V_{DD} = +5$ V and $V_{SS} = -5$ V

TABLE 6.5 Summary of Direct-Coupled, Multistage Amplifier Generic Performance [40,41]

Parameter	Value or Range
Gain at bandwidth	25 dB at dc—2 GHz and 10 dB at dc—5 GHz
Distortion	−30 dBc (average) at 10 dBm P_{out}
Midband noise figure	15 dB (100-μm input FET)
$1/f$ corner frequency	10–100 MHz

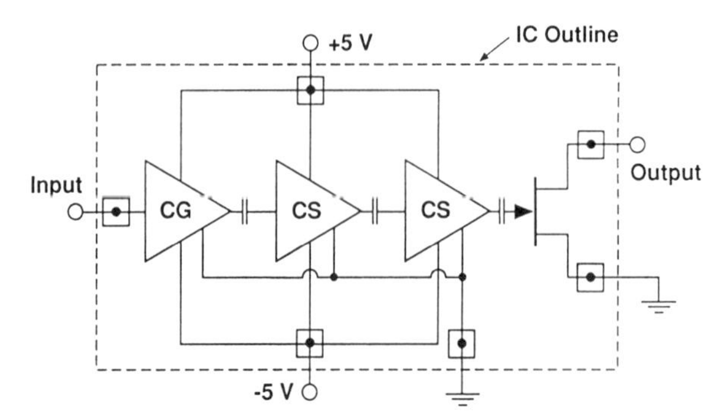

FIGURE 6.44 Block diagram of ac-coupled 3-GHz GaAs IC amplifier.

(compared to $V_{DD} = +6$ V and $V_{SS} = -4$ V for the SA-2). Two goals dictated design changes in the SA-2: (1) reduce the static power dissipation and (2) improve the static bias point stability because of reduced V_{DD}. Lowering V_{DD} to 5 V requires the inverter's internal node voltage V_i to remain dc-centered (i.e., $V_i = \frac{1}{2}V_{DD}$) to avoid premature clipping at the output (especially for stage 3, where the highest signal swings occur). To achieve this degree of stability, feedback is necessary. The BLS network (see Fig. 6.21) dissipates 75%–80% of the total dc power in the SA-2. This is too high to meet the first goal listed above. The dc power dissipation is reduced by splitting the BLS network into a level-shift–feedback network with lower quiescent current and a separate buffer–driver network with a lower supply voltage. Referring to Figure 6.46, the level-shift–feedback network is formed with FET5 and FET6 along with the four-diode stack and series resistor R_D. The size of current source FET6 is chosen such that current I_B is a small fraction of the inverter's drain current (Estreich [45] chose I_B to be 18% of I_D).

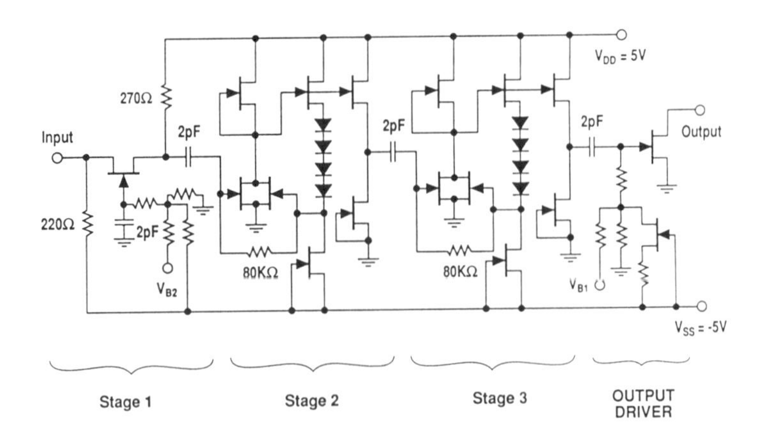

FIGURE 6.45 Complete schematic circuit of AC-coupled 3-GHz GaAs IC amplifier.

TABLE 6.6 FET Widths in AC-Coupled Amplifier

Stage	FET	FET Width (μm)
Input stage	CGF	248
Second stage	Inverter	100
Second stage	Active load	63
Second stage	Source follower (driver)	108
Second stage	Current source (driver)	108
Second stage	Feedback source follower	12
Second stage	Feedback current source	12
Second stage	Feedback FET	27
Third stage	Inverter	150
Third stage	Active load	94
Third stage	Source follower (driver)	164
Third stage	Current source (driver)	164
Third stage	Feedback source follower	18
Third stage	Feedback current source	18
Third stage	Feedback FET	52
Output stage	Output FET	528
Output stage	Bias control	20

Lower dc power dissipation is achieved in the buffer–driver (i.e., source follower as described in Section 5.7.2) because of the lower supply voltage (5 V versus 10 V in the SA-2).

The dc and low-frequency feedback in the "modified SA-2" is applied to both the feedback device FET7 and the inverter FET1. However, RF feedback in the passband is applied only through FET7 because R_{dcf} is large (80 kΩ in this amplifier). The capacitance presented to R_{dcf} places the loop transmission pole (in the loop back to FET1) below the lower frequency of the amplifier's passband. The principal design parameter in establishing

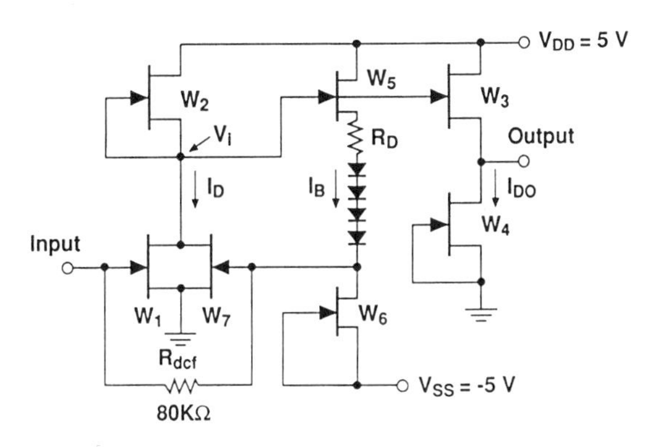

FIGURE 6.46 "Modified SA-2" schematic circuit with feedback path separate from buffer–driver output.

$V_i = \frac{1}{2}V_{DD}$ is the voltage drop across the level-shift network. With the enhanced feedback in the "modified SA-2" the sensitivity of voltage V_i to the FET parameters is greatly reduced.

The nominal gain of the "modified SA-2" gain block is 6.5 dB with a −3-dB bandwidth greater than 3.5 GHz. Its operation is analogous to that described for the SA-2 in Section 6.3.2 with one exception. The buffer–driver has a smaller maximum output signal swing because of the reduced voltage across the combination of FET3 and FET4.

To achieve a lower noise figure (relative to the Van Tuyl–Hornbuckle amplifiers), while still retaining high voltage gain and simultaneously obtain a broadband match to 50 Ω, a common-gate FET forms the input stage. The common-gate FET is known to perform an active matching function as explained in Section 5.3.1.2. Section 6.2.5 presented the superior noise performance of the common-gate FET over a common-source FET of identical gate width (see Figure 6.12). Although a larger common-source FET could be used to achieve a lower noise figure, holding the dc power dissipation to under 200 mW (so as to keep the total amplifier dc power to below 1 W) dictated the selection of the smaller common-gate stage. Figure 6.47 shows the input stage used in this amplifier where passive load and source bias resistors are used, rather than FETs, to reduce the noise figure. The voltage gain of this stage is

$$G_{v1} = \frac{g_m r_{ds} + 1}{1 + (r_{ds}/R_L)} \tag{6.45}$$

where R_L is defined in Figure 6.47. An FET with a width of 220 μm ($g_m = 29.7$ mS and $r_{ds} = 455\ \Omega$) gives $G_{v1} = 5.40$ (14.65 dB) with $R_L = 270\ \Omega$.

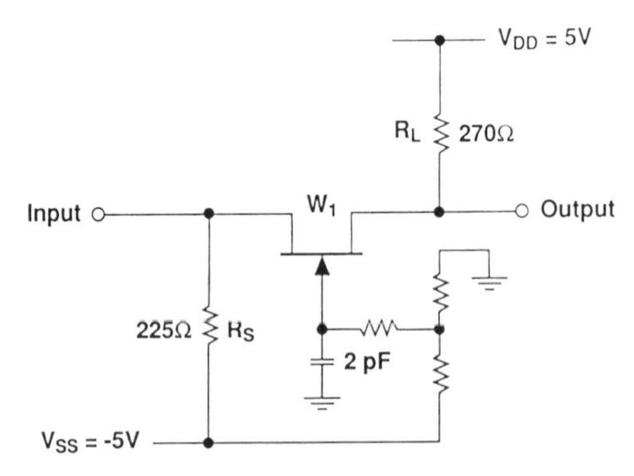

FIGURE 6.47 Common-gate FET input stage of ac-coupled 3-GHz GaAs IC amplifier.

The input resistance R_{in} at low to midrange frequencies is

$$R_{in} = \frac{(r_{ds} + R_L)R_S}{(g_m r_{ds} + 1)R_S + (r_{ds} + R_L)} \tag{6.46}$$

Although the $1/g_m$ dependence dominates the behavior of R_{in}, R_L and R_S weakly influence R_{in} because it monotonically but slowly increases as both R_L and R_S increase [45]. The common-gate stage is a good compromise between reasonable noise figure and high voltage gain, but with the additional advantages of small size, low power dissipation, and excellent control of the input impedance over a wide frequency range.

The nominal midband voltage gain of the complete amplifier is 26 dB with a 50-Ω load presented to the output FET. Figure 6.48 shows the amplifier's gain response over a wide frequency range. Both measured and SPICE simulated [45] data are compared in Figure 6.48. The −3-dB bounds on the measured response fall at 5 MHz and 3.3 GHz. A midband noise figure of approximately 9 dB is achieved with the common-gate input stage, and the $1/f$ noise corner frequency is 70 MHz. Input reflection coefficients from 0.1 to 0.25 are typically observed over the 5 MHz–4 GHz range. Figure 6.49 plots the fundamental output power and second and third harmonic powers versus input power at the fundamental frequency of 500 MHz. A 1-dB compressed output power of 15 dBm is typically measured (as high as 18 dBm for select wafers). The dc power dissipation is approximately 0.9 W when $V_{DD} = +5$ V and $V_{SS} = -5$ V. A photomicrograph of the amplifier appears in Figure 6.50. No on-chip matching elements are included, allowing for a very small die size of 0.44 mm × 0.65 mm. This amplifier was

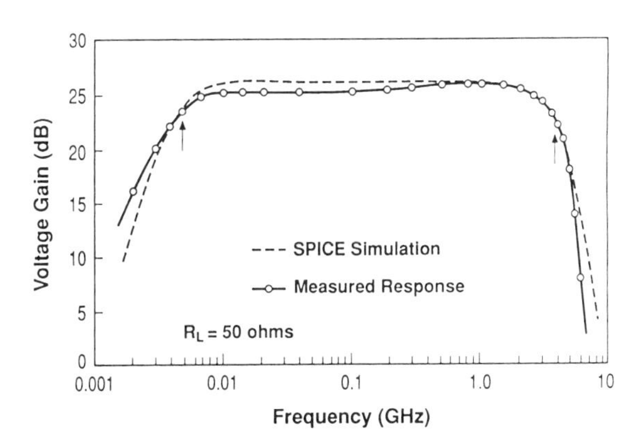

FIGURE 6.48 Voltage gain versus frequency for ac-coupled 3-GHz GaAs IC amplifier. Measured results are compared with SPICE simulation (after Estreich [45], © IEEE; reproduced by permission).

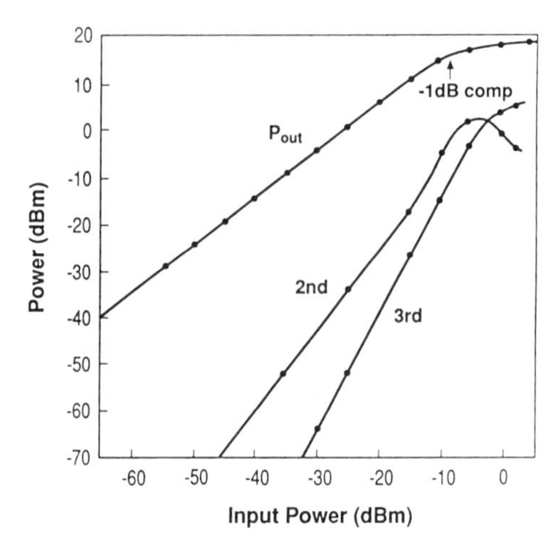

FIGURE 6.49 Fundamental frequency output power, second harmonic and third harmonic powers versus fundamental frequency input power for ac-coupled 3-GHz GaAs IC amplifier (after Estreich [45], © IEEE; reproduced by permission).

used in one of the first electronic instruments (a 3-GHz economy frequency counter introduced in early 1986 by the Hewlett-Packard Company) to utilize GaAs monolithic ICs [46].

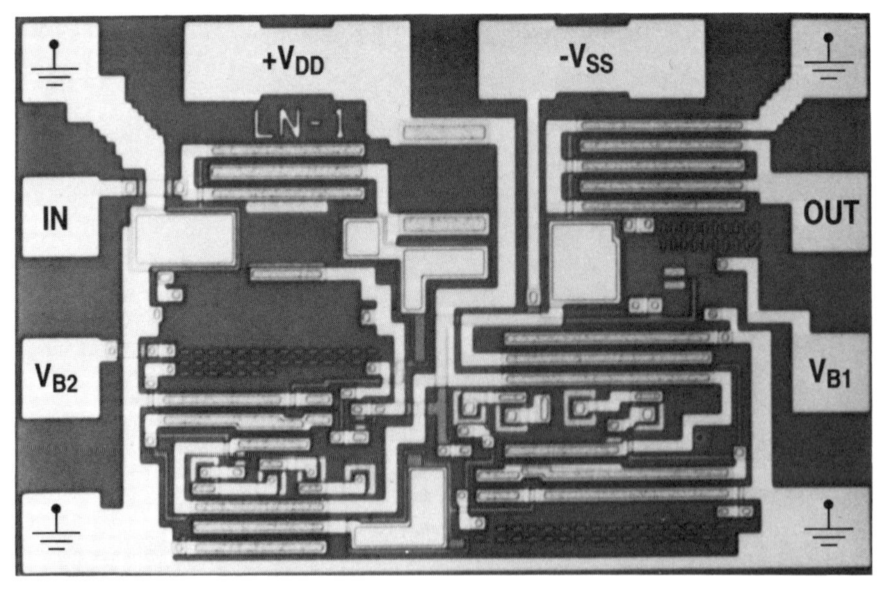

FIGURE 6.50 Photomicrograph of ac-coupled 3-GHz GaAs IC amplifier showing very compact layout (after Estreich [45], © IEEE; reproduced by permission).

6.5 GAIN-CONTROL AMPLIFIERS

In certain applications it is necessary to control or vary the gain of an amplifier. For example, a microwave radio receiver might require a narrow range of RF signal amplitudes input to the mixer for optimum performance. The strength of the input signal from the receiving antenna can vary widely because of changing transmission conditions. A variable-gain amplifier between the antenna and mixer can compensate the signal amplitude to the mixer by using a feedback-leveling loop. In Section 6.5.1 circuit techniques are described to execute the gain-control function in amplifiers. An example of a gain-control amplifier is presented in Section 6.5.2.

6.5.1 Circuit Techniques for Gain Control

Consider the simple common-source FET amplifier cell shown in Figure 6.51. If the drain-to-source resistance r_{ds} is much larger than the load resistance R_L, the voltage gain of the amplifier is

$$G_v = \frac{-g_m R_L}{1 + g_m R_S} \tag{6.47}$$

The voltage gain of this amplifier cell can be changed by either (1) varying the FET's transconductance g_m, (2) varying the load resistance R_L, or (3) changing the magnitude of negative feedback (in this circuit the negative

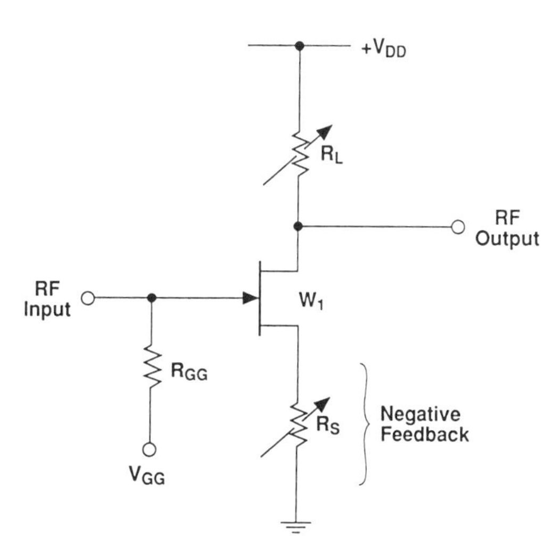

FIGURE 6.51 Common-source FET amplifier stage with variable load resistor and variable source degeneration resistor used for negative feedback. Change in gate voltage provides for changing the FET's transconductance by bias point shift.

feedback is accomplished with a series source resistor R_S, although any form of negative feedback would work in principle). In general, the most practical approach of the three is to vary the strength of the feedback.

Figure 6.13 illustrates the dependence of g_m on the gate-to-source voltage in a typical GaAs FET. Unfortunately, for smaller values of g_m the FET must be biased near pinchoff (i.e., $V_{GS} \cong -|V_p|$). This leads to significant distortion when the input signal level is large (which is often the condition for which the lowest gain setting is required). Furthermore, biasing an FET at very low drain currents (say, $I_D < 0.1\ I_{DSS}$) is not desirable in low-noise applications because the noise figure rises as g_m falls.

Varying either the load resistance or feedback is a better solution. It is difficult to vary the load resistance over a sufficiently wide resistance range and still maintain good linearity under the condition of large signal amplitudes. Greater design flexibility is inherent in varying the magnitude of a feedback signal. In Figure 6.51 the variable source degeneration resistance could be an FET operated in its linear region, where the gate voltage controls the value of R_S. This requires keeping the voltage drop across R_S (i.e., the FET) well below the threshold of saturation (i.e., $V_{D,sat}$) or knee voltage (i.e., $V_{knee} \equiv V_K$). For larger voltage swings more than one FET can be placed in series to divide the voltage among the FETs.

A commonly used method to control the gain of a single gain stage is to use the cascode arrangement[18] with the bias varied on the gate of the common-gate FET [47,48]. This is shown in Figure 6.52*a*, where voltage V_{G2} is the gain-control voltage. For maximum gain both FETs are biased into the saturated region of operation. The control voltage V_{G2} is equal to the sum $(V_{DS1} + V_{GS2})$, subject to the constraint, $I_{D1} = I_{D2}$, assuming negligible gate current in FET2. Figure 6.52*b* illustrates the general behavior of the transconductance of the cascode connection as a function of V_{GS1} and V_{G2}. The maximum transconductance value in Figure 6.52*b* corresponds to $V_{G2} = V_{G2}^{opt}$. As V_{G2} is reduced below V_{G2}^{opt}, drain voltage V_{DS1} decreases. Decreasing V_{DS1} below the saturation threshold of FET1 (i.e., $V_{DS1} < V_{D,sat}$ or V_K) results in a drop in both the transconductance g_{m1} and the output resistance r_{ds1} of FET1. This corresponds to FET1 falling out of saturation. Simultaneously, the dc drain current decreases. An advantage of the cascode arrangement in Figure 6.52*a* is its wide dynamic range (e.g., over 40 dB in power variation demonstrated in Refs. [47 and 48]). The disadvantages are poor distortion performance at the lower gain settings and the careful control required of the microwave impedance presented to the gate node of FET2 [47] Some gate node terminating impedances can lead to instability at the upper edge of the band.

The differential amplifier configuration (see Section 5.3.2.3 of Chapter 5) is especially convenient for implementing gain control because symmetry has distortion cancellation advantages. Consider Figure 6.53, showing the use of

[18] Often a dual-gate FET is used in this application because of its compact layout.

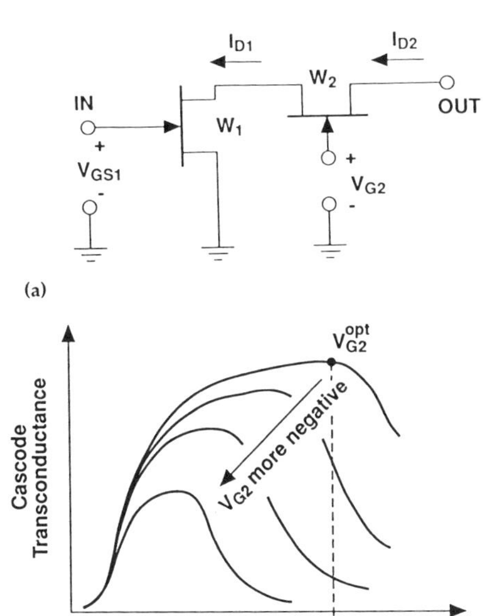

FIGURE 6.52 (*a*) FET cascode configuration (minus bias elements); (*b*) general behavior of the transconductance of the cascode cell as a function of both gate-control voltages.

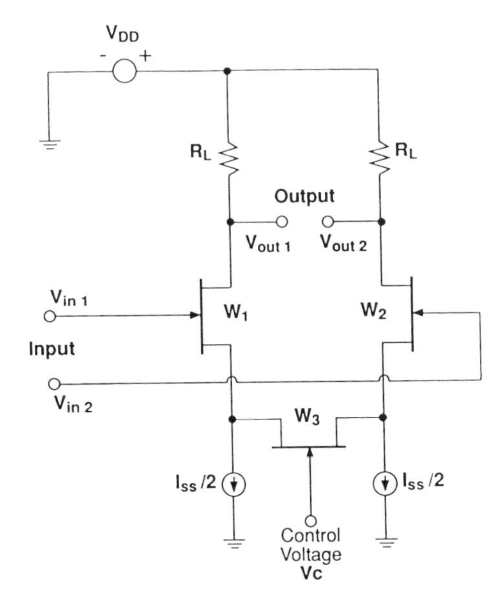

FIGURE 6.53 Differential amplifier with shunt FET3 across source nodes. FET3 is used as a variable source degeneration resistance.

an FET as a variable resistance for controlling the differential gain by means of source degeneration. FET3 is intended to be operated in its linear region of operation. When FET3 is off, the gain is minimum (ideally the gain is zero) because both FET1 and FET2 are constrained to have drain currents equal to $I_{ss}/2$, independent of the input voltages. Raising control voltage V_c turns FET3 on, thereby allowing current to be shared between the source nodes. Assuming the drain-to-source resistance of FET3 to be $2R_S$, where R_S is a function of voltage V_c, the differential gain G_v $(=[(v_{out2} - v_{out1})/(v_{in2} - v_{in1})])$ of the amplifier is

$$G_v = \frac{-g_m R_L}{1 + g_m R_S + [(R_S + R_L)/r_{ds}]} \tag{6.48}$$

In the limit of r_{ds} going to infinity, Equation (6.48) reduces to Equation (6.47). For larger voltage handling capability and better linearity, FET3 can be replaced with multiple FETs in series to divide the signal voltage among the string of FETs.

The same approach can be used to vary the load resistance. The circuit shown in Figure 6.54 uses a drain-to-drain connected FET to vary the load resistance. A disadvantage to this approach is the limited signal range over which the channel resistance of FET3 remains linear (say, of the order of

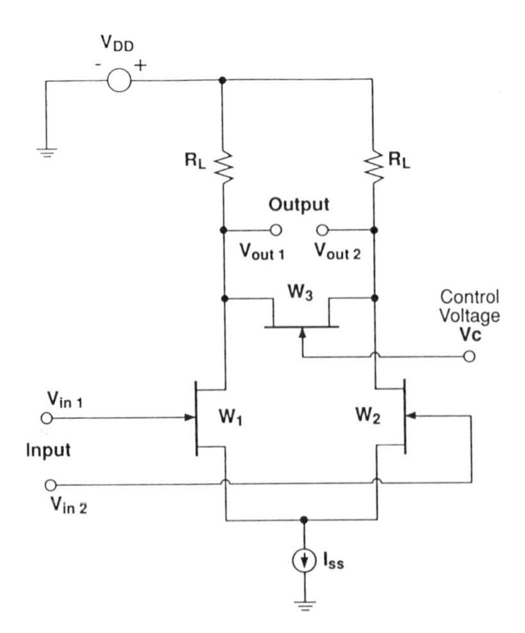

FIGURE 6.54 Differential amplifier with shunt FET3 across drain nodes. The voltage-controlled channel resistance of FET3 allows for a variable load resistance.

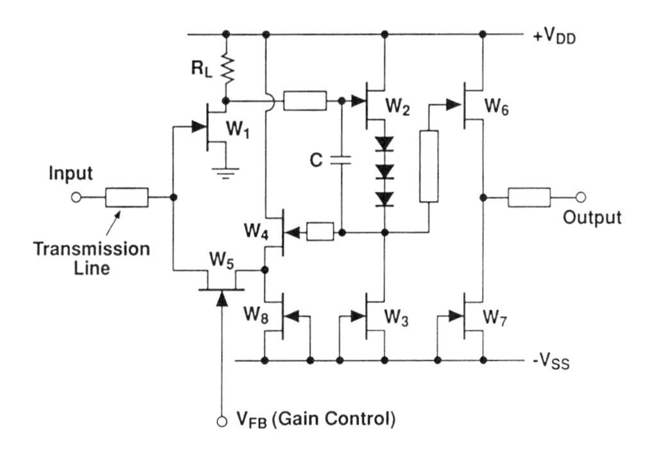

FIGURE 6.55 Variable-gain wideband feedback amplifier stage (after Shigaki et al. [49], © IEEE; reproduced by permission).

±0.2 V). Again, multiple FETs can be used to improve on this limit provided the proper gate-control voltages are supplied. Another approach is to use a cascode arrangement with FET3 imbedded between the internal nodes[19] of the cascode connecting because the voltage swing is usually smaller at the internal node.

6.5.2 A Gain-Control Amplifier Example

The gain-control example [49] presented in this section uses a variable-shunt negative-feedback loop to control the gain. Figure 6.55 shows a gain stage from Shigaki et al. [49]. FET5 is used as a variable resistor to control the magnitude of the feedback. To maintain the high impedance at the level-shift node, a buffer source follower (FET4) must be included. In addition, FET4 improves the input/output isolation. The common-source device (FET1), output buffer FETs (FET6 and FET7), and all level-shift FETs are all 200 μm. The feedback FET is 40 μm. Operation from dc to 6.4 GHz with 11 dB of gain-control range was demonstrated in Shigaki et al. [49]. The maximum gain of the single-gain cell as shown in Figure 6.55 is between 5 and 6 dB. On-chip transmission lines were included to extend the frequency range. A disadvantage to using the FET as a voltage-controlled resistor in a feedback path is its inherent nonlinearity when subjected to larger signal amplitudes. This limits this technique to small-signal applications.

[19] The internal node of the cascode arrangement is the connection of the common-source FET's drain to the common-gate FET's source. This is the node labeled V_b in Figure 5.11 of Chapter 5.

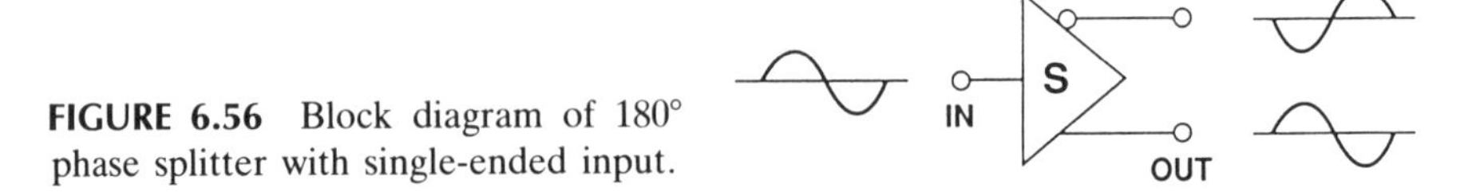

FIGURE 6.56 Block diagram of 180° phase splitter with single-ended input.

6.6 PHASE-SPLITTING AMPLIFIERS

Signal splitters producing a 180° phase shift between two equal-amplitude signals are useful for driving various differential circuits such as amplifiers, mixers, balanced modulators, and frequency dividers. Figure 6.56 is a block diagram of a generic 180° phase splitter with single-ended input. Active phase splitters using the monolithic approach can have greater than unity gain (whereas passive splitters cannot) and excellent isolation between output ports.

The simplest 180° phase splitter can be constructed using a single FET and two resistors as shown in Figure 6.57. The single-FET phase splitter makes use of the inverting property of the common-source FET connection and the non-inverting property of the common-drain FET. It has less than unity voltage gain, which is readily seen using Equation (6.48), where $R_S = R_L \equiv R$ as required for equal signal amplitudes:

$$G_v^{ps} = \frac{\pm g_m R}{1 + g_m R + (2R/r_{ds})} \tag{6.49}$$

The plus sign corresponds to the OUT signal and the minus out to the $\overline{\text{OUT}}$ signal. As an example, for $g_m = 100$ mS, $r_{ds} = 150\ \Omega$, and $R = 50\ \Omega$, $G_v^{ps} =$

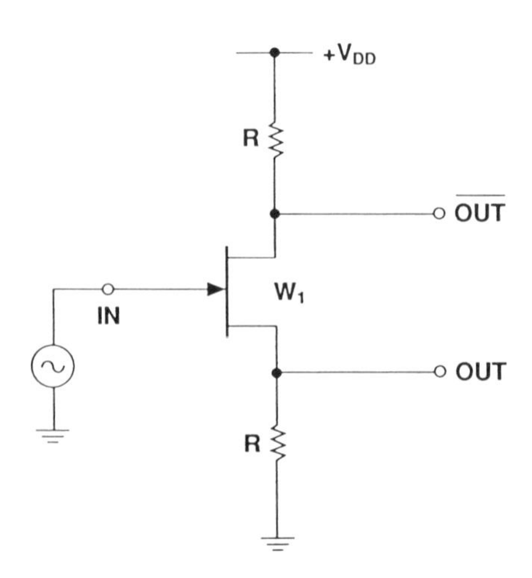

FIGURE 6.57 Simple single-FET phase splitter.

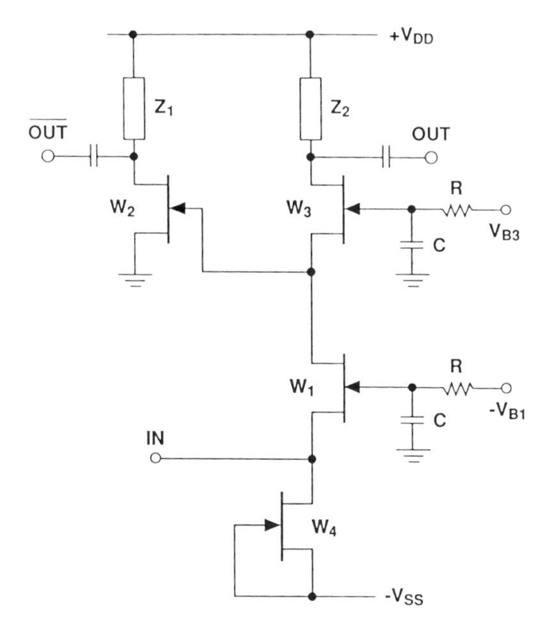

FIGURE 6.58 Wideband FET phase splitter with voltage gain greater than unity.

0.75 or −2.5 dB. The circuit in Figure 6.57 has good linearity because of the large negative feedback (which gives rise to its low gain). The input impedance is capacitive and varies over a wide range for wide bandwidths. The addition of a small amount of compensating capacitance at the source node may be advantageous to improve phase tracking between OUT and $\overline{\text{OUT}}$ at higher frequencies.

An example of an active phase splitter with greater than unity gain is shown in Figure 6.58. FET1 is a common-gate stage (refer to Section 5.3.1.2), which allows for controlling the input impedance over a very wide band and also serves as a buffer for improved isolation from output to input. The output of FET1 drives two independent gain blocks. FET2 is a common-source inverter (refer to Section 5.3.1.1) generating the $\overline{\text{OUT}}$ signal and FET3 is another common-gate stage generating the OUT signal. To match the output amplitudes[20] and phases, the designer can vary the gate widths of both FET2 and FET3 along with load impedances Z_1 and Z_2, respectively. If the overall gain is not too high, excellent bandwidths are achieved with this phase-splitter design.

[20] Recalling the discussions from Chapter 5, it will be realized that identical FET sizes and load impedances will not lead to equal voltage gains. The role of the FET's output resistance affects the gain balance and must be compensated for during the design.

6.7 TRANSIMPEDANCE AMPLIFIERS

The final class of amplifiers considered in this chapter is the *transimpedance amplifier*. It is basically a current-to-voltage converter and is implemented using an inverting amplifier with shunt resistive feedback.

An important application for transimpedance amplifiers is for *optical receivers* (conversion of a modulated optical signal to an electrical signal for information recovery). Figure 6.59 shows three optical receiver front ends: (*a*) low-impedance ($R_d = 50\ \Omega$) preamplifier, (*b*) high-impedance ($R_d \gg 50\ \Omega$) preamplifier, and (*c*) transimpedance preamplifier. For all three the optical to electrical conversion uses a diode (either a PIN diode or avalanche photodiode is used [50]). The low-impedance receiver has the advantage of being able to use a very wide variety of commercially available amplifiers, the widest dynamic range because of the low signal level at the input of the preamplifier, and thus easily achieves the widest bandwidths (e.g., see Ref. [51]). Its wide dynamic range is at the expense of the noise performance. Photodiodes act as current sources and the input signal voltage from the photocurrent can be increased by operating into a higher resistance. To improve the noise performance, higher values of R_d are used (i.e., the best signal-to-noise ratio is achieved as R_d approaches infinity). While higher input impedance levels improve the sensitivity, they limit the bandwidth and lead to earlier saturation in the amplifier because of the large amplitudes at the input of the amplifier. However, high-impedance front ends require equalization to compensate for their lack of bandwidth. Equalization is

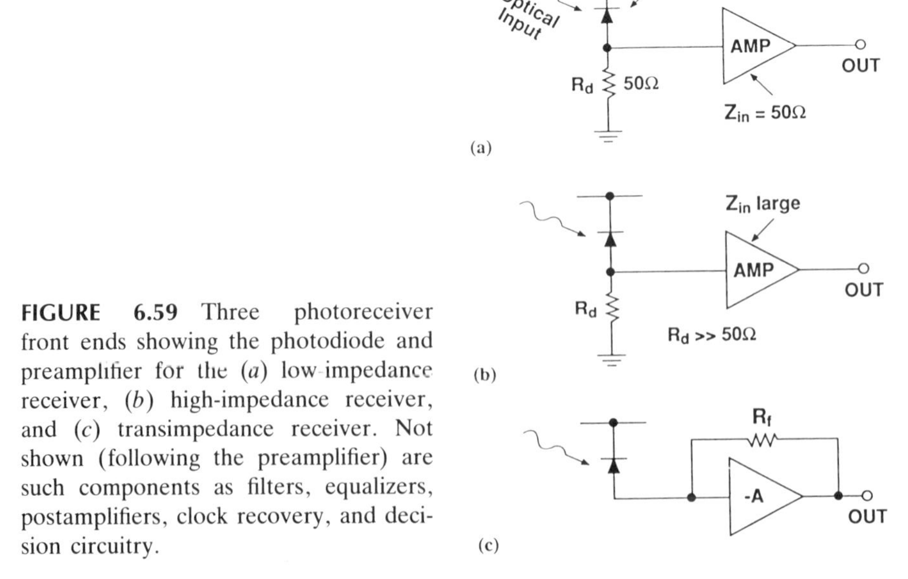

FIGURE 6.59 Three photoreceiver front ends showing the photodiode and preamplifier for the (*a*) low-impedance receiver, (*b*) high-impedance receiver, and (*c*) transimpedance receiver. Not shown (following the preamplifier) are such components as filters, equalizers, postamplifiers, clock recovery, and decision circuitry.

usually provided by the use of simple RC networks. Transimpedance amplifiers typically have better dynamics range than high-impedance receivers and better sensitivity than low-impedance receivers. This is achieved by using negative feedback. The generic transimpedance amplifier is discussed in more detail in the next section.

6.7.1 Principle of Operation

Figure 6.60 shows an equivalent circuit of the transimpedance amplifier driven by a current source I_p with capacitances included. The input capacitance consists of the diode capacitance C_d, the input capacitance of the amplifier C_a ($\cong C_{gs} + C_{gd}$ for a common-source FET used as the input device), stray wiring capacitance at the input C_s, and a small parasitic feedback capacitance C_f that shunts the feedback resistor R_F. The total input node capacitance is

$$C_T = C_d + C_s + C_a + (1 + A)C_f \tag{6.50}$$

where A is the open-loop gain of the inverting amplifier. The dominant pole of the transimpedance amplifier is located at $1/R_F C_T$. The ratio of the output voltage to input current is given by

$$\frac{V_{out}}{I_p} = \frac{-R_F}{1 + j\omega R_F C_T/(1 + A)} \tag{6.51}$$

Equation (6.51) positions the corner frequency ω_c at $(1 + A)/R_F C_T$. Compared to the high-impedance front end the bandwidth is increased by approximately A with the transimpedance approach. At low through midband frequencies $V_{out} = -R_F \cdot I_p$ as expected.

Stability is an important issue in transimpedance amplifiers. The capaci-

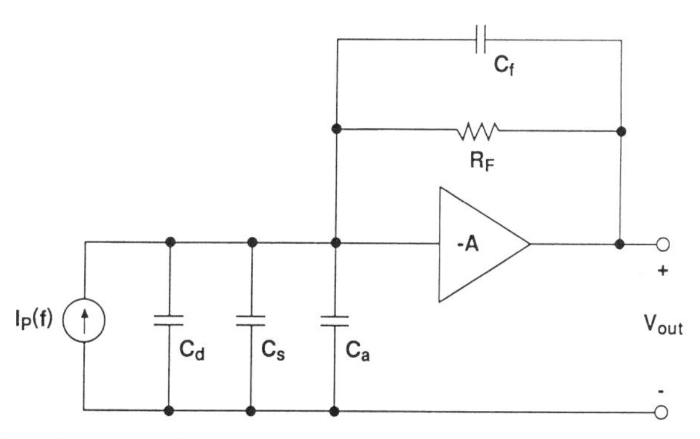

FIGURE 6.60 Transimpedance amplifier equivalent circuit showing input node capacitance components and a current source representing the photocurrent from the diode.

tive input results in a 90° phase shift at high frequencies, which directly adds to the 180° phase shift due to the amplifier's signal inversion. For stability it is common practice to maintain at least a 45° phase margin at the upper edge of the band, as this allows for very little excess phase shift within the amplifier (such as in SA-1- and SA-2-type amplifiers as described in Section 6.3.2). This limits the magnitude of the gain (e.g., fewer stages within the amplifier because each stage contributes excess phase shift[21] at higher frequencies) used in the feedback loop. A smaller feedback resistance is required to maintain a specified bandwidth. Of course, a smaller feedback resistance restricts the transimpedance amplifier's sensitivity and degrades its noise performance.

A commonly used implementation of the transimpedance amplifier using GaAs FETs is shown in Figure 6.61 [53]. This circuit uses a single SA-1 cell with shunt feedback resistor R_F and a source-follower buffer at the output (FET5 and FET6 serving as a current source). In addition, resistor R_A is combined with the current source FET4 to fine-tune the bias point of FET1 by changes in voltage supply V_A.

Noise performance is discussed next. In an FET the principal noise sources are thermal noise in the channel, induced gate noise from charge fluctuations in the channel, shot noise from gate leakage current, and $1/f$ noise. Ignoring the $1/f$ noise, the mean-square noise of the common-source

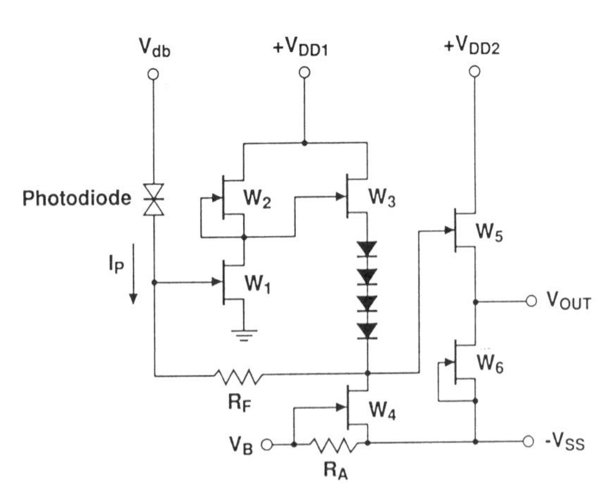

FIGURE 6.61 Simple single-stage transimpedance amplifier based on SA-1 gain block with feedback resistor R_F and output buffer stage (reproduced from Wada [53] by permission of author).

[21] Each stage adds at least an additional pole to the response function. Although the input capacitance and resistance determines the dominant pole, the additional poles all contribute phase shift, which rapidly adds to the phase around the loop.

inverter FET current referred to the input is [52]

$$\langle i_{\mathrm{fet}}^2 \rangle = \frac{4kT\Gamma}{g_{\mathrm{m1}}} (2\pi C_{\mathrm{T}})^2 I_3 B^3 + 2qI_{\mathrm{g}} \cdot I_2 B \tag{6.52}$$

Capacitance C_{T} is defined in Equation (6.50) and g_{m1} is the transconductance of the inverter FET. Table 6.7 defines the parameters used in Equation (6.52) and in the noise equations below. For the active load (i.e., FET2), the mean-square noise current is

$$\langle i_{\mathrm{load}}^2 \rangle = \frac{4kT\Gamma}{g_{\mathrm{m1}}} \cdot \frac{g_{\mathrm{m2}}}{g_{\mathrm{m1}}} (2\pi C_{\mathrm{T}})^2 I_3 B^3 \tag{6.53}$$

where g_{m2} is the transconductance of the active-load FET. Finally, the noise due to the feedback resistor R_{F} is by the expression for thermal noise:

$$\langle i_{\mathrm{r}}^2 \rangle = \frac{4kT}{R_{\mathrm{F}}} \cdot I_2 B \tag{6.54}$$

The total mean-square noise current at the input of the amplifier is the sum of the individual mean-square current fluctuations:

$$\langle i_{\mathrm{amp}}^2 \rangle = \langle i_{\mathrm{fet}}^2 \rangle + \langle i_{\mathrm{load}}^2 \rangle + \langle i_{\mathrm{r}}^2 \rangle \tag{6.55}$$

TABLE 6.7 Definition of Symbols for Noise Equations

kT	= thermal energy (Boltzmann's constant × absolute temperature) = 0.0259 eV = 4.142×10^{-21} J for $T = 300$ K
q	= charge on an electron (1.602×10^{-19} C)
B	= bit rate in reciprocal seconds (s^{-1})
R_{F}	= feedback resistance in ohms
I_{g}	= FET gate leakage current in amperes
C_{T}	= total input node capacitance in farads
g_{m}	= FET transconductance in siemens
I_2 and I_3	= normalized noise–bandwidth integrals [54][a]
Γ	= FET channel noise factor[b]

[a] The normalized noise–bandwidth integrals depend on the input optical pulse shape and the output pulse shape [54]. For rectangular NRZ (non-return-to-zero) input pulses and output pulses with a full raised cosine spectrum, $I_2 = 0.562$ and $I_3 = 0.087$. The equivalent −3-dB filter bandwidth is 58% of the bit rate B. For 50% duty-cycle rectangular RZ (return-to-zero) pulses, $I_2 = 0.403$ and $I_3 = 0.036$, with an equivalent filter bandwidth of 39% of the bit rate B.

[b] The FET channel noise factor Γ accounts for the channel thermal noise, induced gate noise, and the correlation between these two noise sources [55]. Values for Γ can range from approximately 1 to 3; $\Gamma = 1.75$ is a typical value for short-channel GaAs MESFETs.

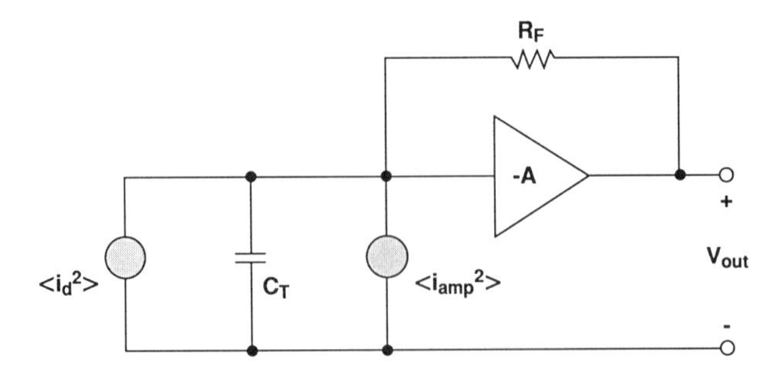

FIGURE 6.62 Transimpedance amplifier noise equivalent circuit.

Figure 6.62 shows the noise equivalent circuit of the transimpedance amplifier. Noise contributions from the other FETs within the amplifier are usually much smaller than those of the input FET provided the first stage has a large enough gain. Noise source $\langle i_d^2 \rangle$ represents the mean-square noise current of the photodiode, which is independent of the amplifier.

For best sensitivity $\langle i_{amp}^2 \rangle$ must be minimized. From Equation (6.54) it is clear that making R_F as large as practical minimizes $\langle i_r^2 \rangle$. To lower the noise contribution from the input FET it is necessary to choose an FET with a low leakage current (typically <100 nA). Of even greater importance, the figure of merit to maximize is g_m/C_T^2. To do this requires an FET with a short gate length for large f_T (i.e., unity current gain frequency) and then choosing the optimum gate width for the amplifier–photodiode combination. As is well known, g_m and C_a $(=C_{gs} + C_{gd})$ are directly proportional to gate width. The lowest noise occurs when

$$C_{gs} + C_{gd} = C_d + C_s + (1 + A)C_f \tag{6.56}$$

Thus, the width of the input FET is chosen to minimize the equivalent input noise current. The use of a passive-load resistor instead of an active load can give better noise performance; however, other tradeoffs must be considered as discussed earlier for the SA-1 and SA-2 amplifiers (e.g., voltage gain, drain supply voltage, and dc power dissipation issues). For the transimpedance amplifier a useful figure of merit [52] is

$$\text{Preamplifier figure of merit} = \frac{f_T}{\Gamma[C_d + C_s + (1 + A)C_f]} \tag{6.57}$$

From Section 5.3.1 the unity current gain frequency f_T is equal to $g_m/[2\pi(C_{gs} + C_{gd})]$.

After the FET width is chosen for best noise performance, the value of the feedback resistor must be determined. The bandwidth varies inversely with R_F (assuming that the dominant pole is set by the $R_F C_T$ product), but is directly proportional to the open-loop gain A. Small values of R_F give large bandwidths, but lead to poor noise performance and degraded sensitivity. A practical lower limit to R_F is of the order of a few hundred ohms; lower values give progressively poorer sensitivity and can also give rise to offset problems. The maximum value of R_F is determined by the peak amplitude allowable at the output of the amplifier before clipping or distortion becomes appreciable. In addition to distortion, the virtual ground at the input can be compromised and bit errors result. Scheinberg et al. [57] have suggested the following set of equations for estimating R_F and A. Define N to be

$$N = \frac{\langle i_r^2 \rangle}{\langle i_{fet}^2 \rangle} \tag{6.58}$$

An acceptable value for N might be 0.5. Resistor R_F is calculated using

$$R_F = \frac{3 f_T}{8\pi\Gamma \mathscr{B}^2 C_d N}, \tag{6.59}$$

where $\mathscr{B}$ is the bandwidth in Hz (not the bit rate as used in the noise equations), and A is determined using

$$A = \frac{3 f_T}{2\Gamma \mathscr{B} N} \tag{6.60}$$

For $C_d = 0.2\,\text{pF}$, $N = 0.5$, $f_T = 25\,\text{GHz}$, $\mathscr{B} = 2\,\text{GHz}$, and $\Gamma = 1.75$, R_F is found to be 4.26 kΩ and A is 21.4 (or 26.6 dB). These are representative of typical design values for GaAs FET transimpedance amplifiers.

6.7.2 Transimpedance Amplifier Example

Figure 6.63 shows a circuit schematic of a recently reported two-stage transimpedance amplifier using a novel active-load element. The input (and first stage of gain) is a cascoded SA-1-type amplifier with inductive peaking added to the active load (15-nH spiral inductor in series with the source of FET3 [56,57]). There are two advantages to this form of active load. First, the inductive peaking extends the bandwidth as discussed in Sections 5.4.3 and 6.2.3. Equation (6.61), giving the effective impedance (Z_{a1}) of the active load, clearly reveals the impedance increase due to the inductor as the frequency increases:

$$Z_{a1} = r_{ds3}(1 + g_{m3}|\omega L_3|) + |\omega L_3| \tag{6.61}$$

In addition, the inductor reduces the noise contribution of the active-load FET at higher frequencies. With the inductor included, the mean-square

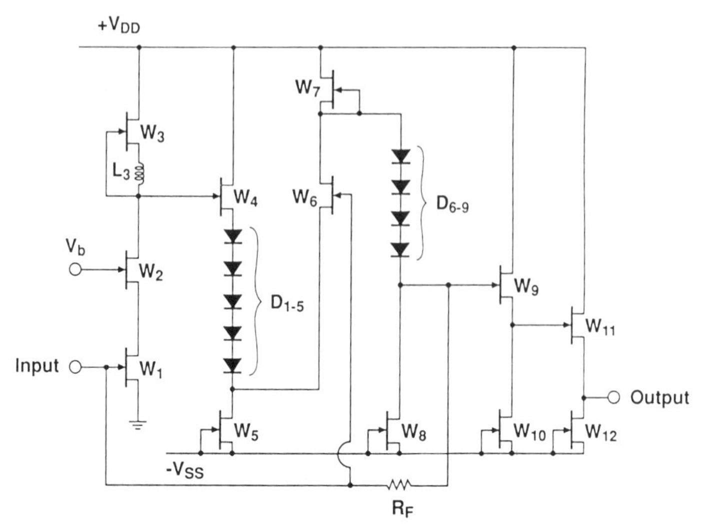

FIGURE 6.63 Anadigics transimpedance amplifier with two gain stages and inductively peaked active load (reproduced from Refs. [56] and [57] by permission of IEEE).

input noise current becomes

$$\langle i_{\text{load}}^2 \rangle = \frac{4kT\Gamma}{g_{\text{m1}}} \frac{g_{\text{m3}}}{g_{\text{m1}}} (2\pi C_{\text{T}})^2 \frac{I_3 B^3}{1 + g_{\text{m3}}|\omega L_3|} \tag{6.62}$$

where $\langle i_{\text{load}}^2 \rangle$ is reduced by the actor $1/(1 + g_{\text{m3}}|\omega L_3|)$ when compared to Equation (6.53) for the case without the inductor. Transconductance g_{m3} corresponds to FET3 (active load).

The second stage is a noninverting common-gate amplifier formed by FET6 with active-load FET7. A unique feature of the second stage is the input is connected directly (feedforward) to the gate of FET6. Scheinberg et al. [57] claim that one of the advantages of this connection is improved stability because the feedback resistor and gate-to-drain capacitance of FET6 form a zero at $1/R_{\text{F}}C_{\text{gd6}}$. This zero can cancel one pole thereby eliminating the phase shift from this pole—this leaves two remaining poles with the dominant pole being at $1/C_{\text{T}}R_{\text{F}}$ and the other pole positioned more distant the $j\omega$ axis. Two source followers, FET9 and FET11, step up the current drive capability at the output with the size of FET11 chosen so that the output impedance is approximately 50 Ω.

The reported performance [56] of the transimpedance amplifier shown in Figure 6.63, where $C_{\text{d}} + C_{\text{s}} = 0.6$ pF, is a transresistance of 2.5 kΩ with a load resistance of 50 Ω and a −3-dB bandwidth of 2.26 GHz. Without the 15-nH peaking inductor the bandwidth was approximately 1.7 GHz (i.e., the inductor extends the bandwidth by 33%). In the frequency band of 1–2 GHz the input referred noise current $\langle i_{amp}^2 \rangle$ was less than 64 pA^2/Hz and from 10 MHz to 600 MHz $\langle i_{amp}^2 \rangle$ was less than 16.8 pA^2/Hz [56].

This performance was achieved using a depletion-mode FET technology with a 0.5-μm gate length and an f_T of 25 GHz. Other FET parameters are 140 mA/mm for I_{DSS}, 170 mS/mm for g_m, and −0.8 V for the pinchoff voltage V_p. The chip was 2 mm^2 and the positive power supply range +5 to +8 V (100 mA current), while the negative supply range covers −3 to −5 V (75 mA current). Scheinberg et al. [57] also discuss the use of an FET shunting R_F to vary the magnitude of the feedback resistance for gain control.

ACKNOWLEDGMENTS

The Van Tuyl–Hornbuckle gain blocks have had a profound influence on GaAs IC amplifier design in the 1980s and into the early 1990s. Many designs have built on the basic topology of the SA-1 and SA-2 gain blocks as illustrated by several examples in this chapter.

The author acknowledges the help of Rory Van Tuyl and Derry Hornbuckle in assembling the material for this chapter. He had the pleasure of working with these two pioneers in GaAs IC design during the beginnings of the development and growth of GaAs ICs. In addition, Tim Shirley, Ron Hogan, Don Cook, and Steve Cochran (Hewlett-Packard, Santa Rosa, CA) provided valuable comments on Chapters 5 and 6 during the writing phase.

REFERENCES

1. R. L. Van Tuyl, "A Monolithic Integrated 4-GHz Amplifier," *1978 IEEE International Solid-State Circuits Conference Digest of Technical Papers*, San Francisco, CA, February 15–17, 1978, pp. 72–73.
2. D. P. Hornbuckle and R. L. Van Tuyl, "Monolithic GaAs Direct-Coupled Amplifiers," *IEEE Trans. Electron Devices*, **ED-28** (2), 175–182 (Feb. 1981).
3. D. J. Mellor, *Computer-Aided Synthesis of Matching Networks for Microwave Amplifiers*, Stanford Electronics Laboratories Report, Stanford University, Stanford, CA, March 1975; See also D. J. Mellor and J. G. Linvill, "Synthesis of Interstage Networks of Prescribed Gain Versus Frequency Slopes," *IEEE Trans. Microwave Theory & Techniques*, **MTT-23**, 1013–1020 (Dec. 1975).
4. T. R. Apel, "GaAs FET Amplifier and MMIC Design Techniques," in *Introduction to Semiconductor Technology: GaAs Related Compounds*, C. T. Wang, ed., Wiley, New York, 1990, pp. 382–435.
5. R. Soares, ed., *GaAs MESFET Circuit Design*, Artech House, Norwood, MA, 1988, pp. 110–134.
6. Y. Ayasli, R. L. Mozzi, J. L. Vorhaus, L. D. Reynolds, and R. A. Pucel, "A

Monolithic GaAs 1–13 GHz Traveling Wave Amplifier," *IEEE Trans. Microwave Theory Techniques*, **MMT-30**, 976–981 (July 1982).

7. E. L. Ginzton, W. R. Hewlett, J. H. Jasberg, and J. D. Noe, "Distributed Amplification," *Proc. IRE*, **36**, 956–969 (Aug. 1948).
8. M. S. Ghausi, *Principles and Design of Linear Active Circuits*, McGraw-Hill, New York, Chapter 13, pp. 334–358.
9. G. D. Vendelin, A. M. Pavio, and U. L. Rohde, *Microwave Circuit Design Using Linear and Nonlinear Techniques*, Wiley, New York, 1990.
10. M. S. Ghausi, *Principles and Design of Linear Active Circuits*, McGraw-Hill, New York, 1965, pp. 300–304.
11. H. W. Bode, *Network Analysis and Feedback Amplifier Design*, Van Nostrand, Princeton, NJ, 1945.
12. R. M. Fano, "Theoretical Limitations on the Broadband Matching of Arbitrary Impedances," *J. Franklin Inst.*, **249**, 57–83, 139–154 (Jan.–Feb. 1950).
13. H. Baher, *Synthesis of Electrical Networks*, Wiley, New York, 1984, Chapter 10.
14. K. Honjo, H. Ogawa, and Y. Konishi, "Microwave Integrated Circuits," in *Microwave Integrated Circuits*, Y. Konishi, ed., Marcel Dekker, New York, 1991, Chapter 5, pp. 340–344.
15. J. M. Pettit and M. M. McWhorter, *Electronic Amplifier Circuits*, McGraw-Hill, New York, 1961, pp. 90–96.
16. M. S. Ghausi, *Principles and Design of Linear Active Circuits*, McGraw-Hill, New York, 1965, pp. 604–610.
17. D. L. Feucht, *Handbook of Analog Circuit Design*, Academic Press, San Diego, 1990, pp. 340–343.
18. P. R. Gray and R. G. Meyer, *Analysis and Design of Analog Integrated Circuits*, 2nd Ed., Wiley, New York, 1984, pp. 466–522, 527–537.
19. K. B. Niclas, W. T. Wilser, R. B. Gold, and W. R. Hitchens, "The Matched Feedback Amplifier: Ultrawide-Band Microwave Amplification with GaAs MESFET's," *IEEE Trans. Microwave Theory Tech.*, **MTT-28**, 285–294 (April 1980).
20. F. D. Waldhauer, *Feedback*, Wiley, New York, 1982.
21. IRE Subcommittee 7.9 on Noise, "Representation of Noise in Linear Two-Ports," *Proc. IRE*, **48**, 69–74 (Jan. 1960).
22. L. D. Nguyen, L. E. Larson, and U. K. Mishra, "Ultra-High-Speed Modulation-Doped Field-Effect Transistors: A Tutorial Review," *Proc. IEEE*, **80**, 494–518 (April 1992).
23. B. Hughes, N. G. Fernandez, and J. M. Gladstone, "GaAs FET's with a Flicker-Noise Corner Below 1 MHz," *IEEE Trans. Electron Devices*, **ED-34** 733–741 (April 1987).
24. D. P. Hornbuckle and R. L. Van Tuyl, "Monolithic GaAs Direct-Coupled Amplifiers," *IEEE Trans. Electron Devices*, **ED-28** (2), 175–182 (Feb. 1981).
25. R. L. Van Tuyl, "A Monolithic Integrated 4-GHz Amplifier," 1978 *IEEE International Solid-State Circuits Conference Digest of Technical Papers*, San Francisco, February 15–17, 1978, pp. 72–73.
26. D. Hornbuckle, "GaAs IC Direct-Coupled Amplifiers," *1980 IEEE MTT-S International Microwave Symposium Digest*, Washington, DC, May 28–30, 1980, pp. 387–389.

27. D. P. Hornbuckle and R. L. Van Tuyl, "Monolithic GaAs Direct-Coupled Amplifiers," *IEEE Trans. Electron Devices*, **ED-28** (2), 175–182 (Feb. 1981).

28. R. S. Pengelly, "Hybrid vs. Monolithic Microwave Circuits: A Matter of Cost," *Microwave System News*, **13**, 77–114 (Jan. 1983).

29. R. S. Pengelly and J. A. Turner, "Monolithic Broadband GaAs FET Amplifiers," *Electron. Lett.* **12** (10) 251–252 (May 1976).

30. R. A. Pucel, ed., *Monolithic Microwave Integrated Circuits*, a volume in the IEEE Press Selected Reprints Series, IEEE Press, The Institute of Electrical and Electronics Engineers, New York, 1985.

31. R. A. Pucel, J. Vorhaus, P. Ng, and W. Fabian, "A Monolithic GaAs X-Band Power Amplifier," *1979 International Electron Device Meeting Technical Digest*, Washington, DC, December 1979, pp. 266–268.

32. R. L. Van Tuyl and C. A. Liechti, "High Speed Integrated Logic with GaAs MESFET's," *IEEE J. Solid-State Circuits*, ***SC*-9**, 269–276 (Oct. 1974).

33. R. L. Van Tuyl, "A Monolithic GaAs IC for Heterodyne Generation of RF Signals," *IEEE Trans. Electron Devices*, **ED-28** (2), 166–170 (Feb. 1981).

34. R. L. Van Tuyl, V. Kumar, D. C. D'Avanzo, T. W. Taylor, V. E. Peterson, D. P. Hornbuckle, R. A. Fisher, and D. B. Estreich, "A Manufacturing Process for Analog and Digital Gallium Arsenide Integrated Circuits," *IEEE Trans. Microwave Theory Tech.*, **MTT-30**, 935–942 (July 1982).

35. C. A. Liechti, "Microwave Field-Effect Transistors—1976," *IEEE Trans. Microwave Theory Tech.*, **MTT-24**, 279–300 (June 1976).

36. P. H. Ladbrooke, *MMIC Design: GaAs FETs and HEMTs*, Artech House, Norwood MA, 1989, pp. 85–87.

37. W. T. Colleran and A. A. Abidi, "A 3.2 GHz, 26 dB Wide-Band Monolithic Matched GaAs MESFET Feedback Amplifier Using Cascodes," *IEEE Trans. Microwave Theory Tech.*, **36**, 1377–1385 (Oct. 1988).

38. A. Mauri, E. M. Bastida, P. A. Chiappa, and M. Feudale, "Very High Performance D. C. Coupled MMIC FET Amplifiers," *Conference Proceedings of 20th European Microwave Conference*, Budapest, Hungary, September 10–13, 1990, Vol. 2, pp. 1761–1765.

39. A. Mauri, E. M. Bastida, P A. Chiappa, and M. Feudale, "Design and Development of High Performance D. C. Coupled MMIC FET Amplifiers," *The 3rd Asia–Pacific Microwave Conference Proceedings*, Tokyo, September 18–21, 1990, pp. 779–782.

40. D. Hornbuckle, "GaAs IC Direct–Coupled Amplifiers," *1980 IEEE MTT-S International Microwave Symposium Digest*, Washington, DC, May 28–30, 1980, pp. 387–389.

41. D. P. Hornbuckle and R. L. Van Tuyl, "Monolithic GaAs Direct-Coupled Amplifiers," *IEEE Trans. Electron Devices*, **ED-28** (2), 175–182 (Feb. 1981).

42. R. L. Van Tuyl, V. Kumar, D. C. D'Avanzo, T. W. Taylor, V. E. Peterson, D. P. Hornbuckle, R. A. Fisher, and D. B. Estreich, "A Manufacturing Process for Analog and Digital Gallium Arsenide Integrated Circuits," *IEEE Trans. Microwave Theory Techniques*, **MTT-30**, 935–942 (July 1982).

43. G. P. Vella-Coleiro, "Symmetric Integrated Amplifier with Controlled DC Offset Voltage," U.S. Patent 4,825,174 (April 25, 1989).
44. D. B. Estreich, "A Wideband Monolithic GaAs IC Amplifier," *ISSCC Digest of Technical Papers*, International Solid-State Circuits Conference, San Francisco, February 1982, pp. 194–195.
45. D. B. Estreich, "A Monolithic Wide-Band GaAs IC Amplifier," *IEEE J. Solid-State Circuits*, **SC-17**, 1166–1173 (Dec. 1982).
46. J. Browne, "GaAs IC Complements Budget 3-GHz Counter," *Microwaves RF*, **25**, 221–223 (Jan. 1986).
47. C. A. Liechti, "Performance of Dual-Gate GaAs MESFET's as Gain-Controlled Low-Noise Amplifiers and High-Speed Modulators," *IEEE Trans. Microwave Theory Tech.*, **MTT-23,** 461–469 (June 1975).
48. B. Kim, H. Q. Tserng, and P. Saunier, "A GaAs Dual-Gate Power FET for Operation up to K Band," *1983 IEEE International Solid-State Circuits Conference Digest of Technical Papers*, February 1983, pp. 200–201.
49. M. Shigaki, S. Yokogawa, H. Kurihara, and K. Yamada, "GaAs Monolithic DC-6.4 GHz Variable-Gain Feedback Amplifier," *IEEE Trans. Microwave Theory Tech.*, **MTT-35**, 923–935 (Oct. 1987).
50. S. R. Forrest, "Optical Detectors for Lightwave Communication," in *Optical Fiber Telecommunications II*, S. E. Miller and I. P. Kaminow, eds., Academic Press, Boston, 1988, pp. 569–599.
51. D. J. Derickson, C. M. Miller, and R. L. Van Tuyl, "A 100 kHz–22 GHz Instrumentation Photoreceiver," *1988 IEEE MTT International Microwave Symposium Digest*, Vol. II, New York, May 25–27, 1988, pp. 1063–1066.
52. B. L. Kasper, "Receiver Design," in S. E. Miller and I. P. Kaminow, eds, *Optical Fiber Telecommunications II*, Academic Press, Boston, 1988, pp. 689–722.
53. O. Wada, T. Hamaguchi, S. Miura, M. Makiuchi, K. Nakai, H. Horimatsu, and T. Sakurai, "AlGaAs/GaAs p–i–n Photodiode/Preamplifier Monolithic Photoreceiver Integrated on a Semi-Insulating GaAs Substrate," *Appl. Phys. Lett.*, **46**, 981–983 (1985).
54. R. G. Smith and S. D. Personick, "Receiver Design for Optical Fiber Communication Systems," in *Semiconductor Devices for Optical Communication*, H. Kressel, ed., Springer-Verlag, Berlin, Germany, 1982, pp. 89–160.
55. K. Ogawa, "Noise Caused by GaAs MESFETs in Optical Receivers," *Bell System Tech. J.* **60**, 923–928 (July–Aug. 1981).
56. R. Bayruns, T. Laverick, N. Scheinberg, and K. Li, "A 5k 2 GHz GaAs Transimpedance Amplifier Using a Low-Noise Active Load," *1991 IEEE International Solid-State Circuits Conference Digest of Technical Papers*, February 1991, pp. 272–273.
57. N. Scheinberg, R. J. Bayruns, and T. M. Laverick, "Monolithic GaAs Transimpedance Amplifiers for Fiber-Optic Receivers," *IEEE J. Solid-State Circuits*, **26**, 1834–1839 (Dec. 1991).

CHAPTER SEVEN

Operational Amplifiers

LARRY E. LARSON
Hughes Research Laboratories
3011 Malibu Canyon Road
Malibu, CA 90265

7.1 INTRODUCTION

Operational amplifiers (op amps) are commonly used analog integrated-circuit (IC) building blocks. A number of GaAs MESFET op amps have been described in the literature that exhibit widely varying characteristics. These circuits exhibit wider bandwidth than their silicon counterparts, but some of their low-frequency characteristics are not as desirable.

The design of these circuits in silicon technology is usually enhanced by the availability of both p- and n-type devices. Unfortunately, only n-type devices are available in typical GaAs MESFET processes, so the flexibility of the designs is somewhat inhibited. Furthermore, since the speed of p-type devices (p-channel MESFETs) is typically much worse than that of n-channel devices, the utility of p-type devices for high-speed analog circuit design is questionable, even if they do eventually become available. As a result, this chapter will concentrate on design techniques that exclusively utilize n-type devices.

High-Frequency Analog Integrated-Circuit Design, Edited by Ravender Goyal
ISBN 0-471-53043-3

7.2 HIGH-SPEED OPERATIONAL-AMPLIFIER DESIGN PRINCIPLES

Operational amplifiers are high-gain linear amplifiers that are useful in a wide range of applications as a result of modifications to their feedback transfer functions. Ideally, they possess a large low-frequency gain and a high unity-gain bandwidth, in addition to a large common-mode rejection ratio and input impedance, and low input offset voltage and output impedance. An ideal operational amplifier is also stable under a wide variety of feedback and output loading conditions. A schematic diagram of a typical two-stage silicon bipolar operational amplifier appears in Figure 7.1. This section initially concentrates on silicon bipolar implementations of high-speed op amps, since most of the GaAs MESFET designs to date have borrowed from these earlier designs.

The design of operational amplifiers is a well-established discipline, having been refined for the last two decades in silicon technology. This section summarizes the basic requirements of successful high-speed operational-amplifier design, and the reader is referred to the excellent review article by Solomon [1], or text by Gray and Meyer [2], for a more detailed treatment of operational-amplifier fundamentals.

The drawbacks of the classic two-stage approach illustrated in Figure 7.1 are well known, and are summarized in Solomon [1]. The first drawback of this configuration is that the unity-gain bandwidth of the op amp is set by the location of the nondominant pole, which in turn is set by the transconductance of the second stage and the load capacitor C_L, namely, $\omega_u \cong s_{p2} \cong g_{m2}/C_L$. The second drawback is that the maximum slew rate is limited to approximately $2\omega_u V_T$, where V_T is the thermal voltage (26 mV at

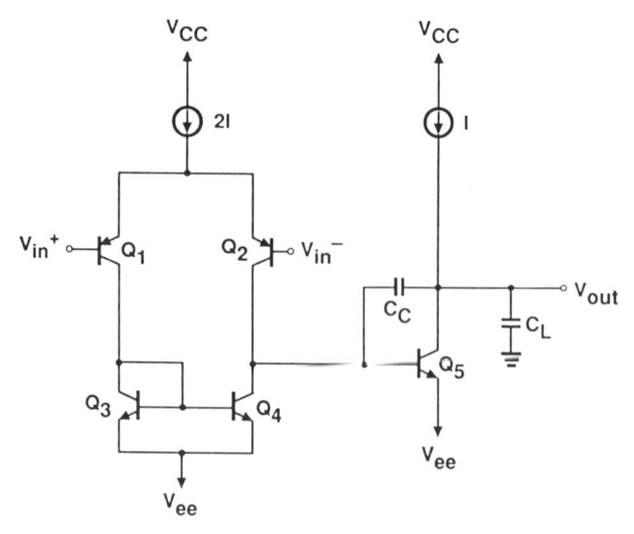

FIGURE 7.1 Simplified model of two-stage bipolar operational amplifier.

room temperature). This limit basically arises because the input stage transconductance saturates for a relatively low differential input voltage, which, in turn, limits the current available to change the voltage across the internal compensation capacitor. This limit is typically circumvcntcd by some form of transconductance reduction technique, which increases the saturation voltage of the transconductance.

The bandwidth limitation can be circumvented by employing a single-stage folded cascode design [3], as shown in Figure 7.2. In this case, the nondominant pole location is the f_T of the pnp devices (Q_3–Q_4). Although these are typically lower than the f_T of the npn devices, the nondominant pole location is still considerably increased.

A second, improved version of the operational amplifier employs "current feedback" to improve the bandwidth of the circuit [4]. A simplified schematic of this version of the operational amplifier is shown in Figure 7.3. In this case, a difference in the input voltages creates a current difference, which is mirrored to the output node. The nondominant pole location is the same as with the folded cascode circuit, but the impedance seen at the inverting input is relatively low, which further improves the bandwidth of the circuit. However, the dc characteristics of this approach, particularly the input offset current and voltage, are somewhat degraded compared with the other two approaches.

All of these designs so far employ pnp as well as npn transistors. This is a serious limitation for most high-speed operational amplifiers, since the f_T of the pnp's is typically at least a factor of 3 less than that of the npn transistors. In GaAs MESFET technology, p-channel devices are simply unavailable. Therefore, it is desirable to develop op-amp designs that employ only fast npn devices, if possible.

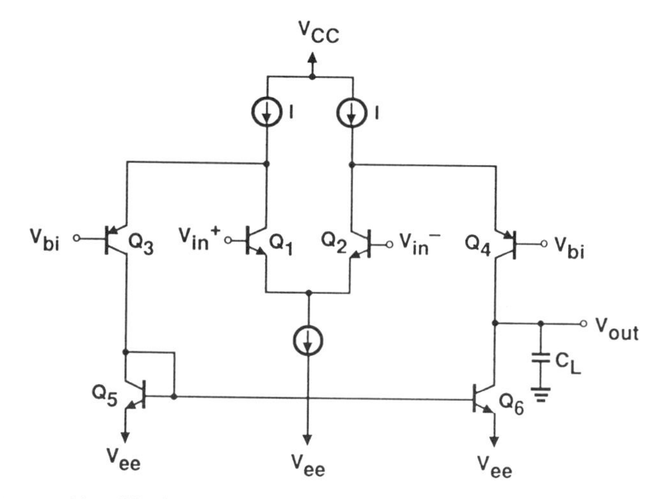

FIGURE 7.2 Simplified model of folded cascode bipolar operational amplifier.

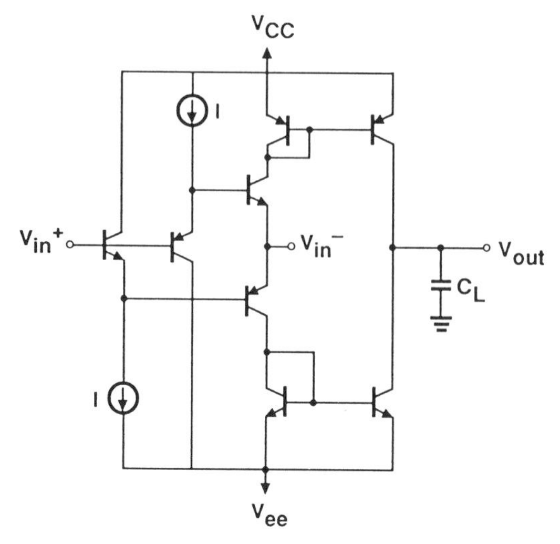

FIGURE 7.3 Simplified model of current feedback bipolar operational amplifier.

The basic topology that most high-speed GaAs MESFET operational amplifiers have adopted was first outlined by Widlar in his classic paper on monolithic silicon bipolar operational amplifiers in 1969 [5]. His proposed high-speed operational amplifier (Fig. 7.4) utilized all npn silicon bipolar transistors in the high-speed signal path. Transistors Q_1–Q_2 formed the differential input circuit. Transistors Q_3–Q_4 performed the level-shifting function along with R_5–R_6, and Q_5–Q_6 perform differential to single-ended

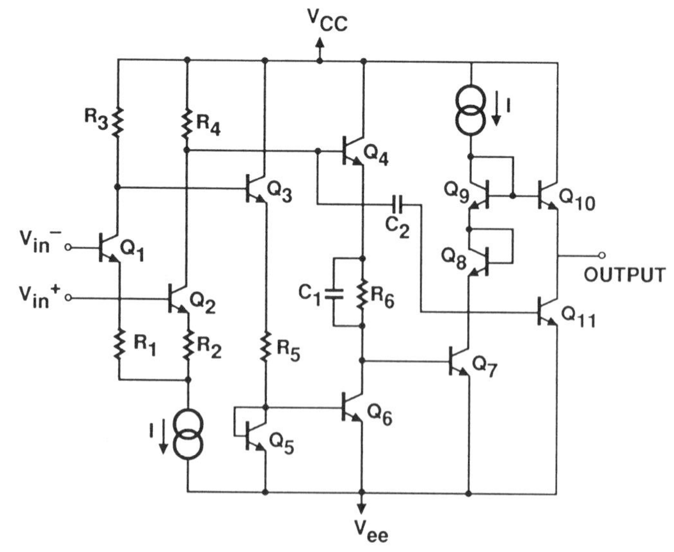

FIGURE 7.4 High-speed silicon bipolar operational amplifier [5].

conversion. The second gain stage is formed by Q_7, which drives a complementary emitter follower. Capacitor C_2 acts as a frequency compensation capacitor, while C_1 minimizes the phase shift through the level-shifting circuitry.

The advantage of this approach is that all the transistors in the signal path were high-speed npn's. Using mid-1960s silicon bipolar technology ($f_T \cong 1$ GHz), the bandwidth of the amplifier was expected to be 50 MHz, with a slew rate of 50 V/μs. The same topology can be applied to GaAs MESFET or HBT devices, which would simply replace the silicon npn transistors of Figure 7.4. In fact, this is the approach that many of the designers of GaAs operational amplifiers have adopted to date.

7.3 BUILDING BLOCKS FOR OPERATIONAL AMPLIFIERS

The equivalent circuit modeling of MESFET devices, which is discussed in Chapter 3, pointed to a number of drawbacks with GaAs FET technology for analog applications:

1. The low intrinsic high-frequency voltage gain ($g_m r_{ds}$) of a short-channel MESFET is typically between 10 and 20, and represents an upper limit on the achievable voltage gain of a simple inverting gain stage. Many precision analog applications require high-gain amplification, so special design techniques must be adopted to increase the achievable gain.
2. The relatively high level of $1/f$ noise and low-frequency oscillations in a typical GaAs MESFET can result in poor dynamic range in amplifier circuits. The magnitude of low-frequency device noise is affected by materials and processing parameters, and has shown dramatic improvement in recent years. However, it is anticipated that this noise will remain relatively high compared to silicon bipolar devices for the foreseeable future. Substrate anomalies also contribute to nonideal behavior, such as hysteresis and long-settling transients, which are especially bothersome in analog integrated circuits.
3. Finally, the degree to which individual MESFETs can be matched to each other is relatively poor compared to that for bipolar transistors, which results in high offset voltage in differential amplifiers and poor uniformity and yield in a variety of matched circuits.

By contrast, GaAs heterojunction bipolar transistor (HBT) technology appears to have few of the drawbacks associated with MESFET technology for analog applications. The intrinsic voltage gain is quite high, there is little $1/f$ noise, and device matching is excellent. The major drawback at this point is the relative immaturity of the technology.

The design of analog ICs in MESFET technology requires the use of "building blocks" (current sources and mirrors, common source and drain amplifiers) that are substantially different from those of silicon bipolar analog integrated circuits. The following sections describe the unique design techniques that have been developed for these circuits in n-channel GaAs MESFET technology.

7.3.1 Current Source Design

Current sources are ubiquitous building blocks in the design of analog ICs in silicon and GaAs technology. GaAs FET technology presents some interesting design constraints that limit the applicability of common silicon current source design practices. This section outlines the basic current source designs in GaAs FET technology and presents a variety of techniques for performance enhancement.

The ideal common-source current source configuration is shown in Figure 7.5. This ideal circuit can be easily approximated by a MESFET whose source is grounded, which is also shown in Figure 7.5. In that case, the current I_{bi} is set by the gate voltage V_{bi} and by the gate width of MESFET, threshold voltage, and full-channel current. The equivalent circuit derived in Chapter 3 can be used to calculate the current through the current source. The current I_{bi} will stay roughly constant as long as the drain-to-source voltage of Q_1 is greater than the drain–source saturation voltage of the FET, $V_{DS,sat}$.

The principal drawback to this configuration is the relatively low small-signal output impedance (r_{ds1}) of Q_1 in the common-source configuration; an ideal current source would have infinite output impedance. Therefore, a variety of feedback techniques have been developed to increase the output impedance of the current source. The use of a cascode arrangement, as shown in Figure 7.6*a*, is a popular technique borrowed from silicon IC design. Here, the negative feedback of Q_2 increases the small-signal output impedance of the circuit to

$$r_{out} \cong \frac{\partial V_{out}}{\partial I_{bi}} \cong r_{ds1} r_{ds2} g_{m1} \qquad (7.1)$$

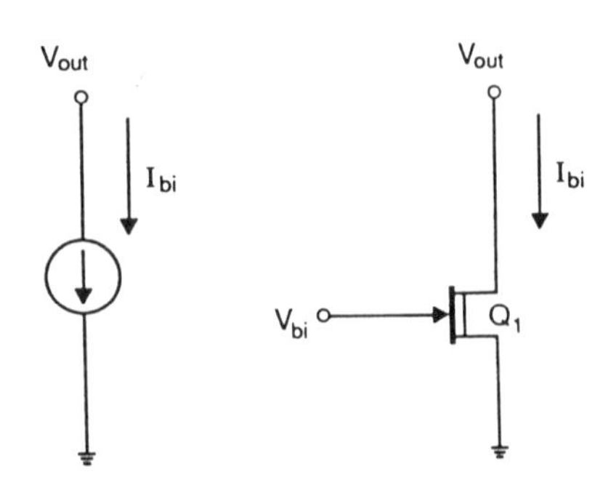

FIGURE 7.5 Ideal common-source current source and FET equivalent.

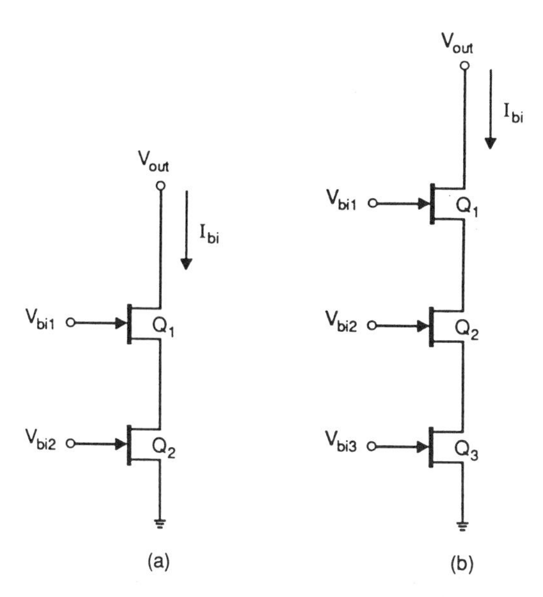

FIGURE 7.6 (*a*) GaAs FET cascode current source; (*b*) three-transistor cascode current source.

Since the $g_m r_{ds}$ product of a typical transistor is at least 10, this technique can lead to a substantial improvement in the effective output impedance of the current source.

If the circuit of Figure 7.6*a* provides insufficient output impedance, this technique can be extended further, as shown in the case of the three-transistor cascode of Figure 7.6*b*. In this case, the small-signal output impedance is further increased to approximately

$$r_{out} \cong r_{ds1} r_{ds2} r_{ds3} g_{m1} g_{m2} \tag{7.2}$$

In the circuits of Figures 7.5 and 7.6, Q_1, Q_2, and Q_3 can be enhancement-mode or depletion-mode FETs.

In theory, the transistors can be stacked even further to realize larger and larger impedance levels. However, the improvement diminishes after a certain point, because second-order effects such as backgating and gate–drain capacitance place an upper limit on the achievable impedance, especially at high frequencies. Furthermore, the dynamic range of the circuit continues to diminish as the number of levels in the cascode increases.

These techniques can also be employed to realize current mirrors, which are the basic building blocks in silicon analog IC design. These circuits act as common-source current amplifiers, with a low input impedance and high output impedance. The design of current mirrors in enhancement-mode FET technology is relatively straightforward, and a typical design is

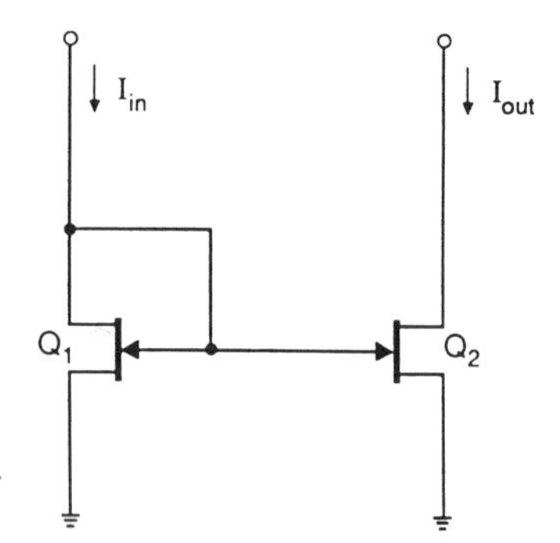

FIGURE 7.7 Enhancement-mode GaAs FET current mirror.

illustrated in Figure 7.7. Since the gate-to-source voltages are identical for transistors Q_1 and Q_2, identically sized devices will have identical currents.

The principal limitation on the operation of a current mirror in an enhancement-mode FET technology is the conduction of the forward-biased Schottky gate under conditions where I_{in} is large; I_{out} will be limited by the turn-on voltage of the Schottky gate. Figures 7.8*a* and 7.8*b* show the designs of the well-known cascode and Wilson current mirrors in enhancement-mode MESFET technology. Once again, these designs are essentially identical to their counterparts in silicon MOS technology, and exhibit improved output impedance and current matching compared to the designs of Figure 7.7.

The design of current mirrors in depletion-mode GaAs FET technology is complicated due to the negative threshold voltage of the device. As a result, design techniques that are slightly different from those previously described are required. Depletion-mode MESFET technology requires that the gate of the FET be maintained at a lower potential than the drain, which was not the case in the previous examples. This requirement arises because the FET

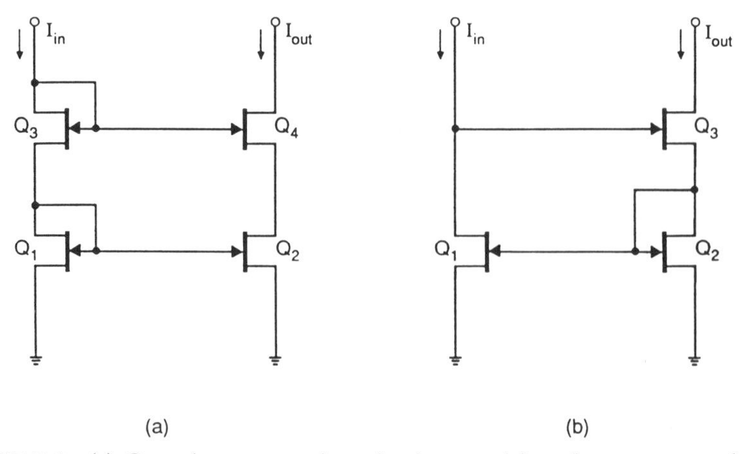

FIGURE 7.8 (*a*) Cascode current mirror implemented in enhancement-mode FET technology; (*b*) Wilson current mirror implemented in HBT technology.

operates in the saturation region only as long as $V_{DG} > -V_T/K > 0$, as discussed in Chapter 3. The value of K is typically between 1 and 2 in most depletion-mode GaAs processes. This condition will not be met if depletion-mode FETs are employed to implement the circuits of Figures 7.7 and 7.8 since $-V_T/K > 0$. Therefore, in a depletion-mode technology, some sort of constant voltage shift is required between the drain and the gate.

This required level-shifting is illustrated schematically in Figure 7.9, which shows the level-shifting voltage realized as an ideal battery. In this case, the level-shift voltage V_{LS} is equal to the V_{DG} of Q_1, and therefore must be larger than $-V_T/K$. Of course, an ideal battery cannot be physically realized, and so forward-biased level-shifting Schottky diodes are usually employed instead.

There are two preferred techniques for the realization of V_{LS}. The first, shown in Figure 7.10*a*, has a source follower combined with forward-biased Schottky diodes. The gate width of transistor Q_3 is normally set equal to that of Q_4, which results in the V_{gs} of Q_3 resting nearly at zero volts. The second approach, shown in Figure 7.10*b* uses forward-biased diodes alone. The advantage of the first approach is that no current is drawn from the reference current by the level-shifting stage. As a result, the dynamic range of the circuit is improved, since I_{in} can drop to zero without affecting overall operation. In both cases, V_{LS} is equal to the voltage across two forward-biased Schottky diodes.

The circuit of Figure 7.10*b* is somewhat simpler than that of Figure 7.10*a*, requiring one less power supply, and dissipates slightly less power. However, the level-shift circuitry of Figure 7.10*b* draws current from I_{in}, and so the minimum value of I_{in} is limited to I_{bi}. In addition, the circuit of Figure 7.10*b* has a somewhat less accurate transfer function than that of Figure 7.10*a* because of the shunting impedance of Q_3.

The linearity and distortion of these current mirrors can be enhanced

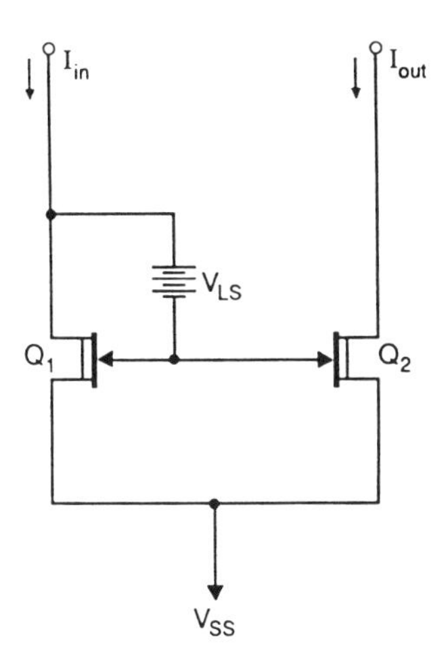

FIGURE 7.9 Depletion-mode FET current mirror.

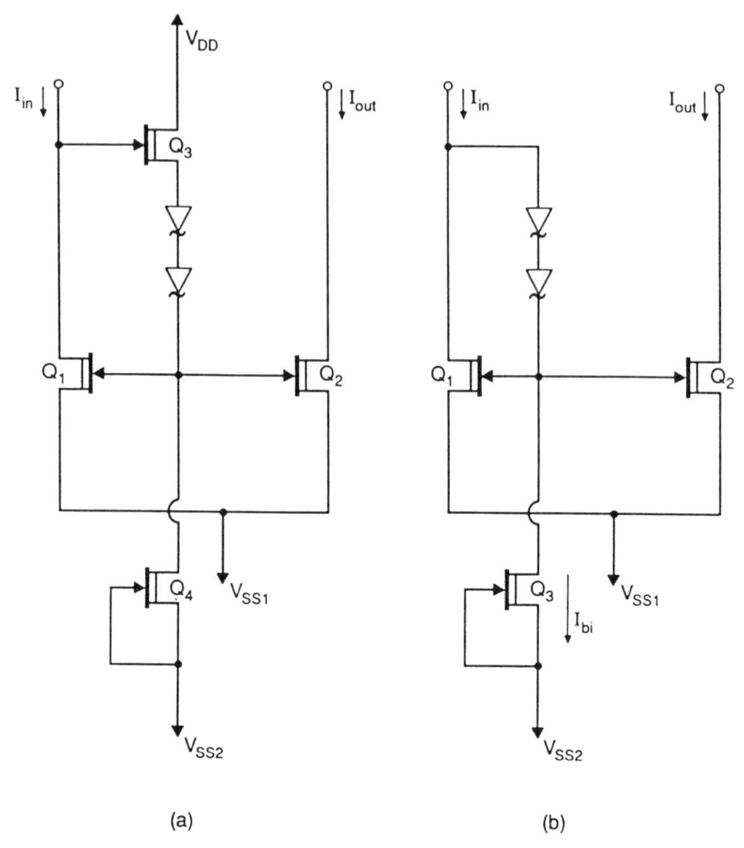

FIGURE 7.10 (*a*) Current mirror employing source follower for level-shift voltage; (*b*) current mirror employing forward-biased diode string for level-shift voltage.

through extensions of well-known silicon techniques, such as the Wilson and cascode current mirrors. These were shown for enhancement-mode MESFET circuits in Figure 7.8.

Current source design in the common-drain configuration is also complicated by the low output impedance of a typical short-channel FET in the saturation region. The most common approach is shown in Figure 7.11, where the gate of a depletion-mode MESFET is connected to its source. The current flowing through the current source is determined by the device gate width, threshold voltage, and current capacity. Its small-signal low-frequency output impedance is given by

$$r_{\text{out}} \cong \frac{\partial V_{\text{out}}}{\partial I_{\text{out}}} \cong r_{\text{ds1}} \tag{7.3}$$

This impedance may be too low for a number of applications, and a

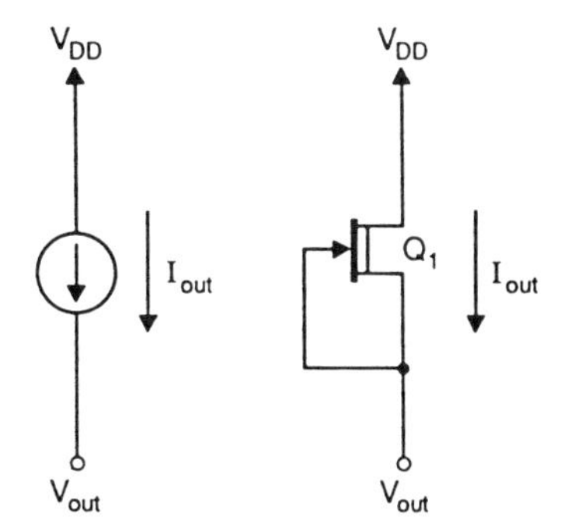

FIGURE 7.11 Common-drain current source.

variety of design techniques have been developed to improve the output impedance. "Bootstrapping" is one approach, which is occasionally employed in silicon bipolar IC designs, to increase the output impedance by employing a buffer to maintain a constant voltage between the drain and source of the transistor. This technique is very effective in GaAs analog IC design. The basic configuration is illustrated in Figures 7.12*a* and 7.12*b* [6]. In the case of the circuit of Figure 7.12*a*, the small-signal output impedance of the current source is increased from r_{ds1} to

$$r_{out} \cong g_{m2} r_{ds1} r_{ds2} \tag{7.4}$$

which is significantly higher than that of Figure 7.11. In addition, the circuit

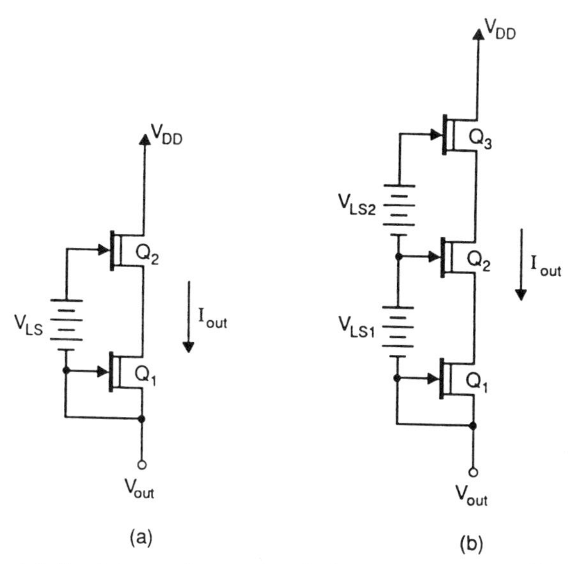

FIGURE 7.12 (*a*) Bootstrapped common-drain current source; (*b*) two-level bootstrapped common-drain current source.

of Figure 7.12*b* has an output impedance given by

$$r_{out} \cong g_{m3} r_{ds3} g_{m2} r_{ds2} r_{ds1} \tag{7.5}$$

In theory, this approach can increase the output impedance to an arbitrary degree, simply by adding more bootstrapping levels. However, just as with the design of a common-source current source, the use of more than two levels of bootstrapping leads to diminishing improvements in output impedance, since backgating and leakage effects begin to limit the achievable gain of the circuit beyond the second level.

Of course, the level-shift voltage V_{LS} is not realized by a battery, but by some circuit that approximates the same function. One approach is shown in Figure 7.13. Here, a source follower and forward-biased Schottky diodes are employed to provide the level-shifting voltage V_{LS}. If the device widths are scaled such that $W_3 = W_4 + W_2$ and $W_5 = W_1 + W_2$, a nominal value of $V_{GS} \cong 0$ will be maintained for all transistors in the circuit over a variety of output voltages. In that case, V_{LS} is the voltage across D_1–D_3, approximately 2.2 V.

Another approach to the realization of the level-shifting voltage for the bootstrapped load is shown in Figure 7.14. In this circuit, the voltage shift is realized by the V_{GS} of transistors Q_3 and Q_4, rather than forward-biased Schottky diodes. If the width of Q_3 is 4 times that of Q_5, and the width of

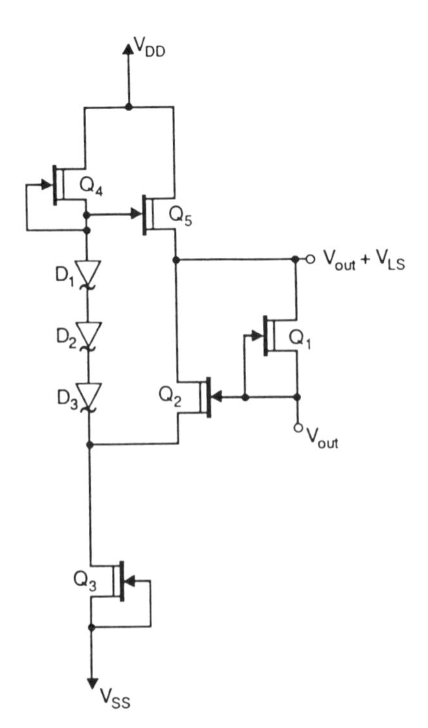

FIGURE 7.13 Two-level bootstrapped current source; $W_5 = W_1 + W_2$; $W_3 = W_4 + W_2$; $V_{LS} = 2.2$ V.

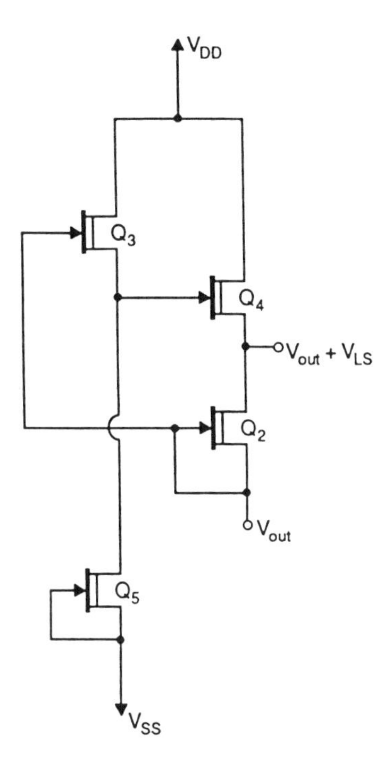

FIGURE 7.14 Alternate realization of two-level bootstrapped active load: $W_3 = W_4 = 4W_2$; $W_5 = W_2$.

Q_4 is 4 times that of Q_2, then, assuming square-law behavior, the V_{GS} of Q_3 and Q_4 will be $V_T/2$. As a result, V_{LS} will be the sum of the two source-to-gate voltages of Q_3 and Q_4, or $-V_T$. This value of V_{LS} is more than adequate to ensure that Q_2 remains in saturation for all bias conditions.

The bootstrapping techniques shown in Figures 7.13 and 7.14 necessitate the use of some kind of circuit to realize the "battery" for level-shifting purposes. However, in some cases, the designer can exploit the "early saturation" phenomenon described in Chapter 3 to realize a bootstrapped load without any level shifting. One example of a "self-bootstrapped" active load appears in Figure 7.15*a*. Since the drain–source saturation voltage of Q_1 is less than $V_{GS} - V_T$, both Q_1 and Q_2 can be biased in the saturation region, as long as

$$\frac{W_2}{W_1} \geq \frac{K^2}{(K-1)^2} \tag{7.6}$$

where W_2 is the width of Q_2, W_1 is the width of Q_1, and K is the "early saturation" factor.

This technique can be extended to a third-level bootstrapped load, as shown in Figure 7.15*b*, where Q_3 has been added. If the width of Q_3 is the same as that of Q_2, then the V_{DS} of Q_1 and Q_2 will be equal, and all three transistors are biased in the saturation region. The equations derived previously for the output impedances of the second and third-level circuits

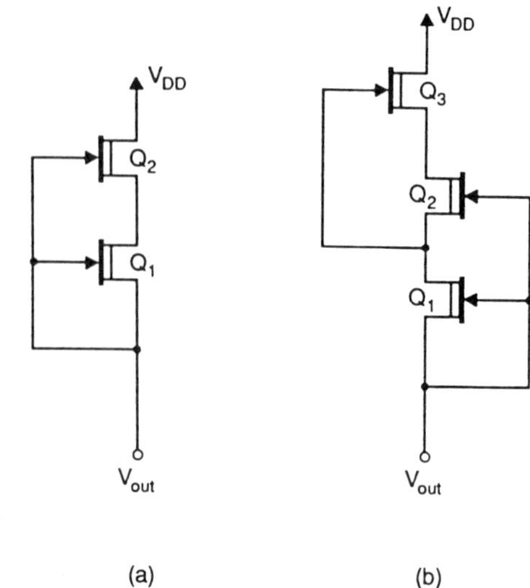

FIGURE 7.15 (*a*) Self-bootstrapped active load; (*b*) two-level self-bootstrapped active load.

[Eqs. (7.4) and (7.5)] also apply to the circuits in Figures 7.15*a* and 7.15*b*, respectively.

The principal disadvantage of this approach is that a large area may be required by Q_2 and/or Q_3, negating the advantage of the simpler circuitry. In fact, in more modern GaAs processes, the value of K is so close to unity that the widths of Q_2 and Q_3 may be unreasonably large. In addition, there is less flexibility in choosing the bias conditions of the load with the approach shown in Figure 7.15.

7.3.2 Single-Stage Amplifier Design

Amplifiers of one kind or another constitute the principal building block of most analog ICs. Optimum design techniques for single-stage amplifiers are well established in silicon technology, and many of the same approaches can be employed in GaAs circuits, especially in the case of HBT technology. However, there are some unique design constraints in the case of GaAs FET technology, which dictate the use of unique approaches. This section outlines the preferred design techniques for common-source and common-drain GaAs amplifier design.

7.3.2.1 Common-Source Amplifier Design The design of high-gain single-stage amplifiers in GaAs technology is complicated by a number of factors. Consider the simple gain stage of Figure 7.16, where the amplifier's load is represented by an ideal current source. In this case, the low-frequency

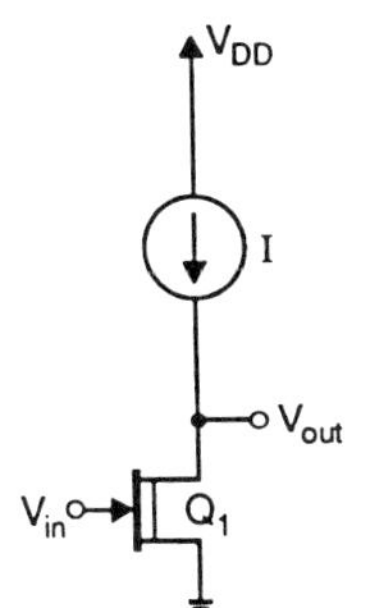

FIGURE 7.16 Single-ended FET amplifier with ideal current source load.

small-signal voltage gain of this circuit, is given by

$$A_V \cong g_{m1} r_{ds1} \tag{7.7}$$

where g_{m1} is the transconductance and r_{ds1} is the drain–source impedance of Q_1. Transistor Q_1 can either be an enhancement-mode or depletion-mode device.

As discussed in the previous section, the $g_m r_{ds}$ product is typically 10–20 at high frequencies, which is too low for the design of high-accuracy comparators and operational amplifiers. Therefore, a variety of design techniques have been developed in an attempt to improve achievable amplifier gain.

Figure 7.17*a* is a schematic of a single-ended GaAs amplifier, where the current source of Figure 7.16*a* has been replaced by a depletion-mode

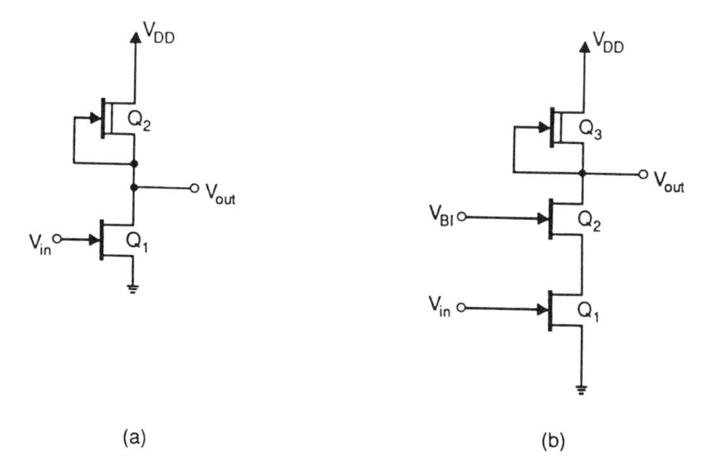

FIGURE 7.17 (*a*) FET amplifier with transistor load; (*b*) FET amplifier with cascode added for increased gain and bandwidth.

transistor. The small-signal gain of this circuit is given by

$$A_{\rm v} \cong -\frac{g_{\rm m1}}{g_{\rm ds1}+g_{\rm ds2}} \tag{7.8}$$

which is generally between 7 and 15. This is often too low for precision analog applications.

The gain and bandwidth of this circuit can be improved through the use of a cascode transistor, which is a common technique in silicon IC design. The cascode transistor can either be an enhancement-mode or a depletion-mode device, depending on the availability. This cascoded gain stage is shown in Figure 7.16*b*, and its small-signal chain can be approximated by

$$A_{\rm v} \cong -\frac{g_{\rm m1}}{g_{\rm ds3}} \tag{7.9}$$

which is a significant improvement over the circuit of Figure 7.17*a*, since the $g_{\rm ds}$ of Q_1 no longer determines the gain of the circuit.

Even higher gains can be achieved from a single stage by another technique borrowed from silicon IC design, which is shown in Figure 7.18 [7]. Here Q_2 acts as a current source, increasing the $g_{\rm m}$ of Q_1, while a smaller current flows through Q_4, increasing its $r_{\rm ds}$. This technique effectively decouples the output impedance of the load transistor from the transconductance of the driver transistor, allowing the overall gain to be enhanced. Both Q_2 and Q_4 must be depletion-mode devices, although Q_1 and Q_3 can be either enhancement- or depletion-mode. The resulting voltage gain is approximately

$$A_{\rm v} \cong -\frac{g_{\rm m1}}{g_{\rm ds4}} \tag{7.10}$$

which is larger than that achievable by the circuit of Figure 7.17*b*, since the $g_{\rm m}$ of Q_1 is increased and the $g_{\rm ds}$ of Q_4 is diminished by this approach.

Finally, the gain of the circuit of Figure 7.18 can be extended even

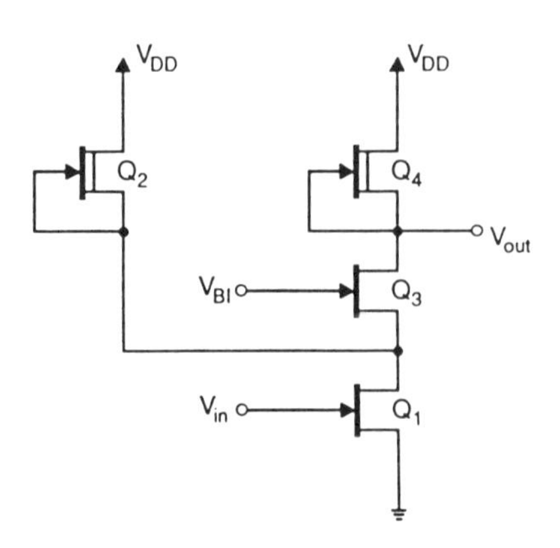

FIGURE 7.18 Enhanced gain FET amplifier.

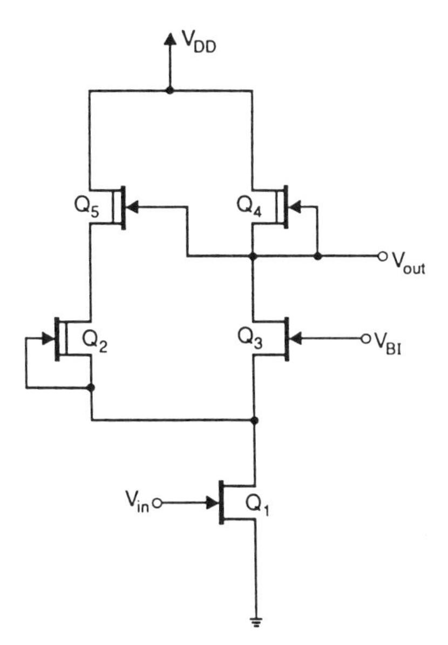

FIGURE 7.19 Enhanced gain FET amplifier with feedback.

further with feedback, as shown in Figure 7.19 [8]. In this case, transistor Q_5 provides positive feedback to the circuit, increasing the achievable gain significantly. The resulting small-signal voltage gain can be approximated by

$$A_v \cong -\frac{g_{m1}}{g_{ds4} - g_{ds2}} \tag{7.11}$$

Therefore, the gain of this circuit can be theoretically infinite, by setting the width of Q_4 equal to that of Q_2 and by setting the bias currents through the two devices equal to each other. This is not advisable, however, because the g_{ds} of the transistor is bias-dependent, and if g_{ds4} falls too low, the circuit will act as a bistable latch and cease to function as a small-signal amplifier. This feedback also increases the output impedance and nonlinearity of the amplifier.

As this discussion has demonstrated, one fundamental difficulty with realizing single-stage GaAs amplifiers that exhibit high gain is the relatively high MESFET output conductance (g_{ds}). The circuits of Figures 7.16–7.19 attempt to circumvent this limitation through a variety of design techniques. These techniques can be combined with the current-source design approaches from the previous section in order to realize amplifier stages that exhibit extremely high gain. An example of a gain stage that incorporates all of these ideas appears in Figure 7.20 [9]. This circuit exhibits a gain that is approximately equal to $(g_m/g_{ds})^3$, which is far larger than that of the circuit of Figure 7.16, which was approximately (g_m/g_{ds}).

The circuits of Figures 7.17*a* and 7.20 were simulated with SPICE, in order to compare their achievable gains and bandwidths. Figure 7.21 is a

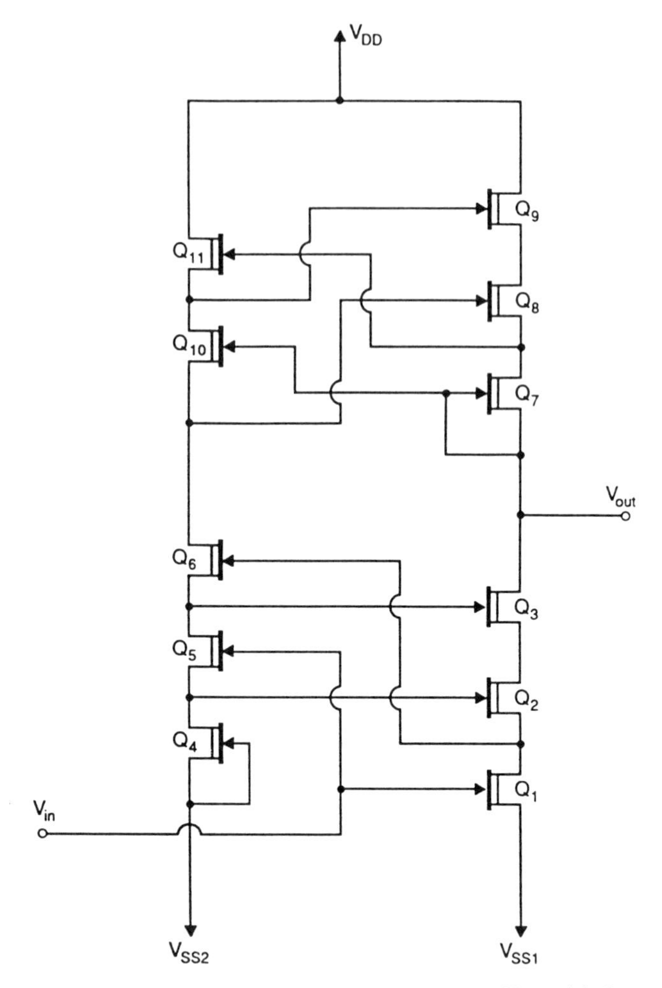

FIGURE 7.20 Fully bootstrapped FET single-stage amplifier with improved gain.

plot of the simulated gain vs. frequency of the two circuits, for a typical 1-μm ion-implanted GaAs MESFET process. It is clear that the fully bootstrapped gain stage of Figure 7.20 will exhibit a far higher gain than that of the simpler circuit of Figure 7.17*a*. Of course, a penalty is paid with the more complex approach in terms of die area and power dissipation. In this simulation, the circuit of Figure 7.17*a* dissipated 5 mW and the circuit of Figure 7.20 dissipated 23 mW. Also note that the bandwidth is compromised slightly with the more complex approach.

The equivalent circuit model employed in this simulation does not account for the frequency dependence of the output impedance at low

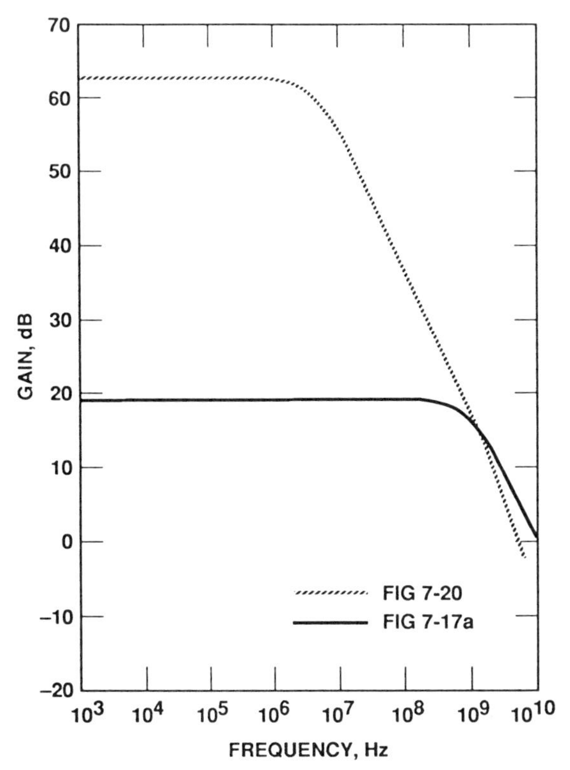

FIGURE 7.21 Comparison of simulated frequency responses of single-stage amplifiers from Figures 7.17*a* and 7.20.

frequencies discussed in Chapter 3, so if the low-frequency value of r_{ds} is used, then the dc gain of these circuits will be even higher than that predicted by these simulations.

7.3.2.2 Common-Drain Amplifier Design Common-drain amplifiers, or source followers, are often employed as buffering, interface, or level-shifting circuits in GaAs analog ICs. Their general importance makes it worthwhile to study their behavior in detail. A useful rule for the design of high-speed circuits is to employ level-shifting circuitry as little as possible, since the extra power dissipation and delay significantly compromised the performance of GaAs analog circuits. However, source followers are often the only choice where level-shifting is necessary, and they can also be used as a buffer between a high-impedance gain stage and a low-impedance load. When carefully designed, they are also suitable for use as moderate accuracy voltage followers in a variety of precision analog applications.

A simple example of a source follower is shown in Figure 7.22. In this

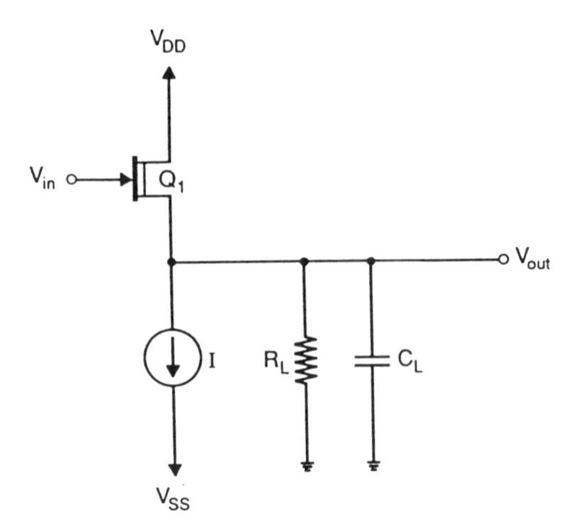

FIGURE 7.22 GaAs FET source follower.

case, ignoring C_{gd}, the small-signal gain is given by the expression

$$A_v \cong \frac{g_{m1}}{g_{m1} + g_L} \frac{1 + (sC_{gs1}/g_{m1})}{1 + [s(C_{gs1} + C_L)/(g_{m1} + g_L)]} \tag{7.12}$$

where g_L is the conductance of the parallel combination of the load resistance R_L and the output impedance r_{ds1} of Q_1, and C_L is the load capacitance.

According to this first-order calculation, the source follower is a single pole, single zero circuit, whose bandwidth can be maximized by setting the pole and the zero to the same frequency. This condition occurs when

$$\frac{C_{gs1}}{C_{gs1} + C_L} = \frac{g_{m1}}{g_{m1} + g_L} \tag{7.13}$$

This condition is not always possible to realize in practice, and often conflicts with other desirable goals, such as maximum gain. For instance, g_L is usually small compared to g_{m1}, but this implies that C_L should be small compared to C_{gs1}, which is often impossible to accomplish.

In cases where the circuit is driving purely capacitive loads, g_L consists entirely of g_{ds1} and the output conductance of the current source I. To maximize the gain of the source follower, it is desirable to minimize g_L, and therefore the output impedance of the current source and transistor Q_1 should be maximized using all the techniques discussed in Section 7.2.1.

Stability or overshoot problems can arise when a source follower (in the case of FET or HEMT technology) or emitter follower (in the case of HBT technology) is driving a purely capacitive load [10]. This can be seen by examining the emitter-follower circuit of Figure 7.23*a*. If C_{cb} (collector-to-

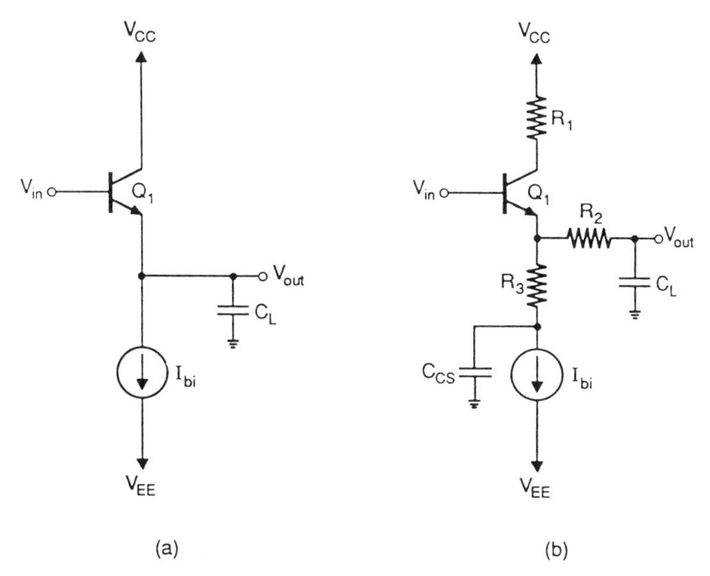

FIGURE 7.23 (*a*) Bipolar emitter follower; (*b*) emitter follower with improved stability and high-frequency behavior.

base capacitor) is assumed to be negligible, the transfer function of this circuit can be approximated by

$$A_v(s) = A_{v0} \frac{(1 + s\tau_t)}{1 + s\tau_t(R_E + R_S)/R_E + s^2\tau_t\tau_e(R_S/R_E)} \tag{7.14}$$

where

$$\tau_t = \frac{1}{2\pi f_T}, \qquad \tau_e = \frac{1}{R_e C_L}, \qquad R_s = R_{in} + r_b, \qquad \text{and}$$

$$A_{v0} = \frac{R_E}{R_E + 1/g_m + R_s/\beta}$$

Equation (7.14) describes a second-order, underdamped, lowpass filter, whose step response can exhibit severe overshoot and ringing.

This result demonstrates that emitter and source follower circuits are prone to poor performance when connected to highly capacitive loads. This limitation can be minimized by the addition of a number of small damping resistors, as shown in Figure 7.23*b*. Resistor R_1 minimizes the injection of current into V_{cc} during a step input. Resistor R_2 minimizes the deleterious effects of C_L on the damping of the circuit. Resistor R_3 minimizes the effect of C_{cs} on the step response of the circuit. The best way to choose the optimum values of these resistors is with a trial-and-error approach, where

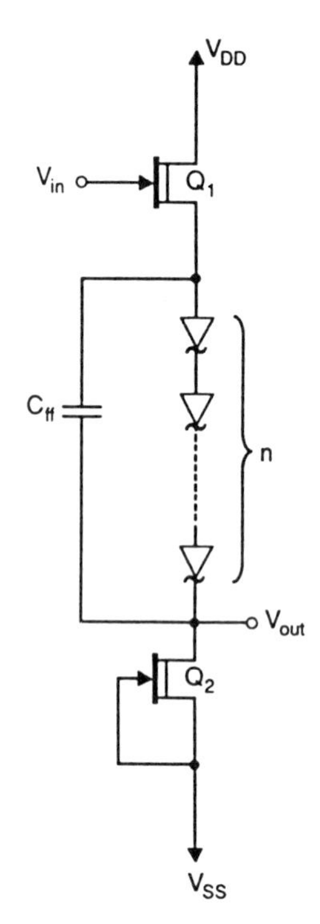

FIGURE 7.24 Source follower used as level shifter, with feedforward capacitor for enhanced speed.

computer simulation is combined with judicious iteration to arrive at a design. Typical values of these resistors should be between 10 and 100 Ω.

If level shifting as well as buffering is desired, the addition of forward-biased Schottky diodes to the source follower, as shown in Figure 7.24, is appropriate. If the device width of a depletion-mode transistor Q_1 is made equal to the width of transistor Q_2, then the V_{GS} of Q_1 will be approximately zero, and the level-shifted voltage will be given by

$$V_{out} = V_{in} - nV_{di} \tag{7.15}$$

where V_{di} is the voltage across a forward-biased Schottky diode.

One drawback of approach is that the dynamic resistance of the forward-biased diodes limits the achievable gain of the circuit, and adds additional delay from the input to the output. This delay can be partially minimized by the addition of the feedforward capacitor C_{ff} in Figure 7.24.

7.4 OPERATIONAL-AMPLIFIER DESIGN

A number of GaAs MESFET op amps have been described in the literature, and their designs illustrate some of the concepts described in this chapter. Generally, op-amp designs can be broken into two classes: those intended as stand-alone discrete circuits, and those intended solely for on-chip applications in an LSI (large-scale IC) circuit. Discrete op amps have far more demanding requirements in terms of output loading and stability, and therefore tend to dissipate more power and be more complex. Op amps designed for on-chip applications are generally simpler and dissipate less power, so that many can be placed on a single IC. This section will discuss the design and implementation of both types.

The Anadigics AOP3510 was the first commercially available general-purpose GaAs MESFET stand-alone monolithic operational amplifier [8]. Like many silicon op amps, the circuit employs two gain stages in order to maximize the overall gain. Miller compensation was employed to ensure stability over a wide range of load conditions and process variations. A schematic of the overall op-amp block diagram appears in Figure 7.25.

The input stage is a differential amplifier, followed by a level shifter, and the combination is shown in Figure 7.26. For the case of a general-purpose op amp, it is important to assure that the input common-mode range rises as the power supply voltages rise. This does not occur with the diode level-shift approaches discussed in Section 7.3.2, because the level-shift voltage is fixed by the forward diode voltage drop.

In this op amp, a new approach for implementation of the level shift was developed, which is also shown in Figure 7.26. The circuit employs a resistor, rather than forward biased diodes to provide the level shifting. The purpose of the level shifter is to shift the voltage at node V2, which should rest approximately 2 V from the positive rail when both inputs are equal, down to the voltage at node V5, which rests approximately 1.4 V from the negative rail, while at the same time providing an ac gain of approximately 1. Transistors Q_7–Q_{11} form a current mirror, so that $I_1 = I_2$ under most bias conditions. Since the currents through the two transistors are the same, the voltage drop across the two resistors, R_1 and R_2, will also be the same.

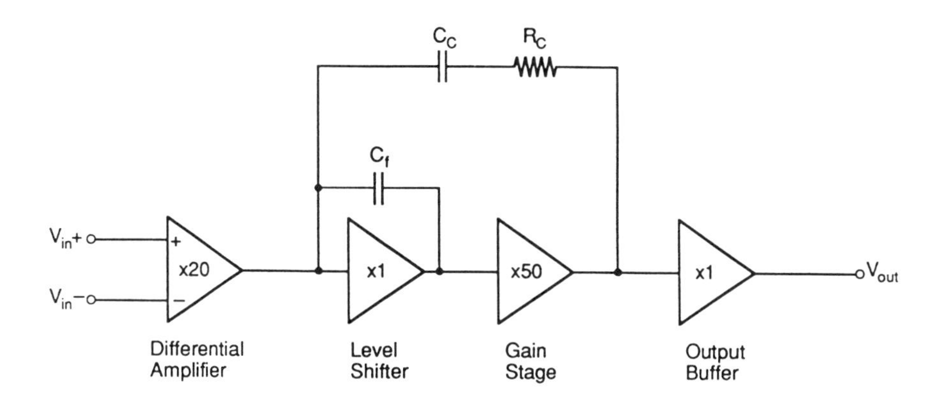

FIGURE 7.25 Block diagram of GaAs MESFET operational amplifier.

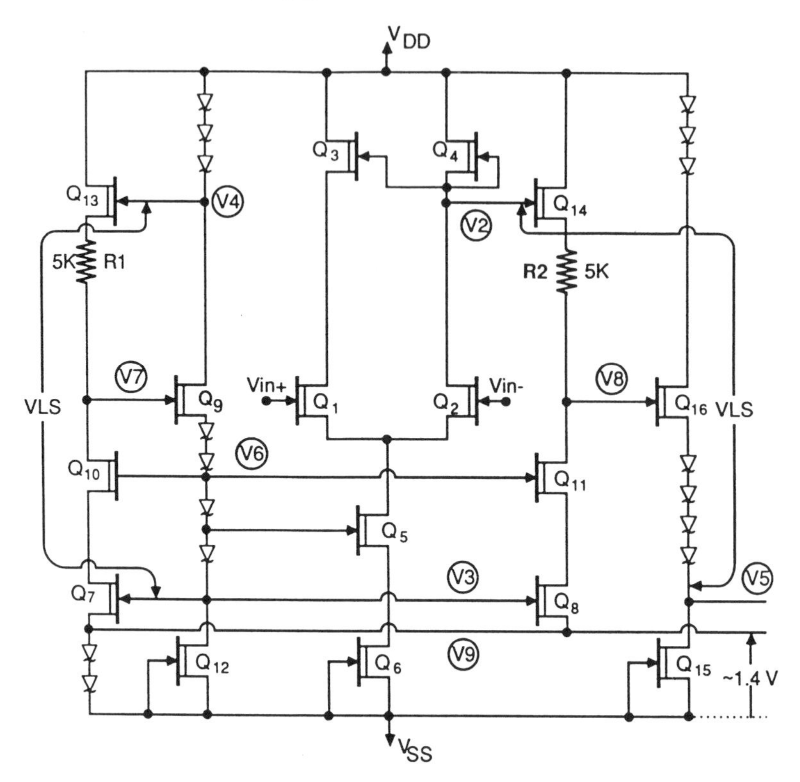

FIGURE 7.26 Complete GaAs MESFET operational amplifier.

Therefore, the voltage drop between nodes V4 and V3 will be $V_{R1} + 4V_{di}$, which is the sum of the voltage across the resistor plus the voltage across the diodes between the source of Q_9 and the drain of Q_{12}. Because the circuit is symmetric, V2–V5 = V4–V3, which means that under quiescent conditions V5 = V3.

The voltage at node V3 is simply $V_{SS} + 2V_{di} + V_{GS7}$, and if the width of Q_7 is picked so that its nominal $V_{GS} = 0$, then the voltage at node V3 is simply $V_{SS} + 2V_{di}$. Of course, as the power supply voltage changes, the current I_1 will change, which forces V_{GS7} to change as well. In order to prevent the change in V_{GS7} from altering the level-shift voltage, the quiescent value of V_{DS4} should track V_{GS7}. This is accomplished by connecting the gate of Q_5 to a node that tracks V_{GS7}. Sincc Q_5 is acting as a source follower, the V_{DS} of Q_6 will track the V_{GS} of Q_7. But when the two differential inputs are at equal potential, the V_{DS} of Q_3 and Q_4 are equal, and any variations in the V_{DS} of Q_6 will be duplicated in the V_{DS} of Q_3 and Q_4.

The result of this technique is that the quiescent voltage at node V5 is maintained at 1.4 V above V_{SS} for a variety of bias and power supply variations. However, one drawback is that the ac performance of the circuit is somewhat limited by the high impedance of resistors R_1 and R_2. In order to minimize this effect, feedforward capacitors have been inserted, which provide a low impedance path through the level-shifter at high frequencies. This creates a low frequency pole and zero in the open-loop transfer function of the amplifier, which is a well-known cause of slow-setting behavior. This has not been observed in the experimental devices, however.

The second gain stage is similar to that of Figure 7.19, where the width of Q_4 was set to one-half that of Q_5. These values increased the gain but ensured that the device did not act as a bistable latch. A push–pull output stage is employed to drive a 50-Ω load while minimizing power dissipation.

In most respects, this GaAs op amp performs like a high-speed equivalent of a standard silicon op amp. Its dc open-loop gain is approximately 60 dB, and its gain–bandwidth product is approximately 500 MHz. Its power dissipation varied from 500 mW to 1 W, depending on the power supplies. The most serious drawback to its use was the presence of low-frequency (~1 kHz) oscillations. These oscillations are not related to the circuit design, but appear to be material- and process-related. In addition, the low-frequency $1/f$ noise of the circuit is significantly higher than that of a typical silicon bipolar op amp, with the corner frequency occuring at approximately 10 MHz.

A second GaAs operational amplifier, with a high common-mode rejection ratio, was described by Katsu et al. in 1987 [11]. This circuit employed few of the gain enhancement techniques described in Section 7.2, but it did employ a novel common-mode feedback technique using a differentially driven common-source feedback amplifier. The open-loop gain of the circuit was 48 dB, with a unity-gain frequency of 150 MHz, and a 1% settling time with a 2-V step of 60 nS. The power dissipation was 225 mW.

When an op amp is intended for use only in on-chip applications, the design philosophy can be slightly different. In high-frequency monolithic applications, where purely capacitive loads predominate, an op amp consisting of a single, high-gain, stage that employs dominant pole compensation at the output is generally preferred over the more common two-stage pole-splitting compensation approach. An explanation for this preference can be seen by examining the single-stage and two-stage op-amp equivalent circuit model of Figure 7.27. Both circuits have approximately the same dc gain, which is $(g_m r_{ds})^2$. However, the nondominant pole location in the two-stage amplifier (Fig. 7.27*a*) is $(-g_{m2}/C_L)$, whereas the nondominant pole of the single-stage op amp (Fig. 7.27*b*) is given by $(-g_{m2}/C_{gs2})$.

Since the nondominant pole determines the achievable bandwidth of the op amp, the single-stage design has the potential for substantially higher bandwidth than the two-stage design. In general, the ratio of C_L/C_{gs} is between 5 and 10 for a typical IC, which means that the ratio of the achievable bandwidths should be of that order, also.

A single-stage GaAs MESFET op amp has been designed and fabricated for monolithic applications, where capacitive loads dominate, and a schematic is shown in Figure 7.28 [12]. This circuit employs all of the gain enhancement techniques that were described in Section 7.2 for a depletion-

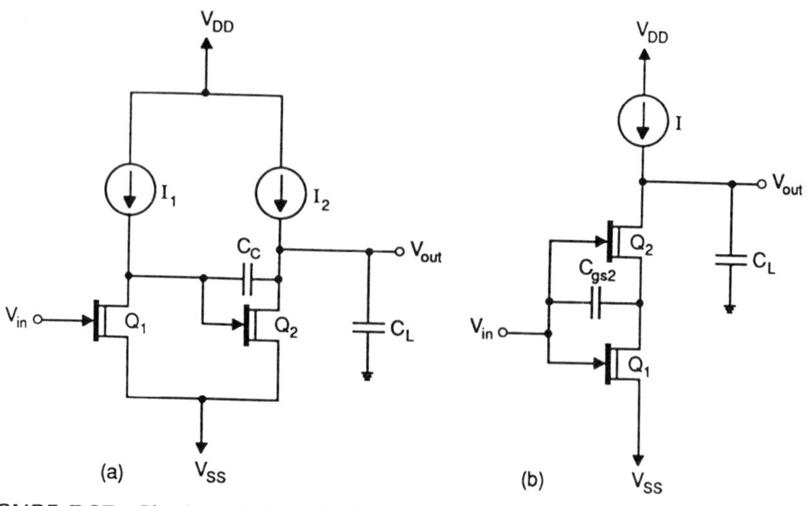

FIGURE 7.27 Single-ended equivalent circuit model of (*a*) two-stage operational amplifier and (*b*) single-stage operational amplifier.

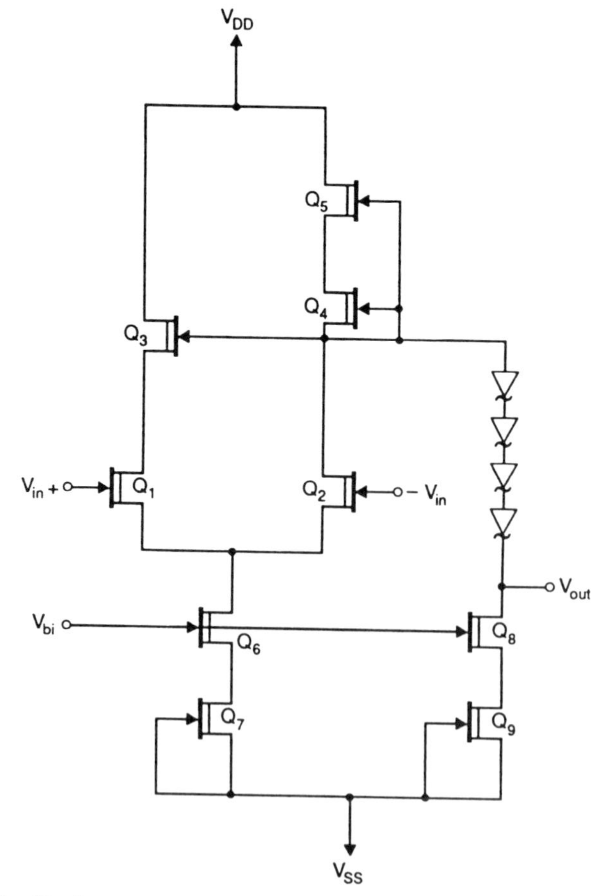

FIGURE 7.28 Single-stage GaAs depletion-mode MESFET operational amplifier [12].

mode technology. The dc gain of the circuit was approximately 40 dB, and the bandwidth, with a 0.4-pF load, was 900 MHz. The circuit dissipated 150 mW. The gain and bandwidth of the circuit are adequate for moderate-precision applications, where the signal frequencies are less than 200 MHz.

Finally, the contributions of C. Toumazou and D. Haigh must be recognized for their pioneering work in the field of broadband analog GaAs design [13–15]. They developed the first GaAs switched-capacitor circuit, originated many new circuit approaches [13,14], and have refined the design techniques to the point where GaAs circuits with 300-MHz sampling rates have recently been demonstrated [15]. Their designs employ many of the techniques developed here for gain enhancement, in addition to some novel approaches for current mirroring and level shifting. These design techniques are described in Chapter 10 of this text.

REFERENCES

1. J. E. Solomon, "The Monolithic Op Amp—A Tutorial Study," *IEEE J. Solid-State Circuits*, **SC-9** (6) (Dec. 1974).
2. P. Gray and R. Meyer, *Analysis and Design of Analog Integrated Circuits*, 3rd ed., Wiley, New York, 1991.
3. T. Choi, R. Kaneshiro, R. Brodersen, P. Gray, M. Wilcox, and B. Jett, "High-Frequency CMOS Switched-Capacitor Filters for Communications Applications," *Digest of Technical Papers, 1983 International Solid-State Circuits Conference*, New York, February 1983.
4. D. Bowers, "A Precision Dual 'Current Feedback' Operational Amplifier," *Proceedings of the 1988 Bipolar Circuits and Technology Meeting*, pp. 68–71.
5. R. J. Widlar, "Design Techniques for Monolithic Operational Amplifiers," *IEEE J. Solid-State Circuits*, **SC-4**, 184–191 (Aug. 1969).
6. A. A. Abidi, "An Analysis of Bootstrapped Gain Enhancement Techniques," *IEEE J. Solid-State Circuits*, **SC-22** (6), 1200–1203 (Dec. 1987).
7. D. Senderowicz, D. A. Hodges, and P. Gray, "High-Performance NMOS Operational Amplifier," *IEEE J. Solid-State Circuits*, **SC-13**, 760–766 (Dec. 1978).
8. N. Scheinberg, "A High-Speed GaAs Operational Amplifier," *IEEE J. Solid-State Circuits*, **SC-22** (4), 522–527 (Aug. 1987).
9. C. Toumazou and D. Haigh, "Analogue Design Techniques for GaAs Operational Amplifiers," *Proceedings of the 1988 International Circuits and Systems and Symposium*, June 1988, pp. 1453–1456.
10. S. Knorr, private communications.
11. S. Katsu, M. Nishiuma, D. Ueda, M. Kazumura, and G. Kano, "A High CMRR GaAs Operational Amplifier IC," *Proceedings of the 1987 IEEE GaAs IC Symposium*, pp. 99–102.
12. L. E. Larson, G. C. Temes, and K. Martin, "GaAs Switched-Capacitor Circuits for High-Speed Signal Processing," *IEEE J. Solid-State Circuits*, **SC-22** (6), 971–981 (Dec. 1987).

13. C. Toumazou and D. Haigh, "Design Optimization and Testing of a GaAs Switched-Capacitor Filter," *IEEE Trans. Circuits Systems*, **38** (8), 825–837 (Aug. 1991).
14. D. Haigh, J. Swanson, J. Taylor, and J. Luck, "High-Frequency GaAs Switched-Capacitor filter Implemented with GaAs Insulated Gate Switches," *Electron. Lett.*, **27** (18), 1619–1620 (Aug. 1991).
15. C. Toumazou and D. Haigh, "Design of GaAs Operational Amplifiers for Sampled-Data Applications," *IEEE Trans. Circuits Systems*, **37** (7), 922–935 (July 1990).

CHAPTER EIGHT

Mixers and Oscillators

PASCAL PHILIPPE,
Laboratoires D'Électronique Philips
94453 Limeil-Brévannes Cédex, France

8.1 MIXERS

8.1.1 Mixer Basics

8.1.1.1 Principle of Operation Mixing or frequency conversion is a process that involves two signals and a nonlinear device. A device is nonlinear when the output current i is no longer simply proportional to the input voltage v. This can be expressed mathematically by taking into account high-order terms in the Taylor series expansion of its transfer function $i(v)$, which has to be written as

$$i = \sum_{n=0}^{\infty} a_n v^n$$

In presence of a two-tone input v at frequencies f_1 and f_2

$$v = V_1 \sin(2\pi f_1 t + \phi_1) + V_2 \sin(2\pi f_2 t + \phi_2)$$

High-Frequency Analog Integrated-Circuit Design, Edited by Ravender Goyal
ISBN 0-471-53043-3

The output current i is of the form

$$i = \sum_{n=0}^{\infty} \sum_{p=0}^{n} C_p^n a_n V_1^p V_2^{n-p} \sin^p(2\pi f_1 t + \phi_1) \sin^{n-p}(2\pi f_2 t + \phi_2)$$

This expression shows that the response contains not only powers of the respective input tones but also crossed products. Using trigonometric formulas, it is possible to express any product of sine functions as a certain sum of sine functions having as argument, a linear combination of the respective arguments. After some tedious calculations, it can thus be demonstrated that the output signal contains harmonics of f_1 and f_2 but also intermodulation products at all possible frequencies that result from a linear combination of f_1 and f_2. In particular, it can be shown that the term of order n in the power series development of $i(v)$ generates signals proportional to

$$a_n V_1^p V_2^q$$

at the frequencies

$$pf_1 \pm qf_2$$

where p and q are integers so that $p \geq q$ related to n by

$$p + q = n$$

This is an important result which shows that a signal multiplication in the time domain results in a frequency addition or substraction in the frequency domain. Mixers take advantage of this property to effect a frequency conversion. Historically, the desired input and output signals of a mixer are, respectively, RF (radiofrequency) and IF (intermediate frequency). The signal that is mixed with the RF signal is supplied by a local oscillator (LO). As the amplitude of the intermodulation products generally decreases as their order increases, one is usually interested at one of the second-order intermodulation products of RF and LO. The IF signal is thus at the difference frequency $|f_{\text{LO}} - f_{\text{RF}}|$ or at the sum frequency $f_{\text{LO}} + f_{\text{RF}}$. It is selected using a filter that eliminates other unwanted signals from the mixer output. It has to be noted that the signal at the frequency f_{IF} may be generated by two different input signals separated by a distance of $2f_{\text{IF}}$ in frequency (Fig. 8.1). Only one of these signals is generally desired. The other, called the image signal (IM), is consequently a potentially disturbing signal. It can be rejected using an input filter.

In order to get a strong IF output signal, two conditions must be met. First, the second-order term a_2 of the transfer function of the mixing device must be large. Mixers are especially designed to enhance this second-order response. Second, the LO signal must be large because the IF signal increases linearly with LO amplitude before reaching saturation. When the

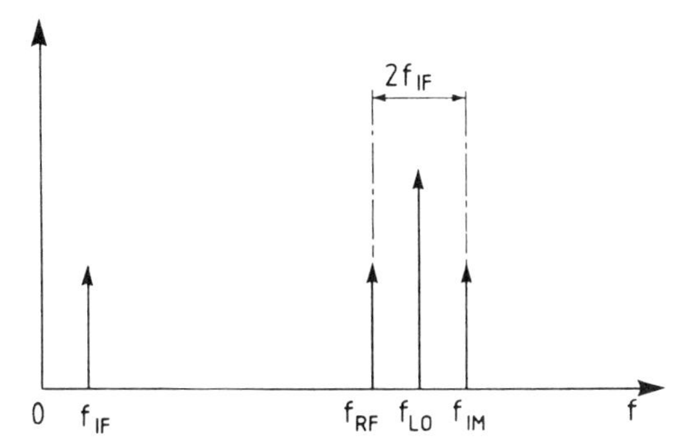

FIGURE 8.1 Image frequency of a mixer. The image gives the same IF as RF. It is the symmetric of RF with respect to LO.

mixer is saturated with respect to the LO signal, the amplitude of the IF signal is maximum and weakly depends on the LO level, which is the typical mixer operation.

The operation of LO-saturated mixers can also be explained by considering the mixer as a switch (Fig. 8.2). The RF signal is turned on and off by a switch operated at the LO frequency. The output signal is the result of the multiplication of the RF signal by the LO switching waveform. In the case of 50% duty-cycle rectangular waveform, a Fourier series expansion of the output shows that the conversion losses from RF to IF are $1/\pi = -9.9$ dB. The output also contains a residual RF signal because the LO waveform has a dc component. Another more sophisticated switch is shown in Figure 8.3. The RF signal is periodically reversed instead of being simply turned off. As the LO waveform has no dc component, the output signal does not contain any RF. The RF is completely transposed, and it can be shown that the conversion losses are now $2/\pi = -3.9$ dB. This figure represents the minimum conversion losses that can be achieved by a passive mixer. Practical mixer structures that approach such an ideal mixer are double balanced diode and passive FET mixers.

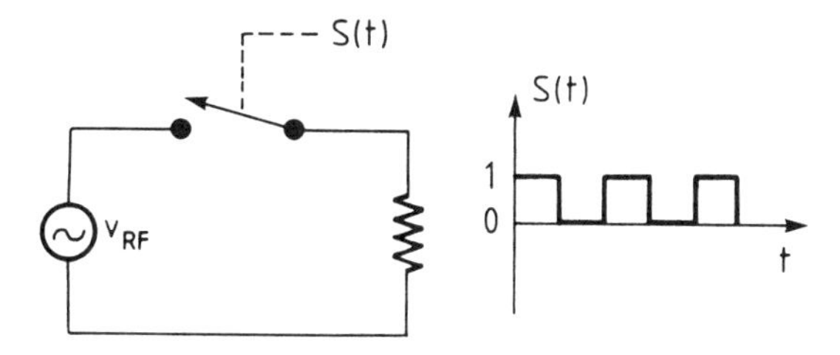

FIGURE 8.2 The mixer as a switch. The output signal is the result of the multiplication of RF by $S(t)$.

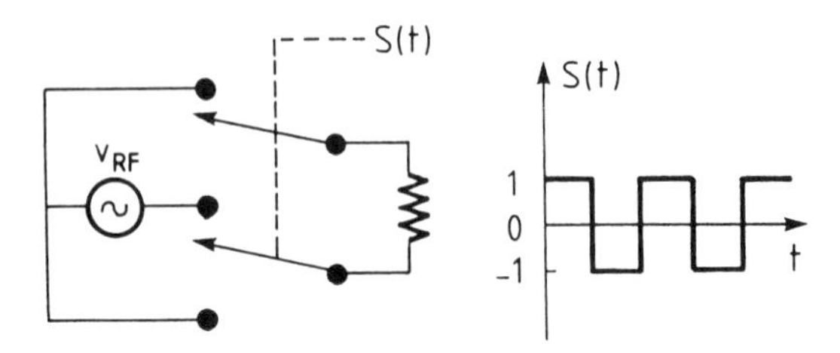

FIGURE 8.3 A more sophisticated switching mixer.

8.1.1.2 Performance Criteria

Conversion Gain The first performance criteria for a mixer is its conversion gain. It characterizes the efficiency of the transposition from RF to IF. Passive mixers have conversion losses, usually in the range -6 to -10 dB, but active mixers may have a conversion gain as high as 10 dB. The conversion gain increases with the LO level but only weakly if it is high enough to saturate the mixer. The required LO level depends on the mixing device, diode or FET, and also on device parameters such as diode threshold voltage or FET pinchoff voltage. Usually, good mixer operation requires a LO power in the range 0–13 dBm.

Isolation Another important feature is the isolation between the three ports of the mixer. As the level is high at the LO port, the LO/RF and LO/IF isolations are the most important features. A good LO isolation is required, for instance, to reduce the LO leakage to the antenna, which may interfere with other receivers or communication services, or to avoid the saturation of circuits following the mixer. These problems can often be addressed by filtering, but it may not be possible to filter LO if it is not widely separated from RF, for instance, or if the LO and RF bands overlap.

Spurious Signal Rejection A good isolation is not sufficient to prevent all undesired signals. Mixers generate a lot of spurious signals that are a combination of RF and LO products that may perturb the system if they are within the IF band of that system. This problem is most serious with high-level low-order products. In down-converters, products such as $2f_{RF} - f_{LO}$ may be in the IF band if the RF band contains signals at twice the IF frequency. This situation occurs, for instance, in satellite TV tuners that have an RF input band ranging from 950 MHz to 2 GHz and an IF frequency of 480 MHz. Such a problem cannot be solved by filtering since it arises from the desired signal itself.

Saturation and Multitone Intermodulation Mixers, like other components using solid-state devices, are subject to saturation. Saturation is the result of odd-order distortion, more particularly third-order distortion [1]. It is

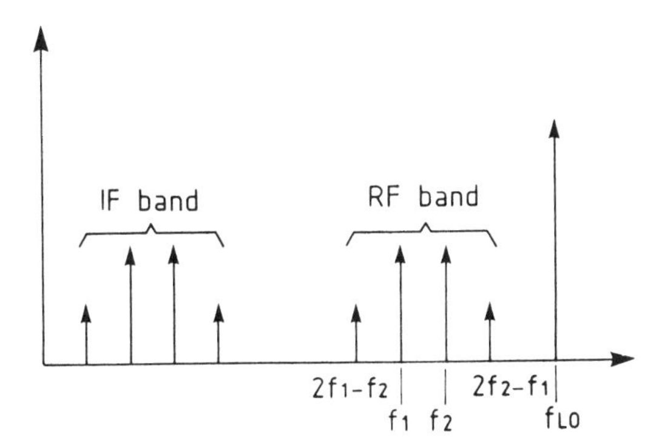

FIGURE 8.4 Third-order intermodulation spectrum. Intermodulation generates spurious signals close to the wanted ones.

usually described by the 1-dB compression point, defined as the power level at which the gain is decreased by 1 dB. Mixers are also subject to intermodulation in presence of multitone RF input signal. Intermodulation products are similar to spurious responses in their method of generation, except that the tones that intermodulate are the input tones, not an input signal and LO. The most serious distortion is usually the third-order intermodulation caused by two closely spaced RF tones at f_{RF1} and f_{RF2}. Intermodulation products are generated at frequencies $2f_{RF1} - f_{RF2}$ and $2f_{RF2} - f_{RF1}$, close below and above the wanted pair (Fig. 8.4). The resulting spectrum is simply converted by the mixer. These products are consequently potentially troublesome signals since they may not be rejected by the IF filter. At levels well below the saturation, their power level increases by 3 dB for a 1-dB increase of the input power, thus more rapidly than the first-order response (Fig. 8.5). The intermodulation performance of mixers is usually measured in terms of the two-tone third-order intercept point,

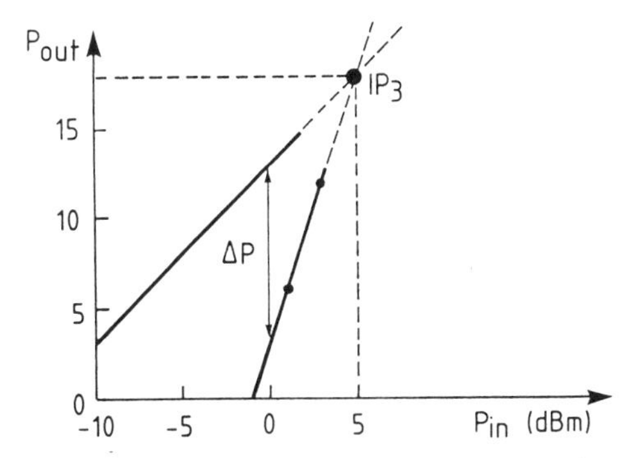

FIGURE 8.5 Intermodulation characteristic. Third-order products increase by 3 dB for 1-dB increase of the first-order wanted term.

defined as the power level at which the output powers of the intermodulation products and of the primary desired response are equal. The intercept point can be referred to the input or the output. The difference in between is simply the gain. This point is fictive because it can never be reached since it is located well above the saturation level. Theoretically, it can be shown that it is 9.6 dB above the 1-dB compression point. The distance ΔP in decibels between the intermodulation products and the desired first-order signal at the output is related to the intercept point (IP3) in decibels by

$$\Delta P = 2(IP_3 - P)$$

where P, in decibels, is the input power if the intercept point is referred to the input, or the output power at first order if the IP3 is referred to the output. The IP3 is an important figure that determines the capability of the mixer to keep the third intermodulation products much below the desired signal. An intercept point can similarly be defined for the second-order intermodulation distortion.

Noise Figure In contrast to large-signal behavior, the noise figure of a mixer is an important figure of merit in small-signal operation. It characterizes the noise power added by the mixer as compared with a noiseless equivalent function. The general noise figure (NF) definition is

$$\mathrm{NF} = \frac{N_{\mathrm{out}}}{GN_{\mathrm{in}}}$$

where N_{in} and N_{out} are respectively the input and output noise powers and G is the gain of the system. The noise figure for a mixer happens to be confusing because of three possible definitions depending on whether the image sideband noise is considered [2]. The most natural definition considers only the frequency conversion from RF to IF. No noise contribution is assumed to come from the image band. In this case, the mixer output noise N_{out} has a similar expression to that of an amplifier:

$$N_{\mathrm{out}} = Gk(T_0 + T_{\mathrm{SSB}})B$$

where G is the mixer power gain, k is the Boltzmann constant, B is the mixer bandwidth, T_0 is the RF source temperature, and T_{SSB} is the single-sideband mixer noise temperature that characterizes the noise added by the mixer when converting RF into IF. If the mixer was noiseless, the output noise would be simply the RF input noise multiplied by the mixer gain:

$$N_{\mathrm{out}}(\text{noiseless}) = Gk(T_0)B$$

A first single sideband (SSB1) noise figure is thus given by

$$NF_{\mathrm{SSB1}} = \frac{T_0 + T_{\mathrm{SSB}}}{T_0}$$

The SSB1 noise figure can be measured by filtering out the image band. If the mixer also responds to the image frequency, the second single-sideband noise figure definition that includes the image contribution has to be used. Assuming identical source impedances and conversion gains for each band, the mixer output noise is now

$$N_{\mathrm{out}} = Gk(2T_0 + T_{\mathrm{SSB}})B$$

In the second single-sideband (SSB2) noise figure definition, the mixer output noise is still referred to the noise power coming from the RF band only, therefore leading to

$$NF_{\mathrm{SSB2}} = \frac{2T_0 + T_{\mathrm{SSB}}}{T_0}$$

In the last noise figure definition, the total output noise power is referred to the sum of noise powers coming from RF and IM. The result is known as the double-sideband (DSB) noise figure given by

$$NF_{\mathrm{DSB}} = \frac{(2T_0 + T_{\mathrm{SSB}})}{2T_0}$$

The DSB noise figure is lower than the SSB2 noise figure by exactly 3 dB. As the SSB1 noise figure, the DSB noise figure is 0 dB for a noiseless mixer, whereas it is 3 dB using the SSB2 definition. SSB1 and SSB2 definitions, however, tend to the same value for infinite noise figure. One of the DSB or SSB2 noise figures is usually given for low-IF mixers. If IF is not very low with respect to RF, the most relevant figure is the SSB1 noise figure.

8.1.1.3 Single and Double-Balanced Mixers Balanced mixers are particular combinations of two (single-balanced) or four (double-balanced) single mixers. Because of the phase relationships between the LO and/or RF signals on each mixer, some unwanted products cancel out at the IF output. Let us consider, for instance, a mixer with a transfer function given by

$$i = \sum_{n=0}^{\infty} a_n v^n$$

If the voltage v is the sum of the RF and LO voltages, we obtain

$$v = v_{\mathrm{RF}} + v_{\mathrm{LO}}$$

The output current i is of the form

$$i = \sum_{n=0}^{\infty} \sum_{p=0}^{n} C_p^n a_n v_{\mathrm{RF}}^p v_{\mathrm{LO}}^{n-p}$$

A single-balanced mixer is built using two single mixers with either RF or LO applied 180° out of phase; the other is applied in phase. In the situation where LO is the signal applied 180° out of phase, the voltages in each mixer are

$$v_1 = v_{\mathrm{RF}} + v_{\mathrm{LO}}$$

$$v_2 = v_{\mathrm{RF}} - v_{\mathrm{LO}}$$

The mixers responses are of the form

$$i_1 = \sum_{n=0}^{\infty} \sum_{p=0}^{n} C_p^n a_n v_{\mathrm{RF}}^p v_{\mathrm{LO}}^{n-p}$$

$$i_2 = \sum_{n=0}^{\infty} \sum_{p=0}^{n} C_p^n a_n v_{\mathrm{RF}}^p (-1)^{n-p} v_{\mathrm{LO}}^{n-p}$$

The second order IF responses are 180° out of phase. They can be added in phase by taking their difference $i_1 - i_2$. In this operation, all products involving an even power of v_{LO} are cancelled. As the result also holds at zero power, it means in particular that the output current does not contain any RF or harmonics of RF. The RF/IF isolation is thus greatly improved. If RF signals are now applied 180° out of phase but LO signals in phase in the mixing devices, a similar result is obtained showing that all products involving an even power of v_{RF} are suppressed from the difference of the output currents. In particular, LO and harmonics of LO are rejected, leading to a high LO/IF isolation.

In a double-balanced mixer, four single mixers are used where the phase relationships between the RF and LO signals applied to each mixer are such that

$$v_1 = v_{\mathrm{RF}} + v_{\mathrm{LO}}$$

$$v_2 = v_{\mathrm{RF}} - v_{\mathrm{LO}}$$

$$v_3 = -v_{\mathrm{RF}} + v_{\mathrm{LO}}$$

$$v_4 = -v_{\mathrm{RF}} - v_{\mathrm{LO}}$$

The individual mixers outputs are combined so that the total output current i is given by

$$i = i_1 - i_2 - (i_3 - i_4)$$

In this operation, the second-order IF responses are added in phase, whereas it can be easily demonstrated that all products involving an even power of v_{RF} or v_{LO} are cancelled out. In particular, RF and LO and their harmonics are rejected, which leads to a high LO/IF and RF/IF isolation.

The capability of balanced mixers to reject some undesired signals make them attractive for applications that require good LO/IF and/or RF/IF isolation and few radiations outside the system bandpass. Some LO rejection capability also exists, usually at the RF port, which yields high LO/RF isolation. Theoretically, the rejection ratio is infinite but it is limited in practice to 20–25 dB because of the nonperfect symmetry of the balanced system. Beside its rejection capabilities, a balanced mixer also has generally a better large-signal performance than does a single-ended mixer, as well as a better noise figure, because some noise contributions are canceled. They are, however, more complex and require 0/180° baluns.

Characteristics of single-ended and balanced FET mixers are summarized and compared in Table 8.1.

8.1.1.4 Image Rejection Mixer It is often desirable to eliminate the mixer response to the image frequency, but this may not be easy with the use of conventional filtering in certain situations. If IF is low, for instance, a sharp filter is required because IM is close to RF. If the RF and IF bands overlap, the image cannot be rejected by a simple band reject filter. A solution to these problems lies in another type of mixer, the image rejection mixer. Figure 8.6 shows the basic circuit principle of the image reject mixer. It consists of two mixers driven by LO signals applied in quadrature out of phase. The mixer responses are added to each other through a quadrature phase combiner. The rejection mechanism can be explained by considering the phase shift undergone by the lower (LSB) and upper (USB) sideband

TABLE 8.1 Comparison of Mixers

Mixer Type	Single-Ended	LO-Balanced	RF-Balanced	Double-Balanced
LO/RF isolation	Poor	Good	Poor	Good
LO/IF isolation	Poor	Poor	Good	Good
RF/IF isolation	Poor	Good	Poor	Good
LO harmonics rejection	None	Even	All	All
RF harmonics rejection	None	All	Even	All
Single-tone spurious rejection	None	$p_{RF} + q_{LO}$ with q even	$p_{RF} + q_{LO}$ with p even	$p_{RF} + q_{LO}$ with p or q even
Two-tone 2nd-order products rejection	No	No	Yes	Yes

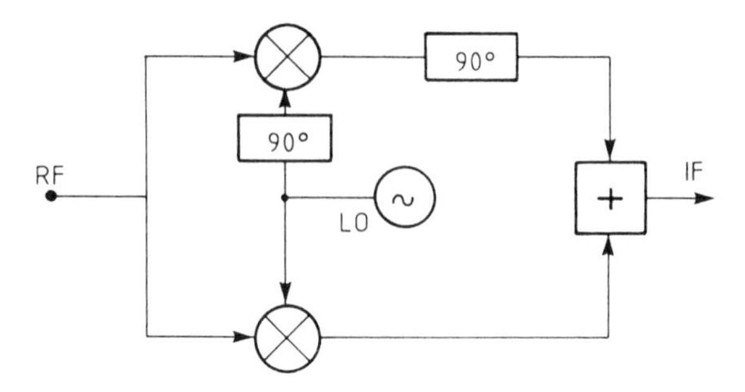

FIGURE 8.6 Image rejection mixer.

signals through the mixer branches. Taking as phase reference the nonshifted LO signal, and remembering that the mixer operates with phase as with frequency, the phases of the LSB and USB signals at the mixer output of branch 1 (nonshifted LO) are

$$\phi(IF_{LSB}) = 0° - \phi(RF_{LSB})$$

$$\phi(IF_{USB}) = \phi(RF_{USB}) - 0°$$

At the mixer output of branch 2 (LO shifted by 90°), the phases are

$$\phi(IF_{LSB}) = -90° - \phi(RF_{LSB})$$

$$\phi(IF_{USB}) = \phi(RF_{USB}) + 90°$$

These relations show that the LSB signal at the mixer output is delayed by 90° in branch 2 with respect to branch 1. However, this in contrast to the USB signal, which is in advance by 90° in branch 2 with respect to branch 1. This discrepancy is due to the fact that the phase of LSB signal is reversed after mixing whereas it is not for the USB. This allows the rejection of one of the sidebands by shifting the IF signal by 90° in one of the branches before adding the signals. For instance, if the signal is delayed by −90° in branch 2, the phases of the respective signals become

$$\phi(IF_{LSB}) = -180° - \phi(RF_{LSB})$$

$$\phi(IF_{USB}) = \phi(RF_{USB}) + 0°$$

The USB signals are in phase in the two branches, whereas the LSB signals are 180° out of phase, so that the lower sideband is rejected when adding the signals of the two branches. With the −90° delay applied in branch 1 instead of branch 2, the upper sideband would have been suppressed.

Image rejection is complete if the gain values in the two mixer branches are identical and if the phase relationships are completely verified. This is, of course, not possible practically; therefore, the rejection R_{IM} is limited by the relative gain imbalance and phase error, denoted respectively by ε and θ according to the expression [2]

$$R_{\mathrm{IM}} = -10 \log \frac{1 - 2\sqrt{\varepsilon} \cos \theta + \varepsilon}{1 + 2\sqrt{\varepsilon} \cos \theta + \varepsilon}$$

Achieving 20-dB image rejection requires a phase error below 10° and a gain imbalance lower than 1 dB. To improve the rejection up to 40 dB, the phase error must be kept below 1° and the gain imbalance below 0.1 dB.

8.1.2 Passive FET Mixers

Passive mixers are based on the modulation of a resistance. The mixing device is not capable of amplification and behaves as a resistor in the absence of modulation. Passive mixers have conversion losses. Their noise figure is readily calculated since it is related to the conversion losses. This may represent an advantage, for the designer, over active mixers whose noise figure is not easy to predict. Indeed, because of the passive nature of the mixer, the noise available N_{avail} at the mixer output is equal to that of a resistor, and hence is given by

$$N_{\mathrm{avail}} = kT_0B$$

The output noise portion due to the conversion of the input noise from RF to IF (as if the mixer was noiseless) is given by

$$N_{\mathrm{avail}}(\text{noiseless}) = kT_0BL_{\mathrm{avail}}$$

where L_{avail} are the available losses of the mixer, given by the ratio of the power available at the output to the power available from the RF source. (One must not mistake the available losses for the voltage losses; these figures are normally not equal.) The single-sideband noise figure (SSB2) is given by the ratio of these two quantities. We obtain

$$\mathrm{NF}_{\mathrm{SSB2}} = \frac{1}{L_{\mathrm{avail}}}$$

The noise figure of a passive mixer is thus directly determined by its available losses. By subtracting the noise contribution of the image band from the output noise, one can calculate the single-sideband noise figure according to the first definition:

$$\mathrm{NF}_{\mathrm{SSB1}} = \frac{1}{L_{\mathrm{avail}}} - 1$$

The relation for the double-sideband noise figure is also readily deduced:

$$\mathrm{NF}_{\mathrm{DSB}} = \frac{1}{2L_{\mathrm{avail}}}$$

It was shown earlier that the minimum voltage conversion losses are −3.9 dB with a square waveform modulation. If the modulation is a sine wave, it can be easily demonstrated using basic trigometry that the losses drop down to −6 dB. These limits are calculated by assuming that the resistance is modulated between short and open circuits. Practical mixers must approach this ideal situation as close as possible to achieve good efficiency. Devices having a wide range of resistance variation are the Schotky diode and the MESFET at zero drain-to-source voltage. The diode mixer is based on the nonlinearity of the diode I–V characteristic. Its operation can be viewed as a modulation of the instantaneous diode conductance by the LO signal superimposed on RF. In the unbiased FET mixer, the RF and LO signals are applied to separate ports. The mixing is based on the variation of the drain-to-source resistance with the gate voltage.

Schottky diode mixers are randomly used at high frequencies, above a few tens of gigahertz, whereas FET mixers are preferred below a few gigahertz. Some reasons are the intrinsic better LO isolation of FET mixers and the capability of active mixers to provide conversion gain. As diode mixers are mostly microwave mixers, they are not within the scope of this chapter; thus FET mixers will be described here in detail.

8.1.2.1 The FET as Voltage-Controlled Resistor At low drain-to-source voltage, the electron velocity is proportional to the electric field and the FET behaves as a simple resistor (Fig. 8.7). Its drain-to-source resistance, which depends on the opening of the conducting channel, can be varied by changing the gate voltage. This part of the $I_{\mathrm{ds}}(V_{\mathrm{ds}})$ characteristic of the transistor is generally called the *linear* or *ohmic region*. The drain-to-source resistance R_{ds} consists of a fixed resistance, including the source and drain resistances R_{s} and R_{d}, respectively, and of a variable channel resistance R_{ch} that depends on the gate-to-source voltage:

$$R_{\mathrm{ds}} = R_{\mathrm{s}} + R_{\mathrm{d}} + R_{\mathrm{ch}}(V_{\mathrm{gs}})$$

The channel resistance can be expressed as

$$R_{\mathrm{ch}} = \frac{V_{\mathrm{satint}}}{I_{\mathrm{sat}}}(V_{\mathrm{gs}})$$

where I_{sat} is the saturated drain current and V_{satint} is the intrinsic saturation voltage, which may be considered as constant in a first-order approximation. It is related to the saturation electric field of the GaAs material and to the

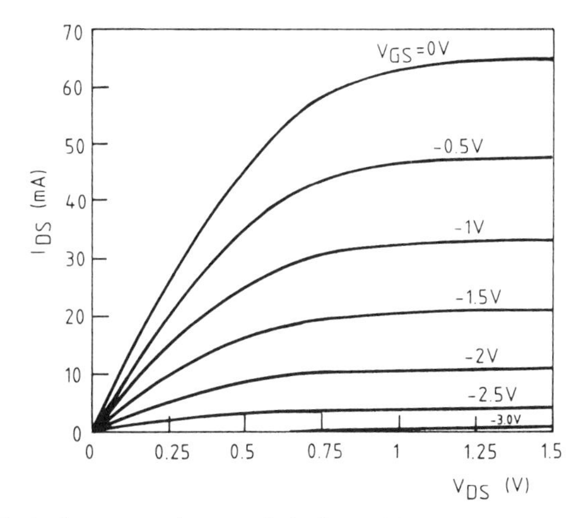

FIGURE 8.7 Typical current characteristic for a high-pinchoff-voltage FET in the linear region.

gate length of the transistor. An order of magnitude for 0.7-μm gate FETs is $V_{\text{satint}} \approx 0.5$ V. An extrinsic saturation voltage V_{satext} can be defined that sets the limit of the linear region of the $I_{\text{ds}}(V_{\text{ds}})$ characteristic. It is related to the saturated current I_{sat} and to the resistances R_{s} and R_{d} by

$$V_{\text{satext}} = V_{\text{satint}} + (R_{\text{s}} + R_{\text{d}})I_{\text{sat}}$$

The voltage drop across R_{s} and R_{d}, causes the linear region to extend to about 1 V at $V_{\text{gs}} = 0$ for FETs with a pinchoff voltage of -3 V.

The range of variation of R_{ds} can be from about 10 Ω at forward gate-to-source voltage to nearly open circuit at pinchoff. High-pinchoff-voltage FETs ($V_{\text{t}} = -3$ V) with a high drain saturation current ($I_{\text{dss}} =$ 300 mA/mm) have the lowest minimum channel resistance ($\approx 5\ \Omega \cdot$ mm). They can therefore achieve a better efficiency than low-pinchoff-voltage devices. As the drain-to-source resistance remains linear up to a V_{ds} of ~1 V, a low intermodulation is expected. Another interesting feature is that the gate is only capacitively coupled to the channel. This coupling is modeled by means of the C_{gs} and C_{gd} capacitances. These capacitances are equal at zero drain-to-source voltage and constitute a quite high impedance, which leads to a good isolation between the gate-control pin and the drain and source ports.

8.1.2.2 Design Considerations The design methodology can be explained by considering the single-transistor mixer in Figure 8.8. Although there is no reason to distinguish between the drain and the source, we assume that the source is the port connected to the load on which the IF signal will be taken. The RF signal is applied to the drain. Direct-current-blocking capacitors

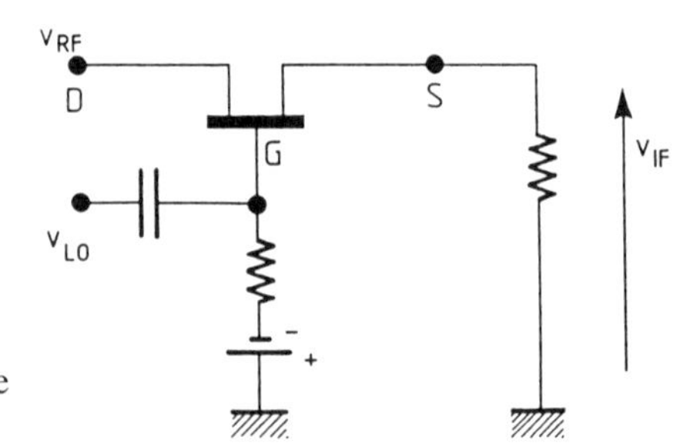

FIGURE 8.8 Single-transistor passive FET mixer.

may be required to avoid any dc current flowing in the transistor. The LO and a negative dc bias are applied to the gate. The gate bias is chosen in order to bias the transistor in a region where the rate of variation of the channel resistance with the gate voltage is maximum. The LO power required to turn on or off the transistor is thus minimum. The optimum bias is therefore close to the pinchoff voltage. To get the maximum conversion efficiency, the LO voltage level must be high enough to get a channel resistance in the on state that is negligible with respect to the load resistor. The order of magnitude of the LO level required is about 2 V peak for -3 V pinchoff-voltage FETs.

The designer then has to consider the constraints due to the mixer environment. If the mixer is an output stage, for example, a 50-Ω load impedance may be imposed. But if the mixer is embedded into an IC, a load has to be chosen, which can be either resistive or capacitive depending on the RF and IF frequencies. In an up-converter application, or in down-converters where the RF and IF frequencies are not widely spaced, a resistive load is more suitable. Its value can be determined, for instance, from the input capacitance of the following stage when a certain output bandwidth is desired, or from considerations regarding power consumption. A low load resistance will require a low RF source impedance and hence, high power consumption in the RF preceding stages. For minimum losses, the transistor width is then selected in order to meet the condition

$$R_{\mathrm{ds-on}} < R_{\mathrm{L}}$$

where $R_{\mathrm{ds-on}}$ is the transistor channel resistance in the on state. For the selection of the RF source resistance R_{S}, two cases can be considered. If output matching to R_{L} is required, R_{S} will be of the same order of magnitude as R_{L}. If power matching is not required, the condition for minimum loading of the RF source will be

$$R_{\mathrm{S}} < R_{\mathrm{L}}$$

In these conditions, the source delivers few current, even during the on state of the mixing transistor. One can therefore consider that the mixer operates as a voltage switch.

If the IF frequency is much lower than RF, a capacitive load can be used. The capacitor behaves as nearly a short circuit for RF, but constitutes a high impedance at the IF frequency. The behavior is somewhat different from that of a resistive load. During the "on" alternate of LO, the RF source delivers a high current due to the low impedance of the load capacitor. The mixer thus acts as current switch rather than a voltage switch.

The designer must also consider constraints related to the LO port. If a large transistor is selected to decrease R_{ds-on}, more LO will be coupled to the drain and to the source through the larger C_{gs} and C_{gd} capacitances. Large capacitances also decrease the switching speed, which may finally degrade the mixer efficiency, and increase the oscillator frequency pushing related to the mixer gate bias. These considerations illustrate the design tradeoff that must be found by the designer in order to meet the specifications.

8.1.2.3 Distortion in Mixers Because of the good linearity of the $I_{ds}(V_{ds})$ characteristic at low drain voltage, a low distortion is expected. This is too simple a reasoning, and deeper insight shows that the distortion is not explained solely by considering the nonlinearity of the current characteristic at constant gate voltage. Another source of distortion is the change in the control voltage caused by the current flowing in the transistor. The control voltage is indeed V_{gs} if $V_{ds} > 0$ or V_{gd} if $V_{ds} < 0$; it is therefore affected by variations of V_d and V_s. The result is a variation of the channel resistance with the input voltage level that may be significant even if the drain-to-source voltage remains low. This phenomenon is thus completely different from the change in the channel resistance caused by the saturation of the transistor. A detailed analysis is given in the case of a mixer loaded by a resistor as shown in Figure 8.8. Its behavior is described in a first-order approximation by the following equation:

$$I_{ds} = \frac{V_{ds}}{R_{ds}} = \frac{V_s}{R_L}$$

The channel resistance can be expressed in a first-order approximation by

$$R_{ds} = R_{ds0}\left(1 - \frac{\Delta V_c}{V_0}\right)$$

where V_0 is a parameter that characterizes the variation of R_{ds} with the control voltage V_c given by

$$V_c = \begin{cases} V_{gs}, & \text{if } V_d > 0 \\ V_{gd}, & \text{if } V_d < 0 \end{cases}$$

If the input drain voltage is positive, the change in the control voltage due to the current in the load resistor is

$$\Delta V_c = \Delta V_{gs} = -V_s$$

As V_s and V_d have the same polarity, ΔV_c is negative. Reporting this value in the preceding equation shows that the channel resistance increases with the incident signal level; therefore the output voltage tends to saturate, or even decrease. This is illustrated by Figure 8.9, which shows the dc transfer characteristic of a high-pinchoff-voltage FET at different gate voltages. With respect to negative input drain voltage, the change in the control voltage is

$$\Delta V_c = \Delta V_{gd} = -V_d$$

where ΔV_c is positive, and the result is now a decrease of the channel resistance with the increasing input level. This effect tends to compensate the saturation due to a large drain-to-source voltage; therefore the transfer characteristic is more linear for negative input as shown in Figure 8.9.

The nonlinearity induced by this mechanism depends on the sensitivity of the channel resistance to the control voltage. High-pinchoff-voltage FETs are less sensitive and therefore more linear than low pinchoff-voltage devices, which are almost unusable in passive mixers because of their excessive distortion.

It is evident from Figure 8.9 that the transfer characteristic is more linear at low reverse gate bias. One may conclude from this observation that a better mixer linearity is obtained if the LO signal applied to the gate is high enough to turn on the transistor is this linear region.

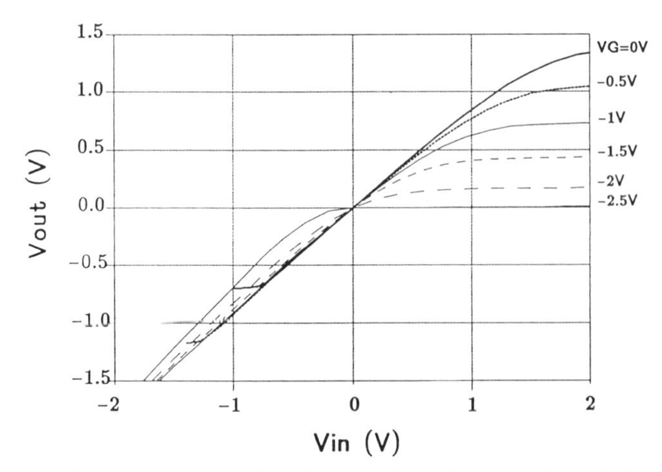

FIGURE 8.9 Direct-current transfer characteristic for a single-ended passive FET switch mixer.

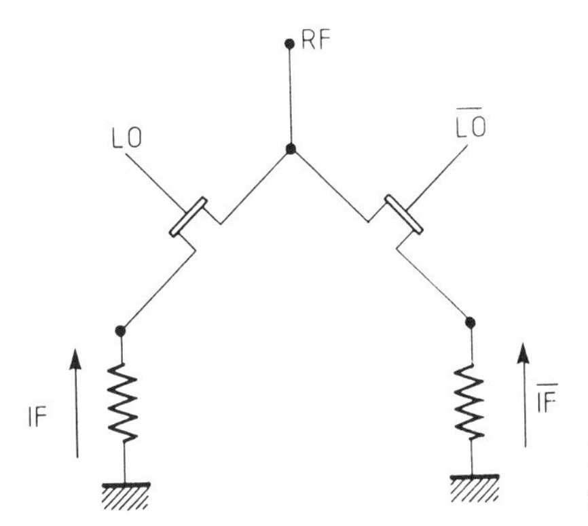

FIGURE 8.10 Single-balanced passive FET mixer.

8.1.2.4 Balanced Mixer Configuration Balanced mixers are often preferred in ICs because of their inherent capability to reject some undesired mixing products. A single-balanced configuration is built by associating two single mixers as in Figure 8.10. RF is applied in common to the mixers, but they are driven by the LO in phase opposition and have separate load resistors on which the IF signals are 180° out of phase. The mixer is completed by an IF 0/180° combiner, which adds the IF signals in phase and effects rejection of the $v_{\text{RF}}^{p} v_{\text{LO}}^{q}$ products wherein v_{LO} is raised at an even power. Some products are also rejected at the RF common input where the mixer currents add to each other in the source impedance. Since LO signals are 180° out of phase, all products $v_{\text{RF}}^{p} v_{\text{LO}}^{q}$ of odd power in LO are rejected, in particular LO, IF, and odd LO harmonics. As a consequence, this single-balanced structure has improved RF/IF and LO/RF isolation as compared to the single-FET mixer. It has also 6 dB more voltage gain if the differential IF voltage is chosen as output voltage.

A double balanced structure can be formed using 4 FETs connected to each other by their drains and sources in a ring arrangement as shown in Figure 8.11. The RF signals are applied 180° out of phase in two opposite nodes of the ring. Two load resistors are placed at the two remaining opposite nodes. The LO signal is applied to the transistors so that any pair

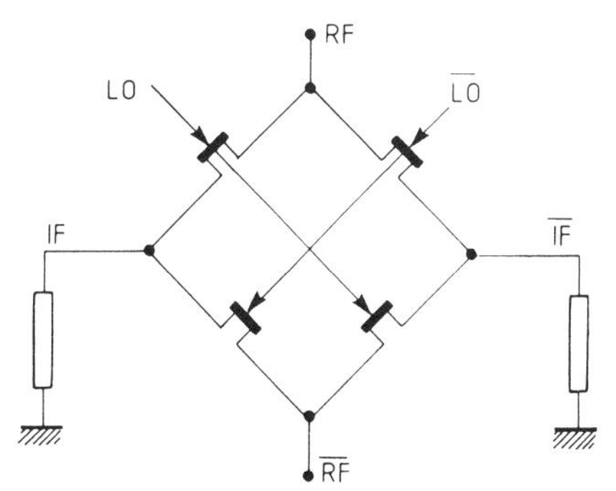

FIGURE 8.11 Double-balanced passive FET "ring" mixer.

of adjacent FETs are driven in phase opposition. If we neglect high-order terms, the voltage v_{IF1} and v_{IF2} on the loads may be written as

$$v_{\mathrm{IF1}} = k[v_{\mathrm{RF}}v_{\mathrm{LO}} + (-v_{\mathrm{RF}})(-v_{\mathrm{LO}})]$$

$$v_{\mathrm{IF2}} = k[v_{\mathrm{RF}}(-v_{\mathrm{LO}}) + (-v_{\mathrm{RF}})v_{\mathrm{LO}}]$$

The current of the transistors connected to the same load resistor add in phase, but the resulting voltages in the loads are 180° out of phase. To obtain an actual double-balanced mixer, the system must be completed by an IF 0/180° combiner. At its output, all intermodulation products $v_{\mathrm{RF}}^{p}v_{\mathrm{LO}}^{q}$ of even power in RF or LO are suppressed. Some unwanted signals are also suppressed at the RF inputs and in the IF load resistors due to the symmetry of the ring. At the RF inputs, which are connected to transistors driven by LO in phase opposition, products of the form $v_{\mathrm{RF}}^{p}v_{\mathrm{LO}}^{q}$ of odd power in LO are rejected. In the IF load resistor, the situation is more complicated. The current components that are 180° out of phase in the pair of transistor connected to the same resistor are canceled. These currents are related to products of the form $v_{\mathrm{RF}}^{p}v_{\mathrm{LO}}^{q}$ where the sum of the exponents $p + q$ is odd. This means in particular that RF, LO, and odd harmonics of RF and LO are rejected at each IF output of the ring. Consequently, even without subtracting the output signals, the ring structure has attractive rejection capabilities, which lead in particular to a high degree of isolation between the different ports of the mixer. In applications where the full spurious rejection capability is not required, only one IF output of the ring mixer is sometimes used.

8.1.2.5 Applications Figure 8.12 shows an example of double-balanced ring mixer applied to an experimental double-frequency conversion integrated TV tuner fabricated using a −3-V pinchoff-voltage GaAs process (Fig. 8.13) [3]. This mixer was operated as an up-converter and was in charge of the conversion of the incoming signal in the VHF–UHF (very- and ultra-high-frequency) bands (50–850 MHz) up to about 2.2 GHz. As the mixer was embedded into the IC, a high load impedance of 600 Ω was chosen. It is indeed preferable to select a high impedance level when possible. This allows to use small transistors in the preceding stages of the mixer, which minimizes the power consumption. Furthermore, it is not necessary to use large transistors to achieve low conversion losses. The gate capacitance can thus be made very low, and a low oscillator frequency pushing due to variations of the mixer gate bias can be obtained. In the present mixer example, 30-μm-gate-width transistors were used. Figure 8.14 shows the simulated conversion losses as a function of the LO level at different gate voltages. The optimum gate voltage is about −2.5 V, which nearly corresponds to the pinchoff voltage. At more negative bias, the losses increase as a result of higher losses during the on state. At less negative

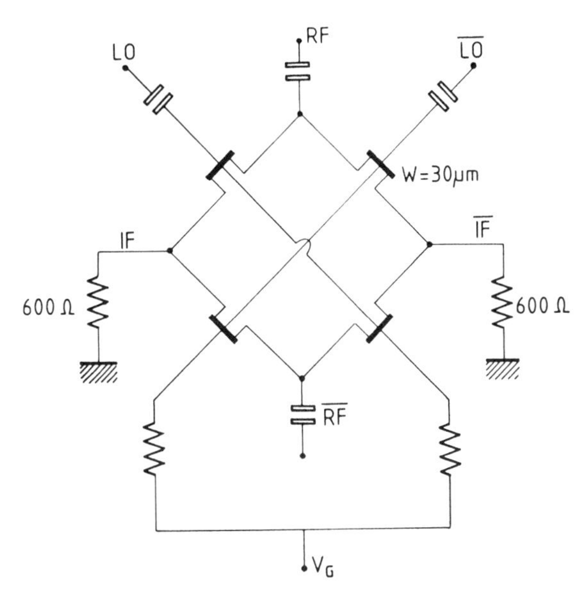

FIGURE 8.12 Double-balanced FET ring mixer for up-conversion in a double-conversion TV tuner.

bias, the efficiency is decreased because the time during which the transistor is switched off is less than half of the LO period. The conversion gain increases with the LO level but saturates above about 2 V peak. Improvement of the third-order intermodulation with LO voltage is also observed as

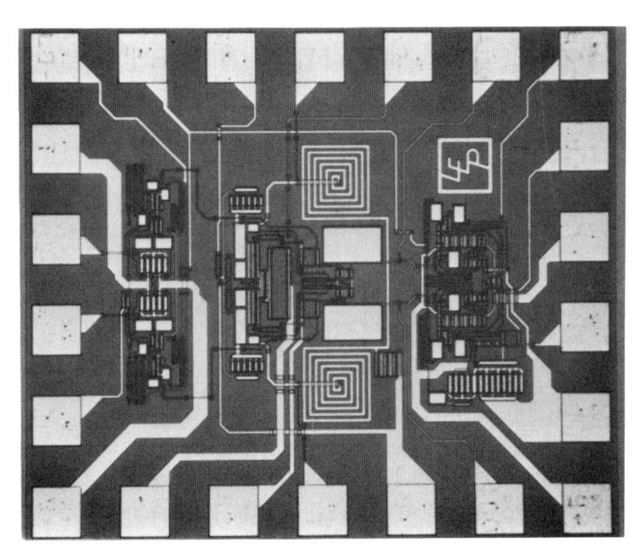

FIGURE 8.13 Up-converter IC of a double-conversion TV tuner system.

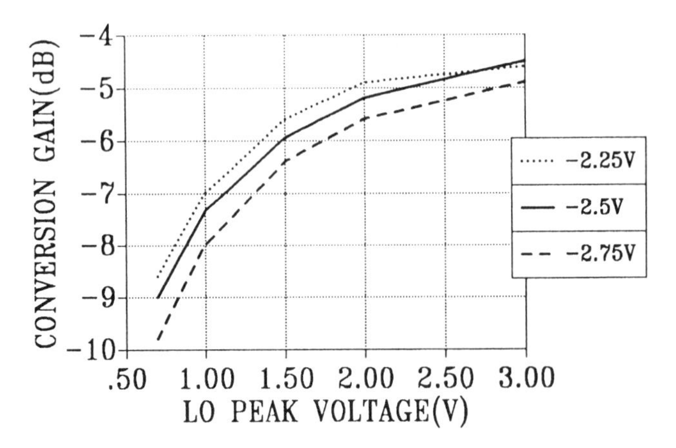

FIGURE 8.14 Conversion losses as a function of LO level.

shown in Figure 8.15, as long as it remains below 2 V. At larger LO voltages, gate conduction occurs, and this has a strong effect on the intermodulation that increases rapidly. An optimum LO level is thus defined that is about 2 V in the present example, but is more generally slightly less than the voltage required for gate current to occur. It is interesting to note that the conversion gain and the intermodulation are optimum almost simultaneously. The optimum IP3 is about 2 V peak (3 dB V) and the corresponding conversion losses are about −5 dB. This last value shows that the resistance modulation waveform is between a sine and a square wave.

The ring mixer is also attractive in down-converter applications, particularly in systems where the intermediate frequency is low, such as in Doppler radars [4]. At an IF below a few tens of megahertz, the noise performance may indeed be degraded by the $1/f$ noise, due to low frequency

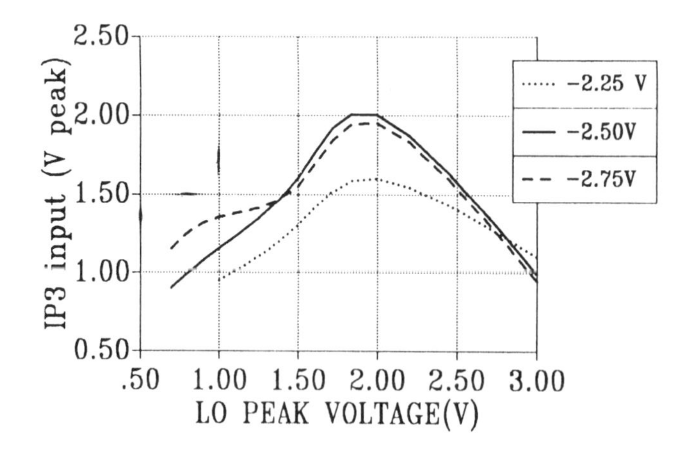

FIGURE 8.15 Intermodulation as a function of LO level.

fluctuations of the dc current flowing in active devices. As its power density decreases as $1/f$, this noise contribution becomes significant at low frequencies. Its magnitude increases with the dc current and depends also on the material. In an active mixer, the $1/f$ noise may cause the noise figure to increase drastically at low IF frequency. Passive FET mixers, however, do not exhibit $1/f$ nose because there is no dc current flowing. Their noise figure remains good down to low frequency; therefore they are often preferred in systems with low IF [5]. Operation down to dc IF is thus possible, and even easy since the mixer output is at zero dc voltage. This is an advantage for applications in direct demodulators. It also allows the operation of the ring mixer as phase detector [6].

8.1.3 Active FET Mixers

The basic active mixer circuit is similar to an amplifier in its topology. The bias conditions are, however, somewhat different. An amplifier is biased in order to get a good linearity, whereas in a mixer a bias is chosen to enhance the quadratic response. In contrast with passive mixers, active mixers are capable of conversion gain. Less amplification is indeed required in front of the mixer, thus reducing the system complexity. Active mixers can offer comparable or even better linearity than their passive counterparts, and their only weak point is their higher noise figure. The noise behavior of active mixers is, moreover, not completely understood, and at least is not easy to predict. The conversion gain is also not easy to calculate. A nonlinear analysis tool is required as well as accurate nonlinear models. This may be why designers have for a long time preferred passive mixers, particularly diode mixers.

Active mixers are built using single-gate or dual-gate FETs. The dual-gate FET mixer offers the possibility for a high LO/RF isolation with a single-ended configuration. Its mechanism of mixing is quite complex, however, because the two transistors interact with each other. The dual-gate FET mixer is not widely used in analog ICs. It is used mostly in microwave circuits at frequencies around the X-band or as discrete component (or nearly discrete, with some biasing or matching elements integrated) for UHF and L-band applications [7]. We will nevertheless devote a paragraph at the end of Section 8.1.3.1 to discussion of its operation principle. Much more popular in IC technology are the single-gate FET mixers. Their operation is based on the modulation of the FET transconductance in the saturated region. Balanced configurations are the most common, in particular the double-balanced Gilbert mixer, which is now used worldwide and may be the most representative active analog mixer.

8.1.3.1 Basic Operation The dominant nonlinearity of a single-gate FET is its transconductance. Other nonlinear circuit elements of the transistor such as C_{gs} and R_{ds} contribute little to the mixing and add only second-order

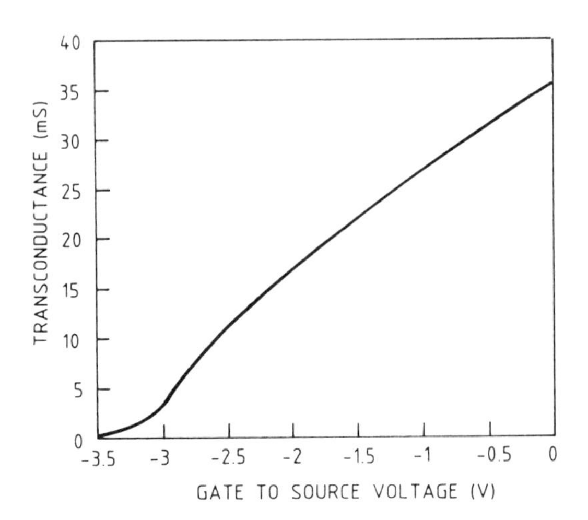

FIGURE 8.16 Transconductance characteristic of a high-pinchoff-voltage FET.

effects to the total performance. Figure 8.16 shows a typical transconductance characteristic as a function of V_{gs} for a high-pinchoff-voltage FET. For maximum conversion efficiency, the operating point must be located in a region where the rate of variation of g_m with V_{gs} is high. It is hence evident that the optimum biasing is close to the pinchoff voltage. Mixing is achieved by combining the LO and RF signals in the transistor in order to have

$$v_{gs} = v_{LO} + v_{RF}$$

For a better isolation, LO and RF are applied to different ports. Usually, LO and RF are respectively applied to the gate and to the source. In order to meet the requirement shown in the preceding relation, the source and the gate must be respectively grounded at the LO and RF frequency. If we denote by Z_s and Z_g respectively the impedance at the source and gate ports, we can express the grounding conditions by

$$Z_s < \frac{1}{g_m} \qquad \text{at LO frequency}$$

$$Z_g < \frac{1}{j\omega C_{gs}} \qquad \text{at RF frequency}$$

The gate grounding requirement is normally met for a well-designed LO driving stage, but the source grounding condition is not easy to meet with a single-ended mixer because it supposes some LO filtering. A filter is not desirable at low microwave frequencies because it is usually a cumbersome circuit to integrate. Filtering may even not be possible if the LO and RF frequencies are not widely spaced. This problem is inherent in single-ended mixers since balanced topologies can take advantage of their symmetry to offer the possibility for virtual grounding. Single-gate FET balanced mixers are discussed in more details in the paragraph dedicated to the analysis of the most widely used topologies.

8.1.3.2 Balanced Mixer Configurations A single-balanced version of the single-gate FET transconductance mixer is realized using a source-coupled FET pair as in Figure 8.17. LO is applied 180° out of phase to the gates of the FET pair so that the source voltage remains unchanged by the LO voltage in a first-order approximation. Thus the sources of the FET pair are virtually grounded with respect to LO regardless of the impedance in the source of the FET pair. In these conditions, RF can be easily injected in the sources for mixing with LO. It can be injected directly. In this case, a resistor to ground is placed in the source of the FET pair. But the injection via a transistor is more often preferred because it provides a superior rejection of the common mode and a higher impedance to the preceding stage (Fig. 8.17). In this case RF is applied to the gate of this third transistor. Regardless of the mode of injection of RF, the gate-to-source voltage of each transistor of the source-coupled pair is determined by the LO and RF voltages according to

$$v_{gs1} = v_{LO} - v_{RF}$$
$$v_{gs2} = -v_{LO} - v_{RF}$$

This results in IF currents 180° out of phase at the drains of the FET pair where a load resistor is usually placed. An actual balanced mixer then uses a differential amplifier to add the IF signals in phase and reject intermodulation products of even LO order.

A double-balanced mixer is built in a similar manner. Two source-coupled FET pairs are now associated to each other as in Figure 8.18. The FET connection is in fact identical to that of the passive ring mixer described earlier, but the difference is that the transistors are now biased in

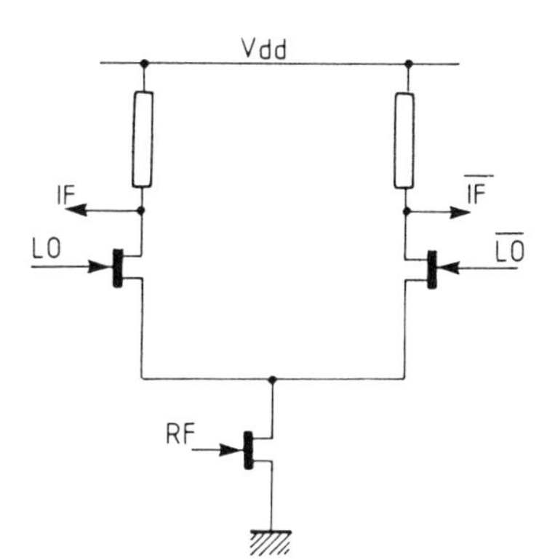

FIGURE 8.17 Single-balanced transconductance mixer.

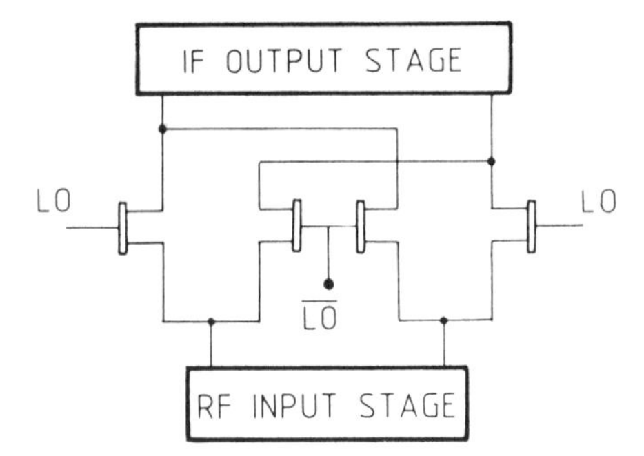

FIGURE 8.18 Double-balanced transconductance mixer.

the saturated region. This arrangement of transistors is known as the *Gilbert cell*. The mixing mechanism is basically the same as in the simple source-coupled FET pair. The respective gate-to-source voltages of the transistor are given by

$$v_{gs1} = v_{LO} - v_{RF}$$

$$v_{gs2} = -v_{LO} - v_{RF}$$

$$v_{gs1} = -v_{LO} + v_{RF}$$

$$v_{gs2} = v_{LO} + v_{RF}$$

Transistors 1 and 3 produce IF currents in phase. They can therefore be connected to the same load resistor. For the same reason, transistors 2 and 4 are also connected to the same resistor, but the resulting voltages in each resistor are 180° out of phase. To realize a true doubled-balanced mixer, an IF differential amplifier is again needed. In some applications, only one IF mixer output is used. Less spurious signals are hence rejected, since only the intermodulation products $v_{RF}^p v_{LO}^q$, where the sum $p + q$ is odd, are rejected. This may be sufficient in some applications where the main interest is in rejecting the LO signal, for example. There is also some rejection capability at the RF inputs (which also exists in the single-balanced version) where intermodulation products of odd order in LO are canceled.

Single- and double-balanced transconductance mixers can also be viewed as current switches. Consider first the single-balanced mixer as in Figure 8.17. The transistor to which RF is applied behaves as a current source and sets the current of the source-coupled gain driven by LO. Under normal operation, the LO voltage is large enough so that the transistors in the source-coupled FET pair are alternatively switched on and off. The RF current is thus alternatively switched from one transistor to the other at the rate of the LO signal. If the LO switching waveform is a square wave, the resulting IF current into each LO transistor is given by

$$i_{IF} = g_m v_{RF} x \frac{1}{\pi}$$

where g_m is the transconductance of the RF transistor and the factor $1/\pi$ comes from the multiplication by LO square function.

The IF currents are 180° out of phase in the LO transistors; if we consider the current as the mixer output, the mixer transconductance g_{m-mix} is given by

$$g_{m-mix} = \frac{1}{\pi} g_m$$

In the double-balanced Gilbert mixer, the source-coupled FET pair are connected in parallel so that their IF currents add to each other. Assuming a RF input stage as in Figure 8.19, the IF output current i_{IF} is given by

$$i_{IF} = \frac{g_m v_{RF}}{\pi} - \frac{g_m(-v_{RF})}{\pi}$$

The Gilbert mixer transconductance g_{m-mix} can now be defined as the IF output current relative to the differential input RF voltage, and is also given by

$$g_{m-mix} = \frac{g_m}{\pi}$$

8.1.3.3 Design Considerations The Gilbert mixer is usually associated with common-source transistors or a differential pair for injection of RF. The IF signals are taken on load resistors placed in the drain of the mixer (Fig. 8.19). As the mixer is basically a transconductance mixer, its voltage

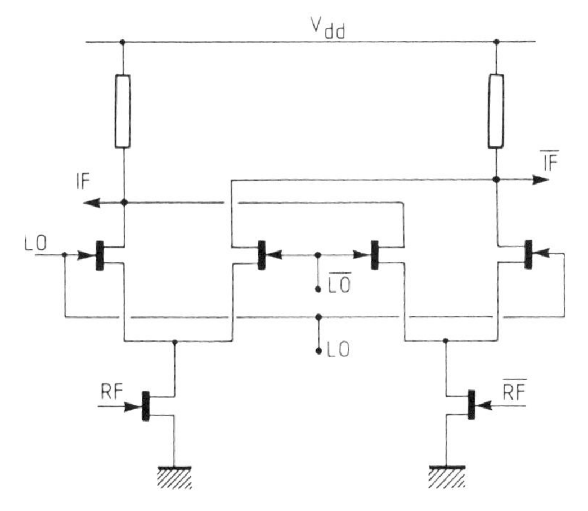

FIGURE 8.19 Double-balanced Gilbert mixer with common-source RF transistors and IF load resistors.

conversion gain G_c can be expressed by

$$G_c = \frac{g_{m-mix} R_L}{1 + g_{d-mix}}$$

where g_{m-mix} and g_{d-mix} are respectively the mixer transconductance and output conductance, and R_L is the load resistance. The load resistance depends on the mixer quiescent current I_{mix} and of the voltage drop V_L that can be accepted in the load resistor according to

$$R_L = \frac{2V_L}{I_{mix}}$$

The maximum conversion gain is thus achieved when the ratio g_{m-mix}/I_{mix} is maximum. This may not necessarily correspond to the optimum noise or distortion performance. For a given load resistor, however, the gain and intermodulation are optimum almost simultaneously, as shown by Figures 8.20 and 8.21, which presents measurement results for a high-pinchoff-voltage FET Gilbert mixer.

The mixer performance depends on the respective sizes of the transistors in the mixer cell and the RF stage. It has to be noted that the RF stage does not provide any voltage gain but rather a few decibel losses because of the low input impedance of the mixer cell. The RF stage is, in addition, not optimally biased for linear operation since it is biased at the same (low) current as the mixer cell. Part of the distortion can therefore be due to the RF stage. It is usual to choose the same V_{gs} biasing voltage for all transistors that leads to a RF transistor twice as large as in the mixer cell. By reducing the size of the RF transistor, its biasing point is displaced toward a less negative gate-to-source voltage, which is more favorable for gain and

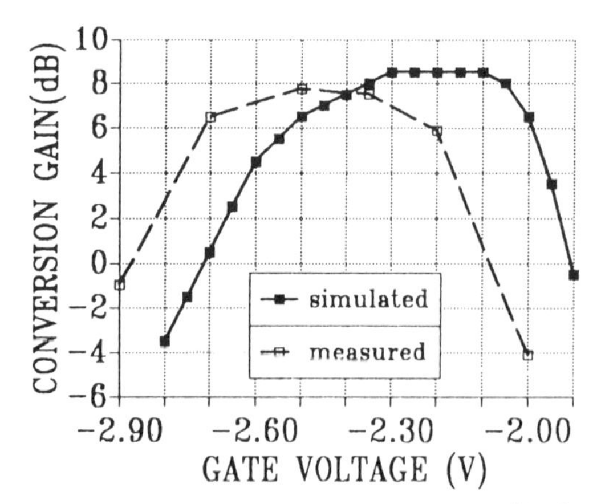

FIGURE 8.20 Voltage conversion gain for a high-pinchoff-voltage FET transconductance mixer.

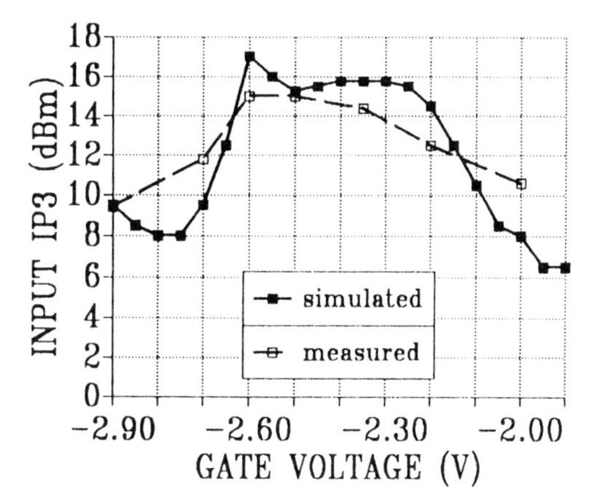

FIGURE 8.21 Input IP3 for a high-pinchoff-voltage FET transconductance mixer.

linearity. But this effect is more than compensated by the reduction in transistor size, which ultimately reduces the gain of the transistor. This is illustrated in the Table 8.2, which compares the wideband unmatched performance of three Gilbert mixers at 2 GHz with different transistor sizes. It is shown that mixer 2, which has a RF transistor half that of the mixer 1 has less gain. This results in a higher noise figure, but the counterpart is a higher input third-order intercept point, due to the lower gain in front of the mixer cell. The table shows in addition that increasing the size of the complete mixer improves its noise figure. This result is due to the fact that the signal-to-noise drain current ratio improves linearly with the square root of the transistor width.

The mixer performance is, of course, dependent on the LO voltage level. As in most mixers, the performance improves with the LO level. The required LO level, as well as the mixer performance, is also strongly dependent on the pinchoff voltage. Table 8.3 compares the wideband performance achieved using high-pinchoff-voltage FETs ($V_t = -3$ V) with those obtained using enhancement-mode FETs ($V_t = 175$ mV). For high-pinchoff-voltage devices, the correct LO level is about 1.5 V peak. Low-pinchoff-voltage mixers require less LO level and have a superior conversion gain, due to their higher g_{m-mix}/I_{mix} ratio. They have also a lower noise

TABLE 8.2 Performance of High-Pinchoff-Voltage Gilbert Mixers at 2 GHz[a]

LO/RF FET Sizes (μm)	Voltage Gain (dB)	Input IP3 (V peak)	NF DSB (dB)	Consumption (mA)
100/200	7.5	2	10	10
100/100	6	2.5	11.75	10
240/480	7	2	8.5	25

[a] IF = 100 MHz, LO voltage = 1.4 V peak.

TABLE 8.3 Comparison of the Gilbert Mixer Performance for Two Threshold Voltages[a]

V_t	−3 V	+175 mV
Supply voltage	9 V	5 V
Voltage gain	6 dB	14 dB
LO voltage	1.4 V peak	0.5 V peak
Input IP3	2.5 V peak	0.25 V peak
NF (DSB)	11.75 dB	7 dB
Current	10 mA	3 mA

[a] IF = 100 MHz, FET size = 100 μm.

figure, but the counterpart is a poorer linearity since they don't accept a high gate-to-source voltage swing.

8.1.3.4 Applications Many applications of the Gilbert mixer have been reported in the field of terrestrial or satellite TV tuners [9–13]. The reasons for using double-balanced mixers in TV tuners are mainly their high isolation and their low second-order distortion. There are indeed drastic specifications that set an upper limit to the power radiated by the tuner that are obviously more easily met using mixers with low inherent radiation. Double-balanced mixers have also the capability to reject second-order products that are potentially troublesome signals for broadband receivers such as TV tuners. The general design target for ICs applied to TV technology is to determine the best tradeoff between dynamic range on one hand, and power consumption on the other hand. This objective dictates the mixer design strategy. Figure 8.22 shows an example of mixer–oscillator IC designed for application in a satellite TV down-converter IC (RF = 950 MHz–2 GHz, IF = 480 MHz) fabricated using an enhancement-mode FET process. A differential pair is used as RF input stage for the Gilbert mixer cell (Fig. 8.23) in order to allow asymmetric input. Series feedback is used in the differential pair to improve the linearity. This, of course, decreases the gain of the RF stage and degrades the noise figure. This technique is however interesting since it allows adjustment of mixer performance to the desired level. The biasing point was chosen in order to maximize the mixer transconductance rather than its voltage gain. The mixer current is thus higher than required for maximum gain because this was found to provide superior linearity. In spite of the series feedback, asymmetric input, low supply voltage (5 V), and nonoptimal biasing conditions for gain, the conversion gain is nevertheless 1 dB. The input third-order intercept point is about 1 V, peak and the noise figure on 50 Ω is 12 dB DSB. Its consumption is about 20 mW.

Other applications have been reported in image rejection receivers [6,15] and phase modulators [14]. The operation of a phase modulator is basically the same as that of a mixer. The Gilbert mixer indeed accomplished a

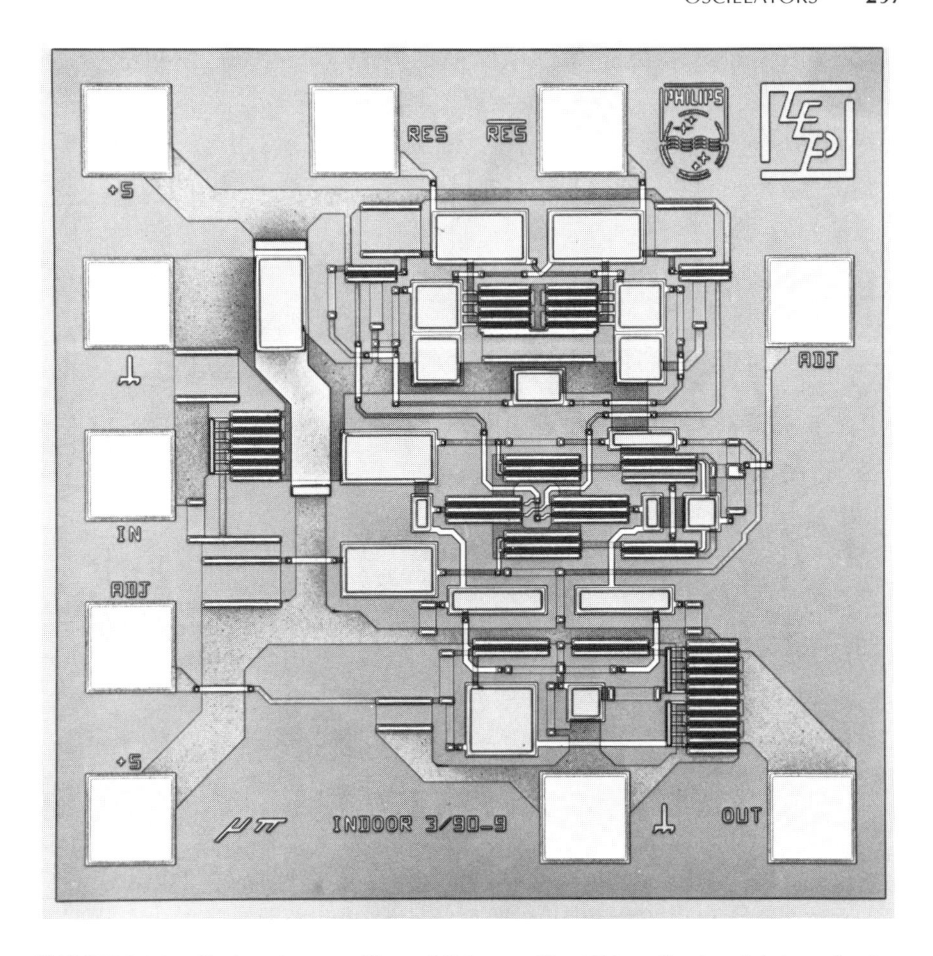

FIGURE 8.22 GaAs mixer–oscillator IC for satellite TV application fabricated using a normally-off process.

biphase modulation since it reverses periodically the phase of the incoming signal at the rate of the LO signal. It can therefore be a basic component for phase-modulated transmission systems.

8.2 OSCILLATORS

8.2.1 Oscillator Basics

8.2.1.1 Oscillation Conditions Basically, an oscillator can be thought of as a closed-loop system composed of an amplifier with gain A and a feedback network with gain β (Fig. 8.24). This system can potentially oscillate at frequencies where the signal phase shift after one trip around the

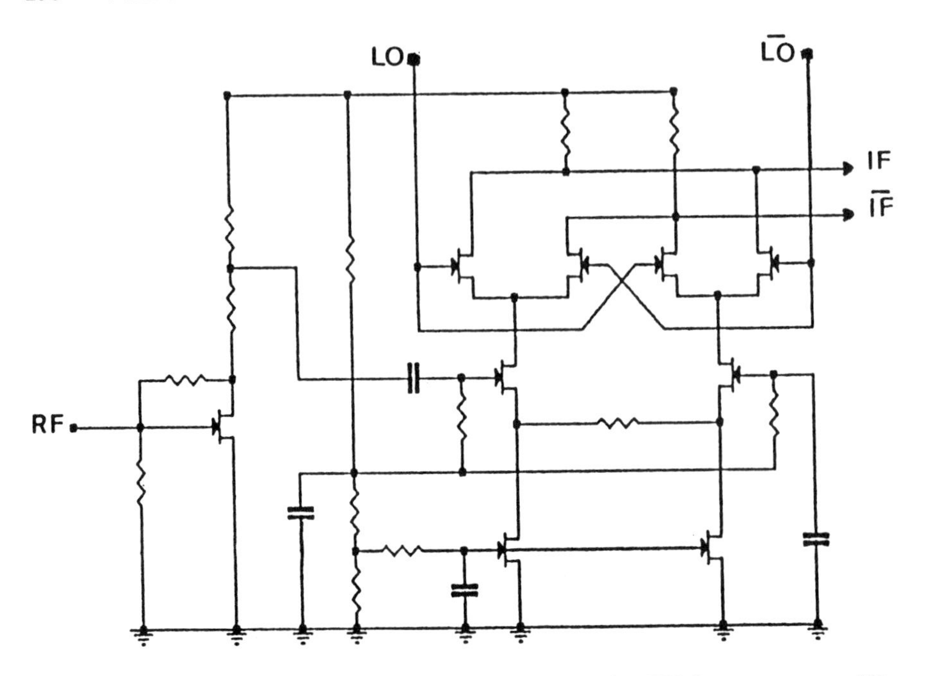

FIGURE 8.23 Gilbert mixer and front-end of the satellite TV down-converter IC.

loop is precisely a multiple of 2π. This condition means, in other words, that the signal must be fed back, through the loop, in phase with the original one. An oscillation can then build up at these frequencies, provided the signal amplifies after each trip around the loop. The loop gain βA must therefore be larger than unity at the oscillation frequency. If we assume this condition to be met, the oscillation level increases rapidly, as long as the amplifier behavior remains linear. As the saturation approaches, the amplifier gain decreases, and hence the loop gain. The oscillation grows less rapidly and finally stabilizes to a level at which the loop gain is exactly equal to unity. This is a stable state because any change in the signal level causes the amplifier gain to vary in a direction that tends to restore the initial level. The stabilization is thus the result of a nonlinear, autolimiting process. It occurs at a level for which the loop gain at the fundamental oscillation

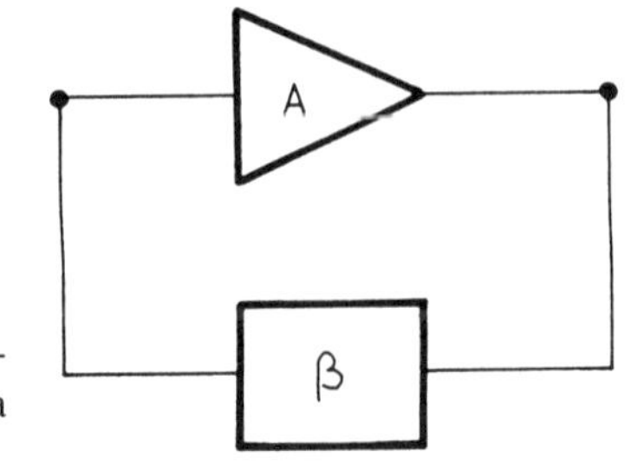

FIGURE 8.24 The oscillator viewed as a feedback system composed of an amplifier and a feedback network.

frequency is precisely decreased to unity. This result is consistent with the well-known formula giving the gain A_f of a feedback amplifier:

$$A_f = \frac{A}{1 - \beta A}$$

It shows that the gain A_f with respect to an external input signal is infinity when the loop gain is unity. This result may be interpreted to mean that a finite output signal can exist even in the absence of input excitation.

The feedback network is very important in the determination of the oscillator waveform. Indeed, the amplifier is usually a broadband circuit; therefore the capability of the closed-loop system to select a particular frequency is due mainly to the feedback circuit. If the phase shift introduced by the feedback network varies slowly with the frequency, the gain β will also be slightly dependent on the frequency. The loop gain and phase shift will remain respectively close to unity and 2π over a wide frequency band, resulting in oscillation spectrum containing a lot of harmonics. However, if the feedback network is essentially a reactive circuit, its phase shift and gain will vary rapidly with frequency. The gain and phase conditions will be verified only in the vicinity of the oscillation frequency, and the oscillation will thus have a sinusoidal waveform. Square waveform oscillators are quite low-frequency devices because of the fundamental need for a frequency much lower than the cutoff frequency of the active components. They become hence more and more difficult to realize as the frequency increases. For applications as local oscillators in microwave emitter–receiver systems, sinusoidal oscillators, which offer superior stability, are more frequently used.

8.2.1.1 Frequency Stability The frequency stability is a measure of the ability of the oscillator to maintain a fixed frequency. Several types of stability can be defined:

- Frequency stability with respect to changes in the environment such as temperature, supply voltage, or load variations
- Long-term stability, which refers to frequency drift over a long time period
- Short term stability, which refers to changes in the frequency within a time period less than a few seconds

The term *frequency stability* is used here to mean short-term stability. On short term, changes in the oscillator frequency result from interactions between the oscillator signal and noise. The interactions can be simple superposition or modulation, the latter resulting from a nonlinear mechanism. The predominant interaction is usually the random phase modulation of the signal by noise. This results in a widening of the oscillation spectrum

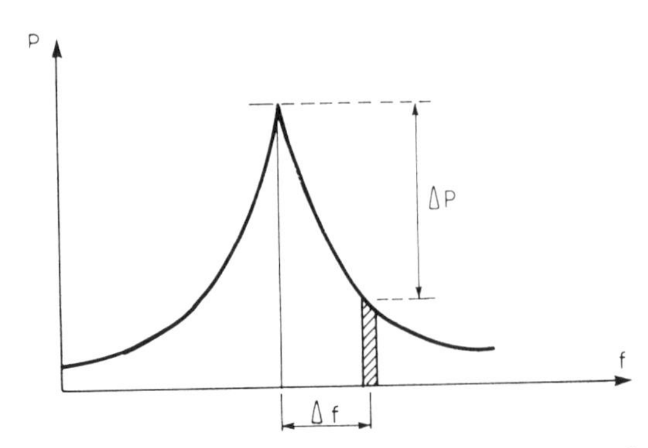

FIGURE 8.25 Oscillation spectrum with phase noise.

as shown symbolically in Figure 8.25. An indication of the phase stability is usually given by the measurement of the sideband noise spectral density at a certain distance of the peak carrier, as referred to the carrier level. An order of magnitude of the phase noise is about −100 dBc/Hz at 100 kHz away from the carrier.

The phase noise is of great concern in most microwave systems where signal levels can span a wide dynamic range. The oscillator noise is indeed transmitted onto the IF signal and may degrade the sensitivity as well as the selectivity of the system. Transmission systems are not equally sensitive to the oscillator phase noise. Digital communication systems, for instance, are sensitive to the noise close to the carrier (less than a few kilohertz offset). It results in phase jitter in the time domain that affects the bit error rate. On the other hand, analog systems are mainly sensitive to the phase noise at large frequency offset. This is because the information is usually placed at several hundred kilohertz away from the carrier.

Although it is important to consider the phase noise in the design of an oscillator, no software tools that are able to predict oscillator phase noise are as yet commercially available. Fortunately, the design guidelines are known. They can be understood from an intuitive description of the oscillator behavior in presence of noise. At the oscillation frequency, the loop phase shift is 2π. Some nonlinear elements are modulated by the noise, resulting in a fluctuating phase shift. The frequency adjusts itself in order to maintain the phase condition. If the loop phase shift varies only slowly with frequency, the frequency variation caused by the phase-shift fluctuation will be important. This is schematically represented by the phase-frequency characteristic shown in Figure 8.26. On the contrary, if there is a linear circuit in the oscillator loop that introduces a large phase variation with frequency, only little change in frequency will result from loop phase fluctuations. A circuit having such a large phase variation is a resonant circuit around its resonant frequency. The phase variation is linear and given

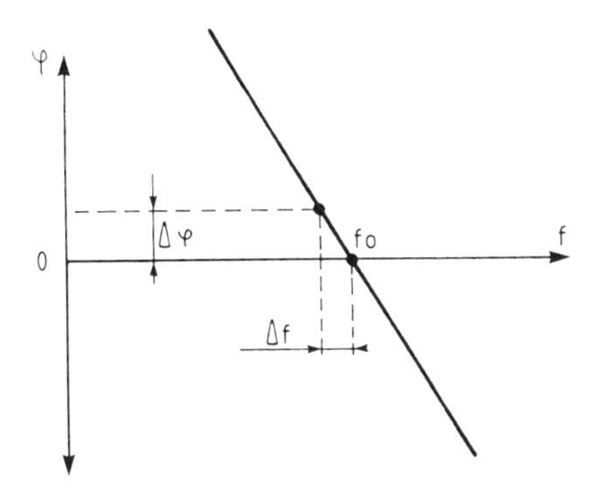

FIGURE 8.26 Conversion of loop phase-shift fluctuations to frequency fluctuations.

by

$$\frac{\Delta\phi}{\Delta f} = \frac{2Q}{f_0}$$

where Q is the quality factor of the resonant circuit. As high Q results in rapid phase variation with frequency, the stability of the oscillator improves with increasing quality factor. Another action point to minimize the phase noise is to try to keep the oscillator operation as linear as possible, or to reduce the contribution of nonlinear capacitance (C_{NL}) to the loop phase shift (low $\Delta f/\Delta C_{NL}$ ratio).

8.2.2 Single-Ended Oscillator Circuits

8.2.2.1 Basic Design Although an oscillator is basically a nonlinear circuit, a linear analysis is extremely useful as a preliminary approach. The oscillation frequency can be predicted rather accurately, and some qualitative information about the oscillation level can also be deduced. In our general analysis, we have considered the oscillator as a feedback loop composed of an amplifier and a parallel feedback network (Fig. 8.27). The output circuit of the amplifier is described by its Y parameters. They are assumed real and therefore reduce to the conductances g_{21} and g_{22}. The feedback network has a gain β and an input impedance Z_{in} when loaded by the amplifier.

It was shown in Section 1.1 that the oscillation conditions can be determined by considering the loop gain. In our simple oscillator model, the loop gain can easily be calculated by hand, but this will probably require some CAD (computer-assisted design) tool in a practical situation. In this context, the oscillator analysis by the loop again method is not convenient because it requires cutting the loop in some places. For the analysis to be valid, then, the circuit must not be left open in the cutting point. It needs to be loaded by the correct impedances, as if the loop were not cut. Although

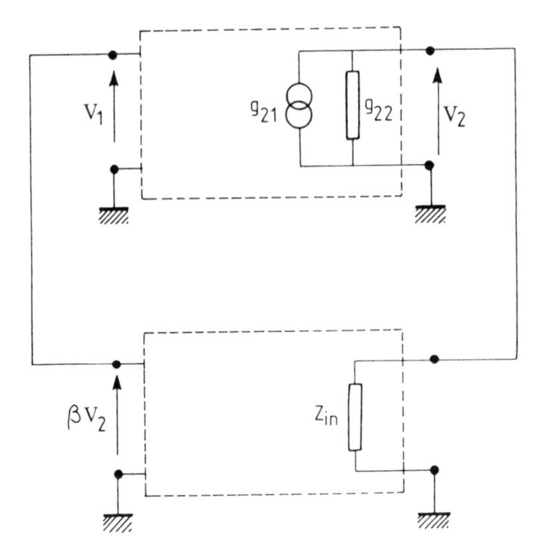

FIGURE 8.27 Basic oscillator model.

possible, this technique is not easy, and a method of analysis that does not necessitate cutting the loop is preferable. The oscillation conditions can indeed be derived from the calculation of the admittance in parallel to one of the oscillator components. This calculation can be readily performed with any software tool. We can calculate, for instance, the admittance Y at the output of the amplifier. It is given, in our simple model, by

$$Y = \frac{1}{Z_{\text{in}}} + g_{22} - g_{21}\beta$$

Taking into account the voltage gain A of the amplifier given by

$$A = \frac{g_{21}Z_{\text{in}}}{1 + g_{22}Z_{\text{in}}}$$

we can write the output admittance of the oscillator as

$$Y = \left(\frac{1}{Z_{\text{in}}} + g_{22}\right)(1 - A\beta)$$

The first term on the right-hand side of this expression is the output admittance in open loop. The multiplication by the term $1 - A\beta$ is thus the effect of feedback. The oscillation frequency is estimated by solving the equation

$$\text{Im}(Y) = 0$$

This equation does not yield exactly the same result as the solution of

$$\mathrm{Im}(1 - A\beta) = 0$$

The two equations are equivalent if the terms $(1/Z_{\mathrm{in}} + g_{22})$ and $(1 - A\beta)$ are simultaneously real. This happens when gain A is real, so that the phase shift through the feedback network is 180° or 0°. This situation is often well verified in practice, so that the oscillation frequency estimated from $\mathrm{Im}(Y) = 0$ is rather accurate. The oscillation will build up in the system if the output conductance of the oscillator is negative. This is expressed by

$$\mathrm{Re}\left[\left(\frac{1}{Z_{\mathrm{in}}} + g_{22}\right)(1 - A\beta)\right] < 0$$

which can be reduced to

$$A\beta > 1$$

The condition for a negative output conductance is thus equivalent to that of loop gain larger than unity. The preceding relations show in addition that the larger is the gain, the more negative is the output conductance. One deduces, in these conditions, that any change in the oscillator circuit that leads to a more negative conductance improves the starting safety margin.

It is also possible to determine the oscillation conditions by looking in series into the oscillator circuit. This requires the insertion of some voltage source in series somewhere in the circuit. If we insert, for instance, a voltage generator in series between the feedback network and the amplifier output, the input impedance Z is given by

$$Z = Z_{\mathrm{in}} + \frac{1}{g_{22}} - \frac{g_{21}\beta Z_{\mathrm{in}}}{g_{22}}$$

This can be rewritten as a function of the loop gain as

$$Z = \left(Z_{\mathrm{in}} + \frac{1}{g_{22}}\right)(1 - A\beta)$$

The expression of Z looks similar to that of Y; the multiplicative factor $1 - A\beta$ is the modification of the open-loop impedance caused by the feedback. The oscillation frequency is estimated by solving the equation

$$\mathrm{Im}(Z) = 0$$

The resulting frequency is not exactly the same as that deduced from the admittance equation because of considerations similar to those discussed above. For the oscillation to start, a negative resistance is required, which is equivalent to a loop gain larger than unity. The higher is the loop gain, the more negative is the resistance. Consequently, the admittance and resistance approaches are equivalent. The design targets are different, however; in a parallel analysis, a more negative conductance is desired when the safety margin is to be improved, whereas it is a more negative resistance in a series approach. The designer then has to consider the safety margin required for starting the oscillation. The absolute value of the negative conductance or resistance, or the loop gain, alone is not a good indicator because it depends on where one is looking into the loop. Information about the safety margin is obtained rather by evaluating the sensitivity of one of these parameters to variations in amplifier or feedback network parameters.

The oscillation frequency determined by linear analysis is not exactly the true one. It is the frequency at which the oscillation will build up. The loop phase shift can indeed be slightly different under large-signal conditions because of nonlinearities. The oscillator will hence finally run at a frequency different from the starting one. As the difference is usually within a few percent, the linear prediction is rather accurate, at least accurate enough in a rough design approach.

The linear analysis, although not able to predict the oscillation level, can nevertheless provide some information about the large-signal behavior of the oscillator. One must not believe that maximizing the negative conductance (or resistance, depending on the method chosen) is the condition for getting the maximum level. This will simply help in starting but, provided there is enough margin, the output level will be determined mostly by the load line of the amplifier. An indication of the large signal operation can therefore be derived from the calculation of the amplifier load impedance Z_l. In our simple oscillator model it is given by

$$Z_l = \frac{Z_{in}}{1 + g_{22} Z_{in}}$$

The load impedance partly depends on the feedback network. One must therefore design a feedback network that does not load the amplifier excessively, even if it is at the expense of a lower loop gain. Accurate prediction of the oscillator level requires the use of some nonlinear CAD tool provided with an accurate nonlinear FET model.

8.2.2.2 Resonant Circuit Oscillator The basic resonant circuit oscillator is shown in Figure 8.28. Depending on where the ground point is located, this

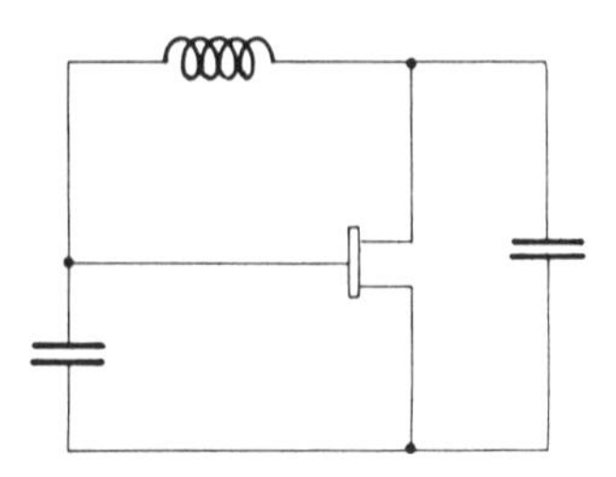

FIGURE 8.28 Basic *LC* resonant circuit oscillator.

oscillator has different names: in the Pierce oscillator, the ac ground is at the source; in the Colpitts oscillator, it is at the drain; in the Clapp, it is at the gate. In a practical circuit, dc-blocking capacitors and load resistors are located at different places for each of the three configurations, making the circuits perform differently. If we restrict our analysis to ideal circuits in a first step, equations can be derived that are valid for the three configurations.

The oscillator circuit of Figure 8.28 can be redrawn in a form that makes the amplifier and the feedback network appear more clearly (Fig. 8.29). The FET is modeled by its transconductance g_m and its output conductance g_d. The C_{gs} and C_{ds} capacitances are included in C_1 and C_2, respectively, whereas C_{gd} is neglected. As capacitors are usually much less lossy than inductors, C_1 and C_2 are considered as ideal capacitors. Ohmic losses in the feedback circuit are accounted for by the series resistance r of the inductor.

If the series resistance is lower than the inductor reactance, $r \ll L\omega$, the oscillation frequency is close to the resonance frequency f_0 of the LC network, given by

$$f_0 = \frac{1}{2\pi}\sqrt{\frac{C_1 + C_2}{LC_1C_2}}$$

At this frequency, and neglecting the series resistance of the inductor, the voltage gain β of the feedback network is given by

$$\beta \approx \frac{1}{1 - LC_2\omega_0^2} = \frac{-C_2}{C_1}$$

The phase shift through the feedback circuit is 180°. As the feedback network is at resonance, its impedance Z_{in} is real and given by

$$Z_{in} \approx \frac{1}{r\omega_0^2 C_2^2}$$

The transistor looks into a resistive load; therefore its phase shift is also 180°, so that the loop phase shift verifies the 360° condition. The condition for loop gain can be found by calculating the FET voltage gain A. If we

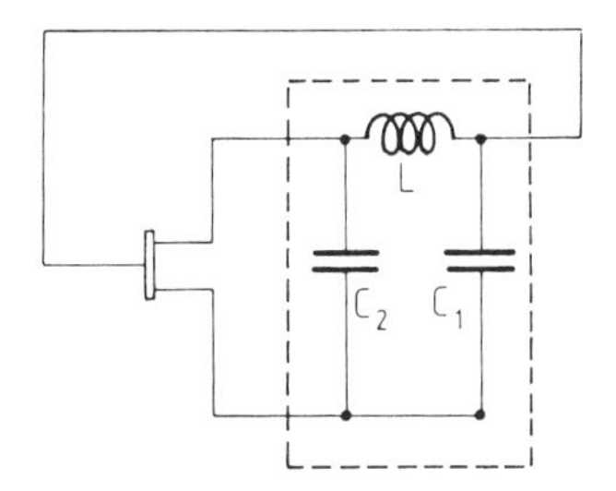

FIGURE 8.29 Basic oscillator circuit making amplifier and feedback circuits appear more clearly.

assume the following inequation is verified

$$g_d Z_{in} \ll 1$$

A is given by

$$A = \frac{-g_m}{r\omega_0^2 C_2^2}$$

The loop gain is

$$A\beta = \frac{g_m}{r\omega_0^2 C_1 C_2}$$

The condition for the oscillation to build up is thus

$$\frac{g_m}{r\omega_0^2 C_1 C_2} > 1$$

The left-hand side of this inequation is maximum, for a given inductor and frequency, when the capacitors C_1 and C_2 are equal. Their values can be calculated from the expression giving the oscillation frequency. The result is

$$C_1 = C_2 = \frac{2}{L\omega_0^2}$$

This condition may not provide, however, optimum transistor loading. If a large drain-to-source voltage swing is desired, for instance, a high load resistance is required that may lead to selection of the capacitance C_2 of lower value than C_1. With the capacitors C_1 and C_2 equal, the oscillation condition can be rewritten as

$$g_m > \frac{4r}{L^2\omega_0^2}$$

This relation shows that the FET transconductance needs to be at least 4 times higher than the equivalent parallel conductance of the inductor equal to $r/L^2\omega_0^2$.

This analysis is somewhat idealized because of the simplifications made, but is has the advantage of allowing a common description of the three basic configurations. Particularities of each topology taking into account biasing resistors and dc blocking or decoupling capacitors are now discussed into more details.

8.2.2.3 Pierce Oscillator A practical Pierce oscillator circuit is shown in Figure 8.30. A gate bias is assumed to be available but is also possible, if the FET is of the depletion type, to self-bias the transistor by adding a source

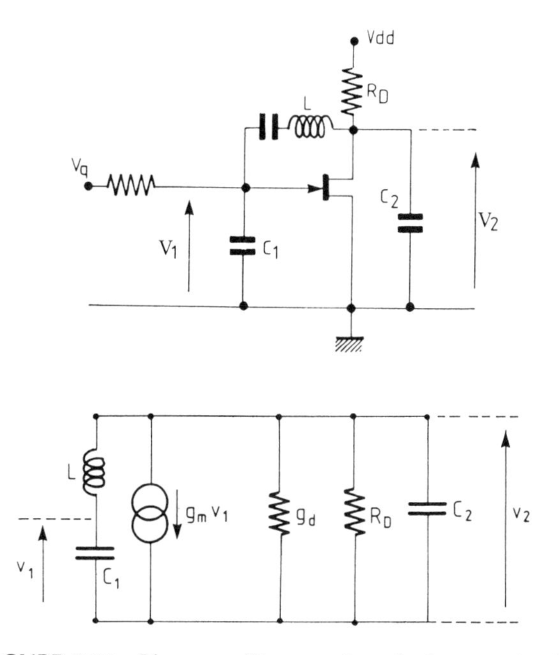

FIGURE 8.30 Pierce oscillator and equivalent ac circuit.

resistor and a shunt capacitor for ac grounding of the source. A dc blocking capacitor is required in series with the inductor, which must have a reactance lower than that of the inductor. A drain load resistor R_D large enough to keep the transistor voltage gain high is also required.

The feedback network gain is

$$\beta = \frac{-C_2}{C_1}$$

At the oscillation frequency, the feedback LC circuit is nearly at resonance and presents a real impedance r_{in} to the transistor. Assuming $r \ll 1/\omega_0^2 C_2$, we can approximate r_{in} by

$$r_{in} \approx \frac{1}{r\omega_0^2 C_2^2}$$

The transistor voltage gain A can be then calculated. It is given by

$$A = \frac{-g_m}{g_d + (1/R_D) + r\omega_0^2 C_2^2}$$

The output conductance g at the drain is given by

$$g = g_d + \frac{1}{R_D} + \frac{1}{r_{in}} + g_m \beta$$

which leads to

$$g = g_d + \frac{1}{R_D} + r\omega_0^2 C_2^2 - g_m \frac{C_2}{C_1}$$

8.2.2.4 Colpitts Oscillator In the Colpitts oscillator, the transistor is connected as a source follower (Fig. 8.31). A source resistor R_S is required to bias the transistor, which must have an impedance higher than that of the capacitor C_2. In Figure 8.31, the gate has its own bias, but when using a depletion-mode FET, the source resistor may be used for self-biasing. In this case, the gate can be dc-grounded through the inductor, thus saving the series dc blocking gate capacitor C. As no drain resistor is required in the Colpitts configuration, one more component is saved as compared to the

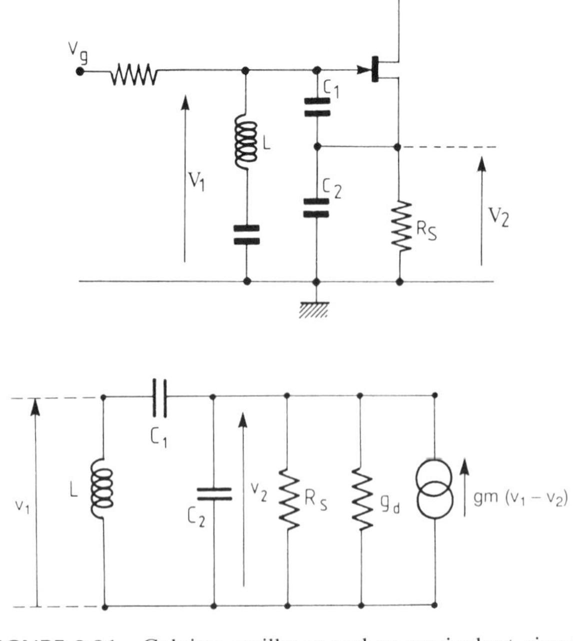

FIGURE 8.31 Colpitts oscillator and ac equivalent circuit.

self-biased Pierce oscillator. Moreover, a lower supply voltage can be used, making the Colpitts oscillator very attractive.

The feedback network gain is

$$\beta = \frac{C_1 + C_2}{C_1}$$

Assuming that the oscillation frequency is the resonant frequency f_0 of the LC feedback circuit and that

$$r \ll \frac{1}{\omega_0 C_2}$$

the impedance presented by the feedback network to the transistor is real and given by

$$r_{in} \approx \frac{1}{r\omega_0^2 C_2^2}$$

The transistor voltage gain A is given by

$$A = \frac{g_m}{g_m + g_d + (1/R_S) + r\omega_0^2 C_2^2}$$

The output conductance g of the oscillator at the source of the transistor is given by

$$g = g_m + g_d + \frac{1}{R_S} + \frac{1}{r_{in}} - g_m \beta$$

leading to a similar expression to that derived for the Pierce oscillator:

$$g = g_d + \frac{1}{R_S} + r\omega_0^2 C_2^2 - g_m \frac{C_2}{C_1}$$

8.2.2.5 Clapp Oscillator In the Clapp oscillator, the transistor is connected as a common-gate amplifier (Fig. 8.32). This configuration requires biasing resistors in the drain (R_D) and in the source (R_S). The source resistor may be used as a self-biasing resistor with depletion-mode FETs, allowing the gate to be dc-grounded. because of the need for R_S and R_D, the Clapp oscillator requires a higher supply voltage than the previous Pierce and Colpitts types. This is true even when compared to the self-biased Pierce oscillator (thus also with a source resistor), because a larger drain resistance is required for the Clapp because of its higher output impedance at the drain.

The Clapp oscillator is not as easily analyzed as the Pierce and Colpitts ones because of the low input impedance of the common-gate transistor.

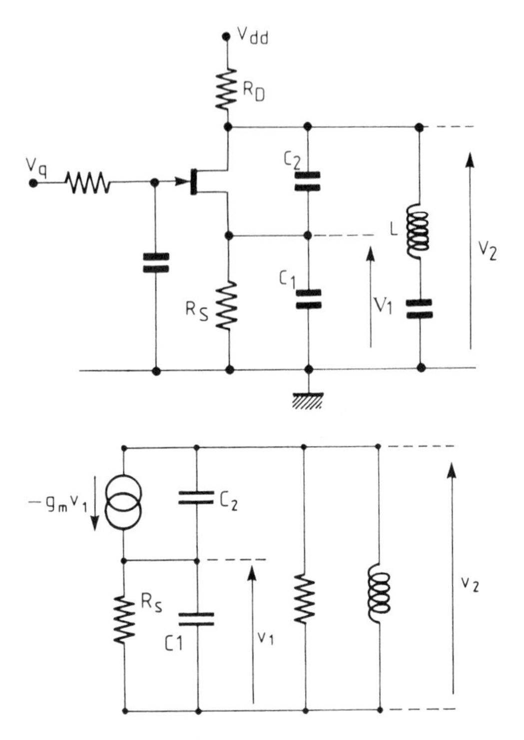

FIGURE 8.32 Clapp oscillator with ac equivalent circuit.

The capacitive voltage divider formed by C_1 and C_2, since looking into a low impedance, does not have a pure real transfer gain. To simplify the calculation, the following assumption must be made:

$$g_m \ll \omega_0(C_1 + C_2)$$

In these conditions, the feedback network gain simplifies to

$$\beta = \frac{C_2}{C_1 + C_2}$$

At the resonant frequency f_0 of the LC circuit, assuming

$$r \ll \omega_0 \frac{C_1 C_2}{C_1 + C_2}$$

and $g_d \ll \omega_0 C_2$, the impedance presented by the feedback network to the transistor is real and is given by

$$r_{in} \approx \frac{(C_1 + C_2)^2}{g_m C_2^2 + r\omega_0 C_1^2 C_2^2}$$

The gain of the transistor is hence real and is given by

$$A = \frac{g_m}{(1/R_D) + [g_m C_2^2/(C_1 + C_2)^2] + r\omega_0^2[C_1 C_2/(C_1 + C_2)]^2}$$

The output conductance g of the oscillator at the drain of the transistor is given by

$$g = \frac{1}{R_D} + \frac{1}{r_{in}} - g_m \beta$$

leading to

$$g = \frac{1}{R_D} + r\omega_0 \left(\frac{C_1 C_2}{C_1 + C_2}\right)^2 - g_m \frac{C_1 C_2}{(C_1 + C_2)^2}$$

Table 8.4 compares the loop gain $A\beta$, the output conductance g, the loss conductance of the feedback network g_{loss} for the three basic oscillators when neglecting the extra components (R_D, R_S), and the drain conductance g_d of the transistor. Although the loop gain condition $A\beta > 1$ is the same for the three oscillators and is given by

$$\frac{g_m}{r\omega_0^2 C_1 C_2} > 1$$

the loop gains are not equal and vary with g_m, although they are equal to unity simultaneously. This may appear surprising, and it means that the absolute loop gain is not a reliable criterion for evaluation of the safety margin.

The second row of the table gives the output conductances. They vary linearly with g_m and are equal to zero simultaneously but have different slopes. As the negative conductance accounts for the energy that is brought to the resonant circuit it must be compared with the conductance g_{loss}

TABLE 8.4 Parameters of Pierce, Colpitts, and Clapp Oscillators

Parameter	Symbol	Pierce	Colpitts	Clapp
Loop gain	$A\beta$	$\frac{g_m}{r\omega_0^2 C_1 C_2}$	$\frac{g_m(C_1 + C_2)}{g_m C_1 + r\omega_0^2 C_2^2 C_1}$	$\frac{g_m(C_1 + C_2)}{g_m C_2 + r\omega_0^2 C_1^2 C_2}$
Output conductance	g	$r\omega_0^2 C_2^2 - g_m \frac{C_2}{C_1}$	$r\omega_0^2 C_2^2 - g_m \frac{C_2}{C_1}$	$\frac{r\omega_0^2 C_1^2 C_2^2 - g_m C_1 C_2}{(C_1 + C_2)^2}$
Resonant circuit loss conductance	g_{loss}	$r\omega_0^2 C_2^2$	$r\omega_0^2 C_2^2$	$\frac{r\omega_0^2 C_1^2 C_2^2}{(C_1 + C_2)^2}$

characterizing the ohmic losses into the resonant circuit. If we calculate the ratio g/g_{loss}, the results are identical for the three oscillators, which is not surprising since the three oscillators are basically the same.

8.2.3 Dual-Phase Oscillators

8.2.3.1 Principle of Operation Dual-phase oscillators consist of two oscillators coupled to each other so that their respective output signals have equal amplitude and are 180° out of phase. Their domain of application is the generation of balanced LO signals for operating balanced mixers, an alternative to the method that consists in using a single-ended oscillator followed by a difference amplifier.

Operation of dual-phase oscillators relies on symmetry. As any symmetric system, dual-phase oscillators can be analyzed by considering successively the differential and common modes. Because of the symmetry, simplifications can be made that allow one to analyze half of the circuit. This method is illustrated in Figure 8.33 showing how a symmetric system can be simplified when considering differential and common modes. The simplications depend on how the circuits are coupled. Two types of coupling can be distinguished:

- Connections perpendicular to the symmetry axis, that is, connecting symmetric nodes. This type of connection is the more common and the more simple to treat.
- Pairs of crossed connections.

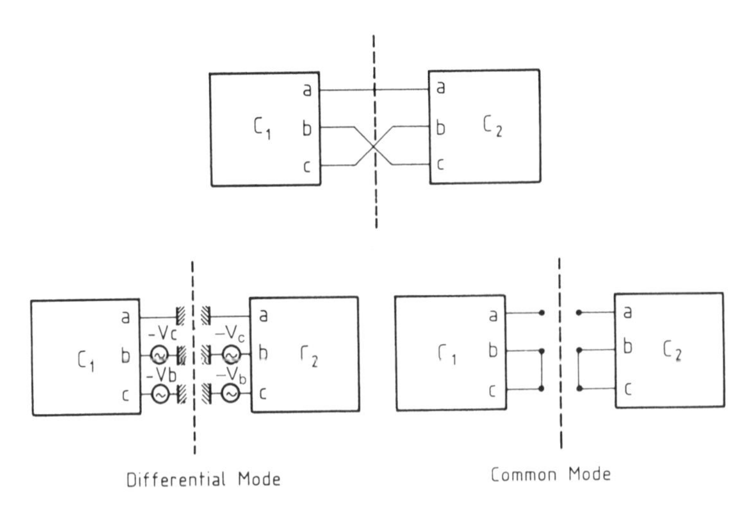

FIGURE 8.33 A pair of coupled circuits and their equivalent circuit in differential and common modes.

In common mode, the signals have equal amplitude and are in phase in the two circuits. Kirchhoff's law stipulates that no current flows in perpendicular coupling connections, and therefore that these leads can be removed. The system behavior is, in addition, unchanged if crossed-coupling connections are not crossed. When this is done, the two circuits are completely independent and can be analyzed separately (Fig. 8.33).

In differential mode, the signals still have equal amplitude but are 180° out of phase in the two circuits. As there is no voltage drop across a short circuit, the voltage needs to be zero at symmetric nodes coupled to each other. These nodes can therefore be grounded without affecting the circuit behavior. For crossed-coupling connections, no such simplifications can be made. The analysis can nevertheless be restricted to half of the system by terminating these connections with a controlled voltage source of equal amplitude but 180° out of phase to the voltage at the symmetric node (Fig. 8.33).

When the electrical circuits of the coupled oscillators pair have been derived for each mode, the corresponding amplifier and feedback network gains can be calculated. This results usually in different potential oscillation frequencies. The oscillation can be forced in a particular mode, if the system is so designed that the loop gain condition fulfilled is only one mode. If the oscillation conditions were verified in the two modes, the oscillation could indeed build up in either the common or differential mode, indifferently. If the differential mode is the one desired, the following conditions must be fulfilled:

$$A_d\beta_d > 1 \quad \text{at} \quad f_{0d}$$

$$A_c\beta_c < 1 \quad \text{at} \quad f_{0c}$$

where f_0 is the oscillation frequency at which the loop phase shift is 2π, and where the subscripts "d" and "c" refer to the differential and common modes, respectively.

In contrast to differential amplifiers, which always have some response to common input, dual-phase oscillators theoretically completely reject the common mode in principle. This is attractive feature for application to balanced mixers wherein LO isolation depends on the degree of balance of LO. The coupled oscillator system is, however, never perfectly symmetric in practice, limiting the rejection to about 25–30 dB.

8.2.3.2 Multivibrator Oscillator One of the most popular dual-phase oscillators has a structure similar to that of a multivibrator (Fig. 8.34). The active part is a differential source-coupled FET pair in which the drains and gates are cross-coupled by means of capacitors C_{dg}, making a positive

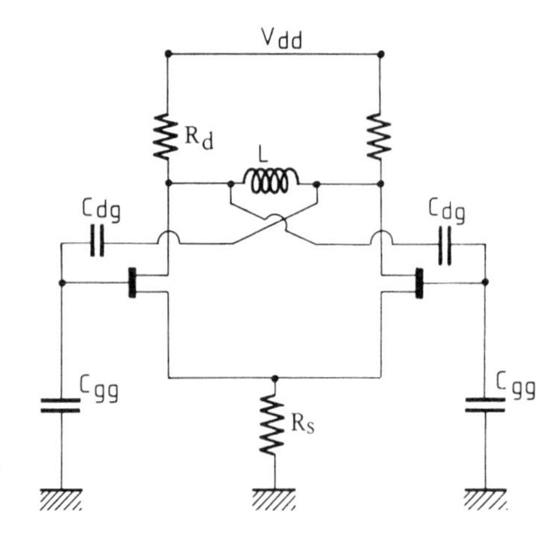

FIGURE 8.34 Multivibrator-like oscillator.

feedback. A reactive circuit is placed in between the drains that looks inductive at the oscillation frequency and resonate with the capacitors C_{dg} and C_{gg}.

The multivibrator is analyzed by separating it into differential and common modes using the method described in the previous paragraph. The equivalent circuit in common mode is shown in Figure 8.35. The transistor is modeled by its transconductance g_m, its output conductance g_d, and its gate-to-source capacitance C_{gs}. The gate-to-drain capacitance is neglected. The source resistance R_S brings series feedback to the transistor, which hence has low gain. The inductor is removed from the circuit, and the capacitor C_{dg} appears as between the gate and the drain of the transistor. The circuit can be further simplified as shown in Figure 8.36, where the new

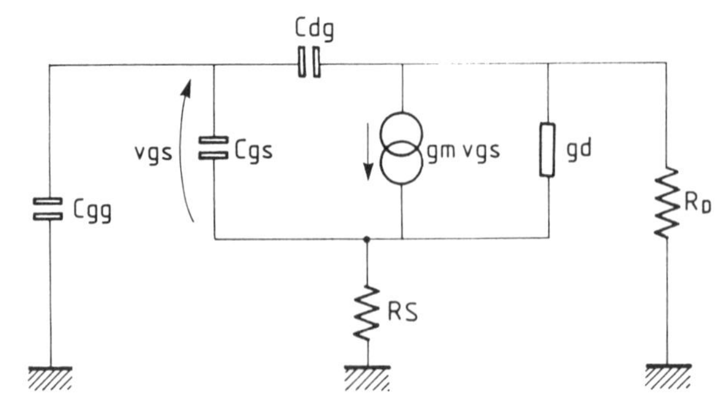

FIGURE 8.35 Equivalent circuit of the multivibrator in common mode.

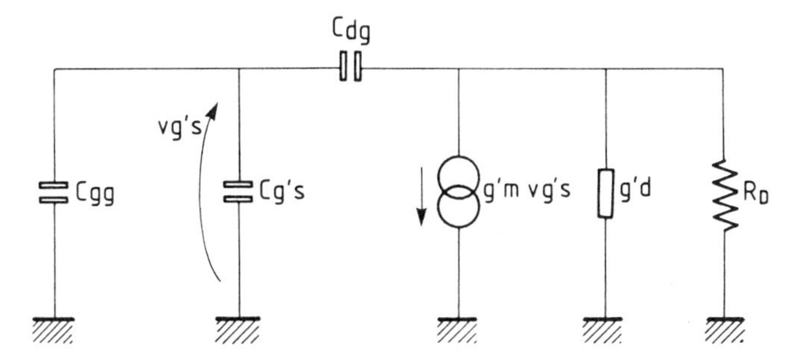

FIGURE 8.36 Simplified circuit in common mode.

transistor parameters are expressed in terms of the old ones according to

$$C'_{gs} = \frac{C_{gs}}{1 + g_m R_S}$$

$$g'_m = \frac{g_m}{1 + g_m R_S}$$

$$g'_d = \frac{g_d}{1 + g_m R_S}$$

The feedback network gain v_1/v_2 is given by

$$\beta = \frac{C_{dg}}{C_{dg} + C_{gg} + C'_{gs}}$$

and the transistor gain is

$$A = \frac{-g'_m + j\omega C_{dg}}{g'_d + (1/R_D) + j\omega C_{dg}}$$

Assuming $g'_d \ll \omega C_{dg}$ and $(1/R_D) \ll \omega C_{dg}$, the loop gain $A\beta$ is given by

$$A\beta = \frac{-g'_m + j\omega C_{dg}}{j\omega(C_{gg} + C'_{gs} + C_{dg})}$$

The loop phase shift tends toward 2π at infinite frequency, but the corresponding loop gain is less than unity. This result shows that no oscillation can start in common mode. The equivalent circuit in the differential mode is shown in Figure 8.37. The resistor R_s is removed and the transistor sources are virtually grounded. The transistor behaves as a

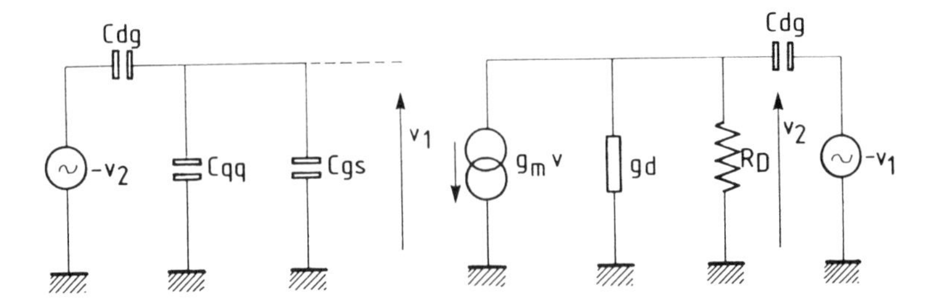

FIGURE 8.37 Equivalent circuit of the multivibrator in differential mode.

common source amplifier and now has high gain. Controlled voltage sources terminate the crossed-coupling leads, which account for the fact that symmetric voltages are 180° out of phase. Defining the C'_{gg} capacitor as

$$C'_{gg} = C_{gg} + C_{gs}$$

the feedback network gain is given by

$$\beta = \frac{v_1}{v_2} = -\frac{C_{dg}}{C_{dg} + C'_{gg}}$$

The oscillation frequency is determined by the resonance of the LC network according to the expression

$$f_0 = \frac{1}{2\pi}\sqrt{\frac{C_{dg} + C'_{gg}}{LC_{dg}C'_{gg}}}$$

At the resonance frequency, the transistor looks into a real impedance. Its phase shift is 180° so that the loop phase shift verify the 360° condition. The input resistance r_{in} of the LC network can be approximated to

$$r_{in} = \frac{(C_{dg} + C'_{gg})^2}{r\omega_0^2 C_{dg}^2 C'^2_{gg}}$$

where r is the series resistance of the inductor. The transistor gain A can then be calculated. It is given by

$$A = -\frac{g_m}{g_d + (1/r_{in}) + (1/R_D)}$$

Reporting the value of r_{in} in the expression of A, and assuming $g_d + (1/$

$R_D) < (1/r_{in})$, the loop gain condition $A\beta > 1$ simplifies into

$$\frac{g_m(C_{dg} + C'_{gg})}{r\omega_0^2 C_{dg} C'^2_{gg}}$$

The left-hand side of this inequation is maximum when

$$C_{dg} \gg C'_{gg}$$

When this condition is fulfilled, the capacitance C'_{gg} is related to the oscillation frequency and the inductance by

$$C'_{gg} = \frac{1}{L\omega_0^2}$$

Reporting this value into the loop gain condition and taking into account that $C_{dg} \gg C'_{gg}$ yields the minimum oscillation condition

$$g_m > \frac{r}{L^2\omega_0^2}$$

This result has to be compared to the similar one derived for basic oscillators earlier. With the same inductor and at the same frequency, the transconductance required for oscillation is 4 times lower with the multivibrator configuration. This property is very attractive and makes the multivibrator oscillator achieve a higher efficiency than Pierce, Colpitts, and Clapp oscillators. It also has implications for the tuning bandwidth. As a transistor of smaller size can be used, thus with lower capacitance, the multivibrator also offers the possibility for a larger tuning range. However, taking full advantage of the oscillation capability of the multivibrator requires very tight cross-coupling of gates and drains ($C_{dg} \gg C_{gg'}$). This has a negative effect on stability and phase noise since the frequency becomes more sensitive to the transistor capacitances [high($\Delta f/\Delta C_{gs}$)].

8.2.3.3 Application Examples Dual-phase oscillators are essentially used as local oscillators for balanced mixers in UHF or L-bands. Many circuits have been reported in the field of terrestrial or satellite TV.

Figure 8.38 shows an example of multivibrator oscillator applicated in a mixer–oscillator IC for satellite TV tuner that was fabricated using a normally-off technology [16]. This type of oscillator has been chosen for its broadband capability and its low power consumption. The cross-coupled differential pair is connected to an external tunable LC resonator that appears inductive at the oscillation frequency. The tuning is performed by means of two silicon varicaps with a capacitance ratio of about 6, allowing spanning of a 1.15-GHz-frequency band from 1.4 to 2.55 GHz. The resonator could not be integrated for basically two reasons: (1) an inte-

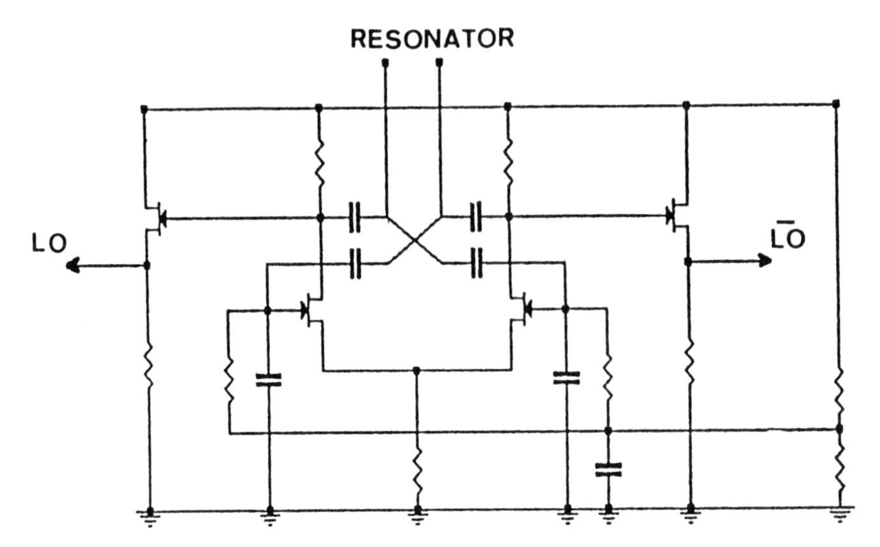

FIGURE 8.38 Multivibrator applied in a mixer–oscillator for satellite TV tuner.

grated resonator would have a lower-quality factor, leading to a poor phase noise performance; and (2) the capacitance ratio of integrated GaAs varicaps is not sufficient to allow such a wide tuning range. The feedback ratio was chosen equal to 0.5 ($C_{dg} = C_{gg} + C_{gs}$) in order to decrease the level of the signal fed back to the gates. Furthermore, a low feedback ratio reduces the sensitivity of the oscillation frequency to variations of the nonlinear capacitances of the transistor, which is known to have a positive impact on phase noise. The capacitor C_{gg} is not absolutely required. It has been added to linearize the transistor input capacitance and make the feedback ratio less dependent on the transistor capacitance. The transistor has been sized for the oscillator to be able to handle a resonator impedance superior to 400 Ω. The oscillation signal is taken from the drains at high impedance and fed to the mixer via a source-follower buffer stage. An example of spectrum measured at 2.5 GHz is shown in Figure 8.39. Taking into account the resolution bandwidth of 10 kHz, the phase noise at 100 kHz away from the carrier is estimated to −100 dBc/Hz.

Dual-phase oscillators derived from single-ended basic topologies have also been reported. Figure 8.40 shows a fully integrated S-band dual-phase broadband Clapp VCO (voltage controlled oscillator) [17]. The varicaps are integrated and consist of Schottky diodes. Their maximum capacitance is about 2.5 pF and equal to the fixed drain to source feedback capacitor. The inductor has an inductance value of about 4 nH. The basic Clapp circuit can be recognized in differential mode where the nodes located on the symmetry axis are virtually grounded. In common mode, the resistors R_D and R_S appear as in series respectively with the inductors and the varicaps. In these

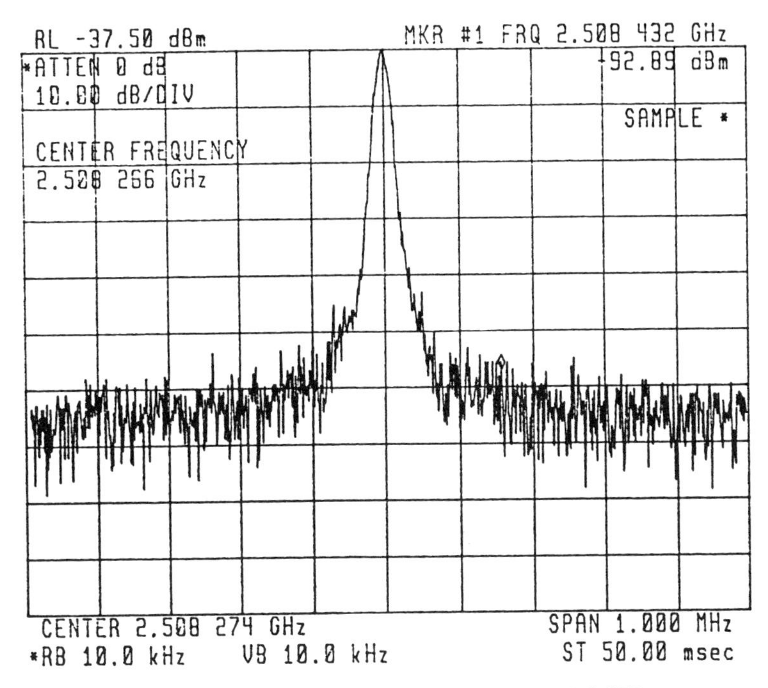

FIGURE 8.39 Multivibrator spectrum at 2.5 GHz.

conditions, the resonant circuit has a very low-quality factor and no oscillations can take place.

The dual-phase Clapp topology offers several advantages as compared to the basic circuit that are related to symmetry. The nodes at which the

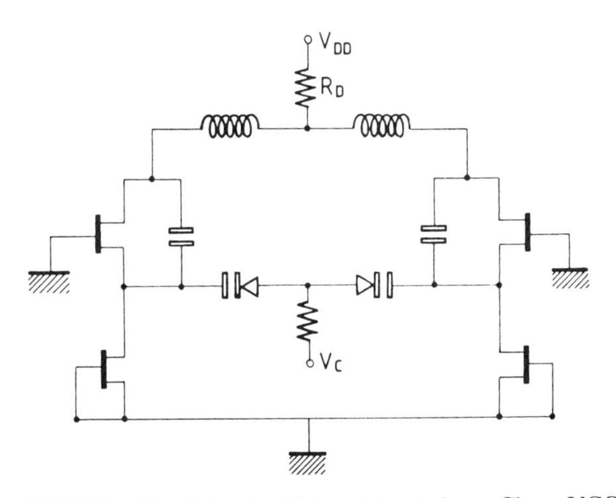

FIGURE 8.40 S-band wideband dual-phase Clapp VCO.

separate oscillators are coupled appear as virtually grounded in differential mode. For instance, no dc blocking capacitor is required as in the single-ended version to ground the inductor placed in the drain of the transistor. As a consequence, there is no need in the dual-phase version for a drain biasing resistor. This is a significant advantage since its resistance must be high for not damping the Q of the resonant circuit. Drain biasing can thus be made directly through the inductor, allowing a lower supply voltage than in the single-ended Clapp oscillator. Another advantage related to virtual grounding is that the oscillator frequency is not dependent as in single-ended oscillators to parasitics ground or drain inductances. This makes the oscillator much less sensitive to its environment.

Figure 8.41 shows the frequency and phase noise characteristics of the S-band dual-phase Clapp oscillator. The frequency can be tuned from about 2.2 GHz up to 3.6 GHz when sweeping the $C(V)$ characteristic of the varicap from 0 V (tuning voltage at 2 V since the anode is not referred to zero) to beyond pinchoff. The phase noise is about −90 dBc/Hz at 100 kHz away from the carrier but exhibits a peak up to −75 dBc/Hz around 0 V of tuning voltage. At this tuning voltage, the varicap is almost pinched off. In this biasing region it has a very low Q [18], which has, as consequence, a higher oscillator phase noise. This illustrates the difficulty of integrating wideband VCOs. A large tuning bandwidth requires indeed high capacitance variation. This can be achieved using integrated Schottky diodes provided their $C(V)$ characteristic is used up to beyond pinchoff. However, around pinchoff, the varicap has a very low Q, leading to a high phase noise when the varicap is biased in this region. The solution to achieve a good phase noise on the tuning range is to restrict to the portion of the $C(V)$ characteristic that extend between 0 V and a few hundred millivolts before pinchoff. The corresponding capacitance ratio, however, is rather small, about 2 or slightly more, and it is then very difficult to get a wide tuning bandwidth.

A similar dual-phase Clapp oscillator but with a narrower bandwidth has

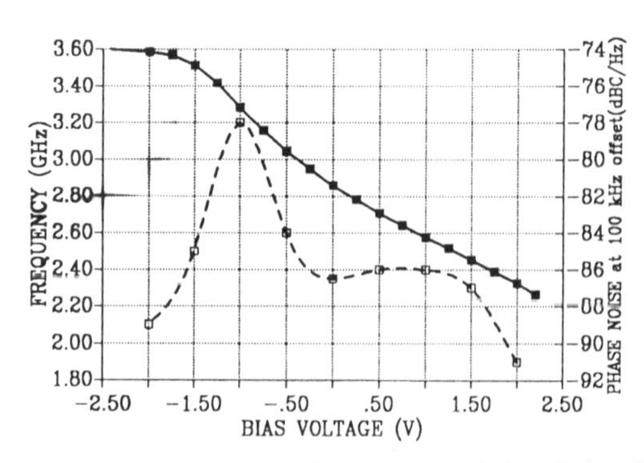

FIGURE 8.41 Frequency and phase noise characteristic of the S-band dual-phase Clapp VCO.

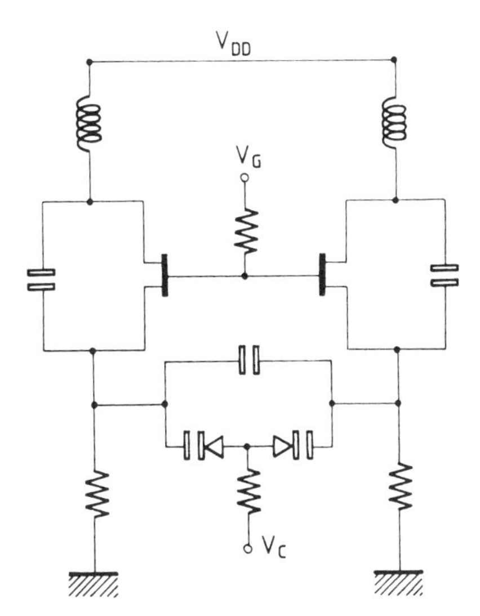

FIGURE 8.42 L-band dual-phase Clapp oscillator.

been designed using an enhancement-mode process for an L-band application at 1.3 GHz [19]. Its electrical circuit is shown in Figure 8.42. The capacitance C_1 consists of a fixed capacitor in parallel to an integrated varicap, which allows tuning of the oscillation frequency over about a 100-MHz bandwidth. As compared to the S-band dual-phase Clapp oscillator described above, the gates of the transistors here are not truly grounded but only virtually in differential mode. True grounding was not possible since the gate of enhancement-mode FETs must be at a positive voltage with respect to the source. Figure 8.43 shows a photograph of this oscillator IC where spiral inductors are clearly seen. The chip also integrates

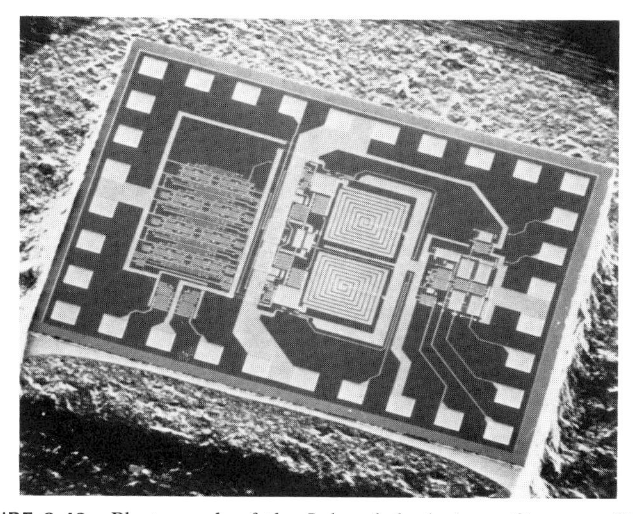

FIGURE 8.43 Photograph of the L-band dual-phase Clapp oscillator.

a frequency divider for phase locking onto a low-frequency crystal oscillator. The oscillator phase noise is about −102 dBc at 100 kHz offset in open loop, which is a rather good figure for a fully integrated oscillator. The phase noise being lower than that of the S-band depletion-mode oscillator described above is due to the narrower band and hence, higher Q of the resonant circuit.

REFERENCES

1. K. A. Simons, "The Decibel Relationships Between Amplifier Distortion Products," *Proc. IEEE*, **58**, 1071 (July 1970).
2. S. A. Maas, *Microwave Mixers*, Artech House, Norwood, MA, 1986.
3. T. Ducourant and P. Philippe, et al., "A 3 Chip GaAs Double Conversion TV Tuner System with 70 dB Image Rejection," *1989 Microwave and Millimeter Wave Monolithic Circuits Symposium Digest*, June 1989, p. 1987.
4. R. Leblanc et al., "GaAs Monolithic Circuit for FMCW Radars," *1988 Microwave and Millimeter Wave Monolithic Circuits Symposium Digest*, May 1988, p. 109.
5. P. O'Sullivan et al., "Highly Integrated I/Q Converter for 900 MHz Applications," *1990 GaAs IC Symposium Digest*, October 1990, p. 287.
6. P. Philippe et al., "A Multi-Octave Active GaAs MMIC Quadrature Phase Shifter," *1989 Microwave and Millimeter Wave Monolithic Circuits Symposium Digest*, June 1989, p. 75.
7. Y. Imai et al., "Design and Performance of Low-Current GaAs MMIC's for L-Band Front-End Applications," *IEEE Trans. Microwave Theory Tech.*, **MTT-39**, 209 (Feb. 1991).
8. R. Meierer et al., "Dual Gate FET Mixers," *IEEE Trans. Microwave Theory Tech.*, **MTT-32**, 248 (March 1984).
9. U. Ablassmeier et al., "GaAs FET Up-Converter for TV Tuner," *IEEE Trans. Electron Devices*, **ED-27**, 1156 (June 1980).
10. H. Mizukami et al., "A High Quality GaAs IC Tuner for TV/VCR Receivers," *IEEE Trans. Consumer Electron.*, **CE-34**, 649 (Aug. 1988).
11. T. Nakatsuka et al., "Low Distortion and Low Noise Osillator Mixer IC for CATV Converters," *1988 GaAs Symposium Digest*, 1988, p. 161.
12. H. Yagita et al., "Low Noise and Low Distortion GaAs Mixer Oscillator IC for Broadcasting Satellite TV Tuncr," *1989 GaAs IC Symposium Digest*, 1989, p. 75.
13. A. Yamamoto et al., "A Compact Satellite 1 GHz Tuner with GaAs ICs," *IEEE Trans. Consumer Electron.*, **CE-35** (Aug. 1989).
14. R. Pyndiah et al., "GaAs Monolithic Direct Linear 1-2.8GHz QPSK Modulator," *19th European Microwave Conference*, London, Sept. 1989, pp. 1201–1204.

15. P. Jean et al., "Wideband Monolithic GaAs Phase Detector for Homodyne Reception," MTT-Symposium, vol. 1, Las Vegas, June 1987, pp. 169–171.
16. P. Philippe and M. Pertus, "A 2 GHz Enhancement Mode GaAs Down-Converter IC for Satellite TV Tuner," MTT-Symposium, Boston, 1991, *Digest*, p. 73.
17. P. Philippe et al., "GaAs MMIC Building Blocks for TV Applications," GaAs Applications Symposium, Rome, 1990, *Conference Proceedings*, p. 128.
18. P. Philippe et al., "Physical Equivalent Circuit Model for Planar Schottky Varactor Diode," *IEEE Trans. Microwave Theory Tech.*, **MTT-36**, 250 (Feb. 1988).
19. D. Meignant et al., "GaAs Normally-off Mixed Analog/Digital Technology for GPS Receiver Front-End," 3rd Asia-Pacific Microwave Conference, Tokyo, 1990, *Conference Proceedings*, p. 763.

CHAPTER NINE

Data Conversion Circuits

P. E. ALLEN
Georgia Institute of Technology
School of Electrical and Computer Engineering
Atlanta, GA 30332-0250

C. M. BREEVOORT
Consultant

9.1 INTRODUCTION

While progress in digital and microwave GaAs circuits has been significant over the years, problems in GaAs MESFET technology have limited the design efforts in analog signal processing circuitry [1–3]. Factors that have frustrated GaAs MESFET analog circuit design include hysteresis, backgating, poor device matching, low transconductance, and poor yield. Although many different analog-to-digital (A/D) and digital-to-analog (D/A) converter architectures have been implemented in silicon, GaAs MESFET technology has found only limited use in data conversion circuits because of these problems. As the understanding of GaAs MESFET technology increases, new circuits and converter architectures will be developed [4]. The applications for GaAs converters are aimed at radar, communications, electronic warfare, instrumentation, and video processing.

The material presented in this chapter has been selected to be representative of the present state-of-the-art techniques in GaAs MESFET analog signal processing. In this chapter, we will discuss D/A and A/D conversion

High-Frequency Analog Integrated-Circuit Design, Edited by Ravender Goyal
ISBN 0-471-53043-3

architectures. The main emphasis will be on the topologies that are most suitable for implementation in the GaAs MESFET technology. This is the current-scaling or current-steering concept for the D/A converter. A/D converter architectures that will be discussed include the parallel (flash), the feedforward (subranging and pipelined), and oversampled ($\Delta\Sigma$) architectures. The major elements in these A/D converter architectures, such as the comparator and sample-and-hold (S/H) circuit, will be presented.

9.2 D/A CONVERTERS

The ability to convert digital signals to analog is very important in applications such as direct digital synthesis (DDS) techniques. It allows the creation of a carrier frequency using a digital accumulator, a sine or cosine lookup table (a ready-only memory) and a high-speed D/A converter. The resulting output signal has fine frequency resolution and phase control, a broad bandwidth of operation, fast switching speed between output frequencies for frequency hopping, and excellent phase noise performance. These qualities are hard to meet with present analog circuitry. The only analog component remaining is the D/A converter. The speed and accuracy of the D/A converter determine the performance of the DDS system.

Figure 9.1*a* shows a conceptual block diagram of a D/A converter. The inputs are a digital word of N bits $(b_1, b_2, b_3, \ldots, b_N)$ and a reference voltage V_{REF}. The output can be a current I_{OUT}, which creates a voltage drop across a termination resistor, or a direct voltage output V_{OUT}. This voltage output can be expressed as

$$V_{\text{OUT}} = KV_{\text{REF}}D \tag{9.1}$$

where K is a scaling factor and the digital word D is given as

$$D = b_1 2^{-1} + b_2 2^{-2} + b_3 2^{-3} + \cdots + b_N 2^{-N} \tag{9.2}$$

where N is the total number of bits of the digital word and b_i is the ith bit coefficient and is either 0 or 1. Thus, the output of the D/A converter can be expressed by combining Equations (9.1) and (9.2) to get

$$\begin{aligned} V_{\text{OUT}} &= KV_{\text{REF}}(b_1 2^{-1} + b_2 2^{-2} + b_3 2^{-3} + \cdots + b_N 2^{-N}) \\ &= KV_{\text{REF}} \sum_{i=1}^{\text{N}} b_i 2^{-i} \end{aligned} \tag{9.3}$$

In high-speed D/A converters the digital word is synchronously clocked. In this case it is necessary to use latches to hold the word for conversion and to provide a S/H circuit at the output, as shown in Figure 9.1*b*. The output of the S/H circuit, denoted by an asterisk, is held constant while the latch

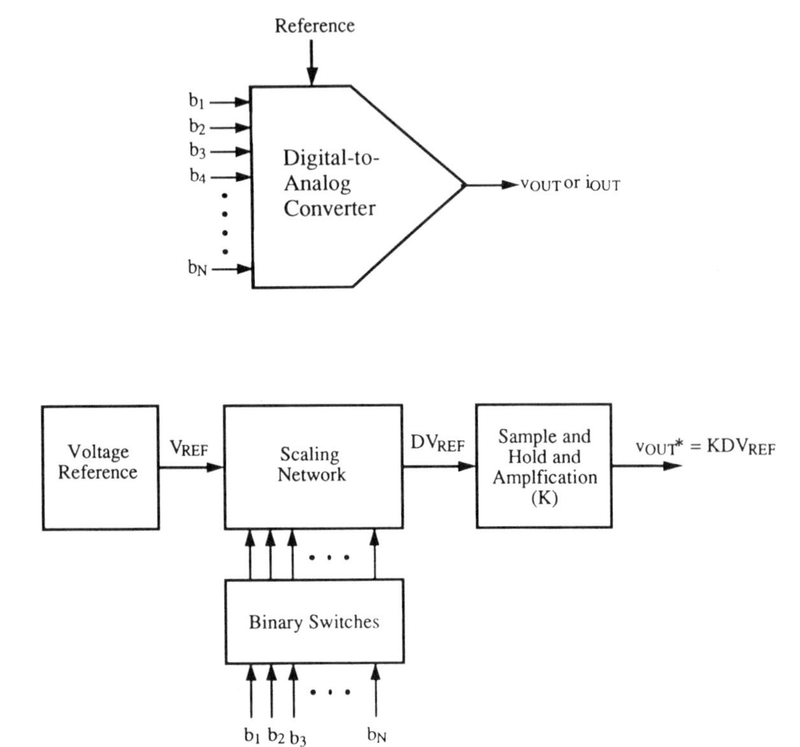

FIGURE 9.1 (*a*) Conceptual block diagram of a D/A converter; (*b*) high-speed realization.

receives its new data. The S/H circuit switches to the sample mode after the output of the D/A converter has been stabilized.

The characterization of the D/A converter is very important in understanding its use and application. Static and dynamic properties determine the performance of the D/A converter. As one might expect, the static properties are independent of time and include the converter transfer characteristics, quantization noise, dynamic range, gain, offset, and nonlinearity.

The transfer characteristics of the ideal D/A converter are shown in Figure 9.2. With each increase in the digital code the analog output voltage increases by 1 least significant bit (LSB), which corresponds to 12.5% of full-scale (FS) for this 3-bit D/A converter. The change in the output is an analog quantity but is often referred to as an LSB change; that is, it is due to an LSB change in the digital input code. An offset has been added to the lowest digital code so that the analog output voltage corresponds to $\frac{1}{16}$ times FS.

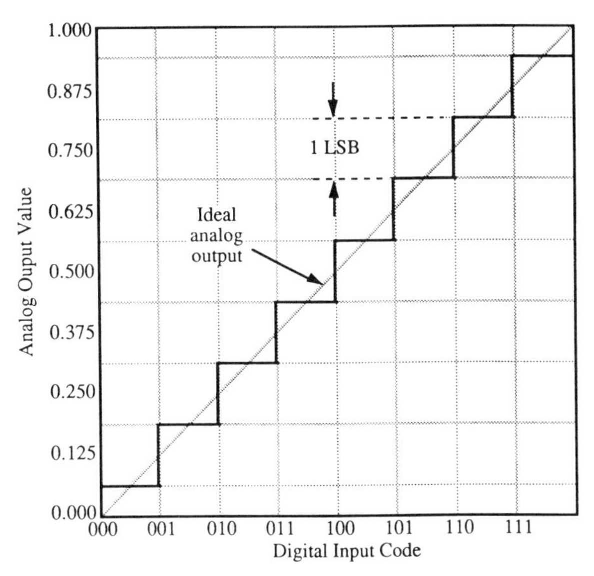

FIGURE 9.2 D/A converter transfer characteristic.

The resolution of a converter is the smallest analog change that can be resolved by an A/D converter or produced by a D/A converter. Resolution is commonly expressed in number of bits N, where the converter has 2^N possible states. Since the converters have a finite resolution, an error is always present in digitizing an analog value. This error is called the *quantization noise* and has a value of up to ± 0.5 LSB. As can be seen in Figure 9.2, the quantization noise is ± 0.5 LSB about each multiple of FS/8. The straight line in Figure 9.2 through the midpoint of each analog step change represents the transfer curve when the resolution of the D/A converter approaches infinity.

The signal-to-noise (S/N) ratio is an important parameter in high-speed converters. The S/N ratio assumes a ramp input applied to an ideal A/D converter that drives an ideal D/A converter. A sawtooth waveform of $\pm \mathrm{FS}/2^{N+1}$ results after subtraction of the D/A converter output from the original input signal. It represents the ideal quantization noise and has an rms value of $(\mathrm{FS}/2^N)/\sqrt{12}$. The S/N ratio can be expressed in decibels as a power ratio as

$$\begin{aligned} \mathrm{S/N(dB)} &= 10 \log_{10}\left[\frac{\mathrm{FS}}{\mathrm{FS}/2^N \sqrt{12}}\right]^2 \\ &= 20 \log_{10}(2^N) + 20 \log_{10}\sqrt{12} = 6.02N + 10.8 \qquad (9.4) \end{aligned}$$

Equation (9.4) shows that the S/N ratio increases by a factor of approximately 6 dB for each additional bit of resolution.

Other static characteristics include offset error, gain error, nonlinearity, and nonmonotonicity. The offset error, shown in Figure 9.3*a*, is seen to be a vertical shift in the D/A transfer characteristic in relation to the ideal D/A transfer curve. The offset error is defined as the deviation from zero in the transfer characteristics. It is normally expressed in millivolts or percent of FS.

A gain error, illustrated in Figure 9.1*b*, is defined as the difference between the ideal and actual transfer characteristics at the FS point when the offset has been set to zero. If may be expressed in percent of FS.

Nonlinearity represents the deviation from the straight line as shown in Figure 9.1*c*. Integral nonlinearity (INL) is defined as the maximum deviation of the actual transfer characteristics from the straight line. The complete transfer characteristics plays a role in the determination of the

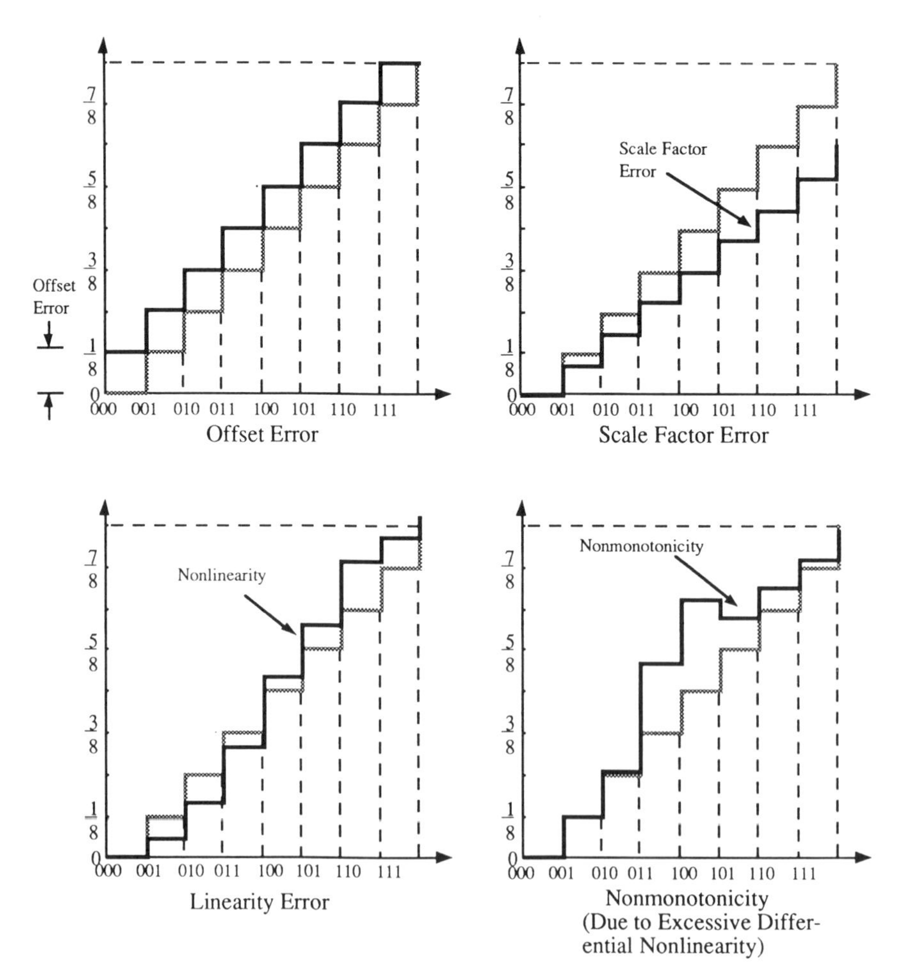

FIGURE 9.3 D/A converter offset error (*a*), gain error (*b*), nonlinearity (*c*), and nonmonotonicity due to excessive differential nonlinearity (*d*).

integral nonlinearity. The differential nonlinearity (DNL), as shown in Figure 9.3*d*, is the maximum deviation of any of the analog output changes caused by an LSB change from an ideal 1-LSB change. Nonlinearity is expressed in fractions of LSBs.

Monotonicity in a D/A converter refers to the property that causes a continuously increasing output value with an increasing digital input code. Nonmonotonicity can occur if the differential nonlinearity error exceeds ± 1 LSB.

Since GaAs MESFET D/A converters operate at high frequencies, the dynamic characterization of GaAs MESFET D/A converters is important. Factors other than the resistor network, current switches, and the stability of the reference voltage become crucial.

The D/A converter settling time is the worst-case time between application of a digital input and the time the output remains within an error band around its final value. Normally, a FS change in the output is used for this measurement, which takes all internal factors, such as turning switches on and off and redirecting currents, into account. The result is expressed in percentage of FS or LSB.

Glitches, which show up as output spikes caused by different delay times in switch turnon and turnoff, represent a potential major error source in high-speed D/A converters. The worst-case glitch occurs when the most significant bit (MSB) changes state, followed by the times when the next lower bit changes. Glitches occur any time when skew exist among the input bits causing distortion and generally higher noise levels.

The glitch impulse or glitch energy represents the net area under the voltage–time curve of the glitch and is expressed in picovolt-seconds. In order to minimize this product, symmetry in the design and layout of the D/A converter is critical. The use of nonsaturating logic accomplishes that turnon and turnoff times will be more symmetric than with saturating logic. The use of a sample-and-hold (S/H) circuit connected to the D/A converter output is an improvement. The S/H circuit operates in the hold mode until the effect of the glitch has diminished, after which it changes to the track mode. It returns to the hold mode just before any glitch can be expected at the output. The S/H circuit masks the glitches produced by the D/A converter, but introduces its own errors, thus limiting its usefulness at high speed.

The remainder of this section discusses several very high-speed D/A converter architectures suitable for implementation in a GaAs MESFET process. All architectures take advantage of the fast switching capabilities of the MESFET and are based on the current-scaling or current-steering architecture [5–8]. It is the logical choice for the high-speed GaAs MESFET technology, since it switches a number of stationary currents into two complementary load resistors. Although not necessarily required, the accuracy of the converters can be increased by the use of precision nickel–chromium (NiCr) resistors.

Since stand-alone GaAs D/A converters are normally operating in a 50 Ω environment, that is, where the value of the load resistors equals 50 Ω, the magnitude of internal currents and voltage drops across internal resistors are determined. As a consequence, current scaling factors and resistor ratios are somewhat constrained.

9.2.1 Current-Scaling D/A Converters

A general block diagram of the current-scaling D/A converter is shown in Figure 9.4. A reference voltage V_{REF} determines binary-weighted currents $I_1, I_2, I_3, \ldots, I_N$. The currents are directed through the switches S_i to one of two output resistors, R_{L1} and R_{L2}. It is seen that the output voltages can be expressed as

$$V_{\mathrm{OUT}} = -I_{\mathrm{OUT}} R_{\mathrm{L1}} \tag{9.5a}$$

$$V^*_{\mathrm{OUT}} = -I^*_{\mathrm{OUT}} R_{\mathrm{L2}} \tag{9.5b}$$

whereby the sum of I_{OUT} and I^*_{OUT} equals the total current I_0 generated by the current sinks. The total current I_0 can be expressed as

$$I_0 = I_{\mathrm{LSB}} + 2I_{\mathrm{LSB}} + 4I_{\mathrm{LSB}} + \cdots + 2^{N-1} I_{\mathrm{LSB}} \tag{9.6}$$

whereby I_{LSB} is the smallest current corresponding to the branch controlled by the least significant bit in the D/A converter. The current I_{LSB} is defined as

$$I_{\mathrm{LSB}} = \frac{V_{\mathrm{REF}}}{R_{\mathrm{S}} 2^N} \tag{9.7}$$

whereby R_{S} represents a precision resistor in the current sinks. Substitution of binary bits $b_1, b_2, \ldots, b_N$ and Equation (9.7) into Equation (9.5a) results

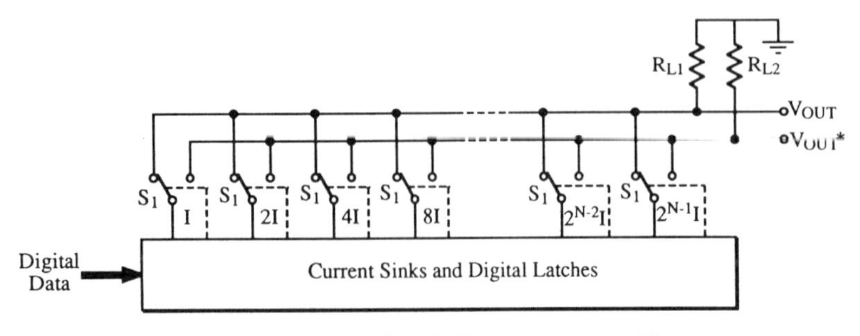

FIGURE 9.4 Current-scaling D/A converter architecture.

in

$$V_{\text{OUT}} = \frac{R_{\text{L1}}}{R_{\text{S}}} [b_1 2^N + b_2 2^{N-1} + \cdots + b_N 2] V_{\text{REF}} \tag{9.7a}$$

Equation (9.5b) can be substituted in a similar way with the realization that the complementary values of b_i have to be used.

$$V^*_{\text{OUT}} = -\frac{R_{\text{L2}}}{R_{\text{S}}} [b^*_1 2^N + b^*_2 2^{N-1} + \cdots + b^*_N 2] V_{\text{REF}} \tag{9.7b}$$

The output voltage V_{OUT} and V^*_{OUT} are determined by the binary bits b_i and their complement b^*_i.

The disadvantage of this architecture is the large ratio of resistor values in the binary-weighted current sinks. This ratio can be expressed as

$$\frac{R_{\text{MSB}}}{R_{\text{LSB}}} = \frac{1}{2^{N-1}} \tag{9.8}$$

whereby R_{MSB} represents the resistor for the MSB and R_{LSB}, the resistor for the LSB. A ratio of 1:128 is required for an 8-bit D/A converter based on this architecture. The currents set by the current sinks will also be binary-scaled according to this ratio, resulting in a 1:128 MESFET width differential between the MSB current and the LSB current. The accuracy requirements for the resistors and MESFET devices in the current sinks are difficult to achieve without expensive trimming techniques.

Two architectural modifications can be applied to decrease the resistor and the MESFET width ratios. The first one, known as *cascading*, uses current dividers between two binary-weighted D/A converter sections. A possible implementation is shown in Figure 9.5*a*. The accuracy of the current dividers needs to be within the magnitude of the LSB of the overall D/A converter resolution. Since the differential topology of this high-speed D/A architecture requires a current divider in each branch, matching between current dividers is also required.

A second approach, which has been successfully implemented in GaAs MESFET technology, uses a master–slave ladder. This approach, shown in Figure 9.5*b*, uses a master ladder, which determines the currents for the most significant bits, and a slave ladder, which sets the currents for the least significant bits. The $I/16$ current source for the slave ladder requires an accuracy of better than ± 0.5 LSB. The exact number of bits in the master and slave sections is determined by the matching properties of the MESFETs and resistors in the current sinks. The binary-weighted current sinks in the master section can be modified by employing digital decoding of some of the most significant bits. For instance, three of these bits could be decoded into eight signals that control eight identical current sinks in the master section. The master section of this D/A converter is depicted in Figure 9.6.

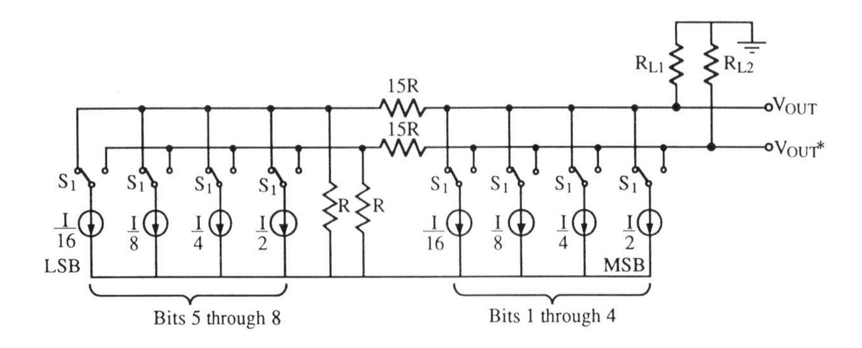

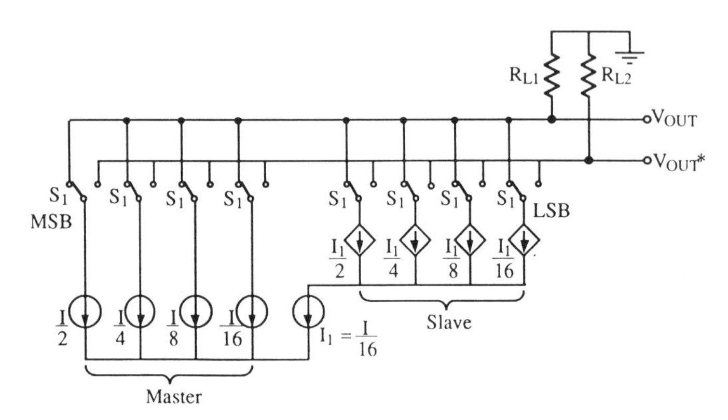

FIGURE 9.5 D/A converter with a current divider (*a*), and with a master–slave ladder (*b*).

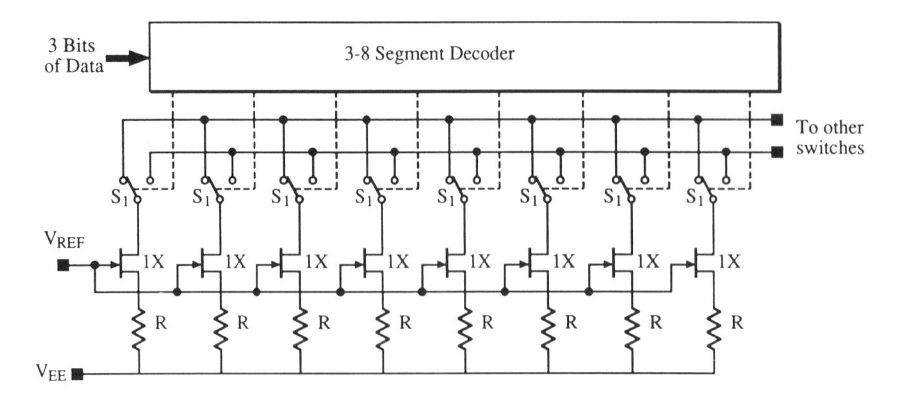

FIGURE 9.6 Master section of master–slave D/A converter.

9.2.2 *R*–2*R* Ladder D/A Converters

The D/A converter architectures discussed here depend to a large extent on the matching of resistor ratios and MESFETs that set the currents in the current sinks. To scale the currents in the current sinks, it is evident from equation (9.7*b*) that the resistor ratio is exponentionally proportional to the resolution. The use of an *R*–2*R* ladder structure circumvents this problem. The disadvantage of the *R*–2*R* ladder is that there are floating nodes present with associated parasitic capacitances. The resistances and parasitic capacitances form an *RC* network that requires charging and discharging, causing a delay in the response of the converter.

Two possible implementations of D/A converters with *R*–2*R* ladder structures are shown in Figure 9.7. More resistors are used than in a

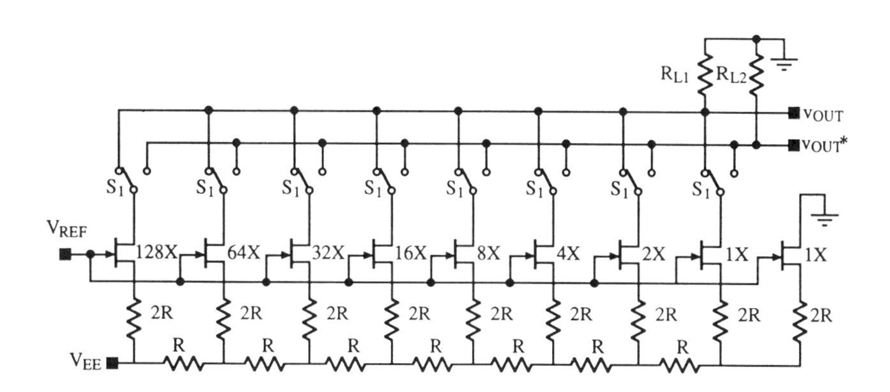

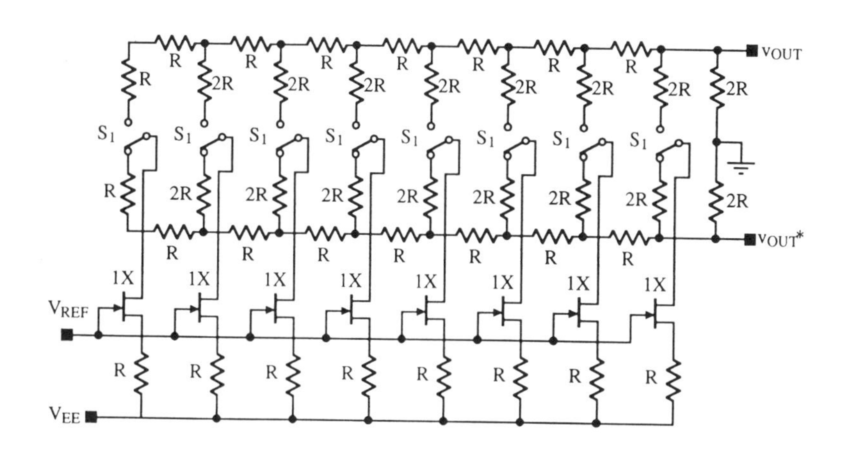

FIGURE 9.7 R-2R ladder structures using binary-weighted MESFET widths (*a*) and identical MESFETs (*b*).

binary-weighted D/A converter, but their ratio has been scaled down to 1:2. Both converters in Figure 9.7 are described by the same equation. The difference is that the converter in Figure 9.7*a* uses binary weighted MESFET widths (all gate lengths are identical). The currents through the MESFETs are proportional to the widths of the devices.

The D/A converter architecture in Figure 9.7*b* uses identical MESFETs in the current sinks resulting in identical currents through all of them.

Since one objective of GaAs MESFET D/A converters is to obtain a high conversion rate, it is obvious that the circuitry around the converter has to operate at a similar speed. The digital interface will therefore be compatible with the emitter-coupled-logic (ECL) voltage levels.

An accurate voltage reference is required for D/A converters in order to establish a basis for absolute measurement accuracy. Reference circuits in Si use the bandgap principle; specifically, the V_{BE} of a transistor with a negative temperature coefficient of 2 mV/°C can be extrapolated to approximately 1.2 V at absolute zero. It is possible to match two transistors with different current densities and produce a constant, temperature-invariant voltage, which can be amplified and buffered. Other Si references can be derived from a buried zener with laser trimming of thin-film resistors.

Unfortunately, the design of accurate voltage references in GaAs MESFET technology is a very challenging task. Effects such as device matching, hysteresis, and backgating make it virtually impossible to design precision op amps. A convenient solution is then to use one of the excellent Si-based references externally.

9.3 A/D CONVERTERS

Many more challenges will be encountered in the design of an A/D converter in comparison to the design of a D/A converter. The A/D converter determines the output digital word corresponding to an analog input signal. Most A/D converters require a S/H circuit at the input because it is not possible to convert a changing analog input signal with enough accuracy. The types of A/D converters that will be considered take advantage of the high-speed operation of the GaAs MESFET technology. They include the parallel (flash), feedforward (subranging, pipelined), and oversampled ($\Delta\Sigma$) architecture. The first two architectures excel in speed with a low to medium resolution, while the last architecture distinguishes itself with a medium to high resolution and a minimum amount of analog circuitry.

Figure 9.8 shows a basic block diagram of a parallel A/D converter. The input to the A/D converter is V_{IN} and could be an output signal from a S/H circuit. The voltages V_0 through V_{N-1} represent reference voltages that are proportional between two applied reference voltages, V_{REF+} and V_{REF-}. The input, along with the reference voltages, is used to determine the digital

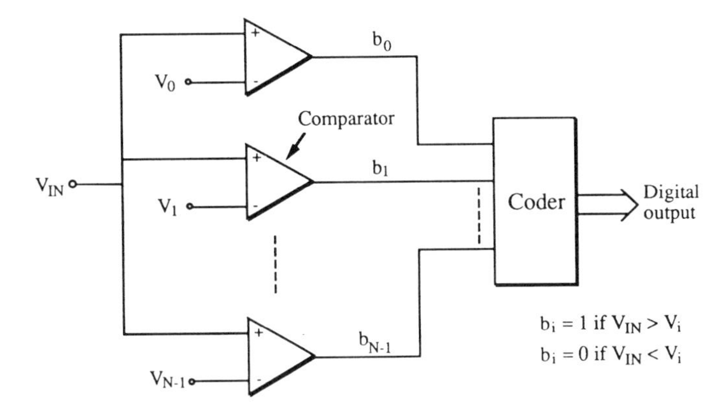

FIGURE 9.8 Block diagram of a parallel A/D converter.

word that best represents the sampled analog input signal. The comparator outputs X_i are then encoded into an output digital word. The encoding can result in one of several binary codes, such as offset-binary code or the Gray code. The result is that the A/D converter converts a continuous range of input amplitude levels into a discrete, finite set of digital words at discrete time intervals.

The characterization of the A/D converter is almost identical to that of the D/A converter if the input and output definitions are interchanged. The static A/D converter properties will be considered first. Figure 9.9 shows the transfer characteristic of an ideal 3-bit A/D converter. The analog input voltage normalized to full scale (FS) is shown on the horizontal axis. The digital code output is depicted on the vertical axis. The A/D converter has been designed so that the output digital word changes when the analog input is at odd multiples of FS/16. This way the output code is stable when the analog input signal is at the halfway point, FS/2. The LSB of the digital output code changes each time the analog input changes by $FS/2^N$, where N is equal to the number of digital bits. The length of each horizontal part of the staircase is equal to 1 LSB, or FS/8. The ranges are centered about even multiples of FS/16 except for the rightmost and leftmost, which have no right or left limits, respectively.

Most static definitions for the D/A converter apply also to the A/D converter. Exceptions are offset error, gain error, nonlinearity, and monotonicity. Figure 9.10 illustrates these errors for a 3-bit A/D converter. The ideal characteristic is shown by dashed lines for comparison. Figure 9.10*a* illustrates the offset error in a 3-bit A/D converter. Figure 9.10*b* shows the gain error, while Figure 9.10*c* depicts the nonlinearity characteristics with the same definitions as in the D/A converter case. Figure 9.10*d* illustrates excessive differential nonlinearity, which causes missing codes in A/D converters. In Figure 9.10*d* the digital codes 010, 011, and 110 are skipped.

The dynamic characteristics of the A/D converter relate primarily to the

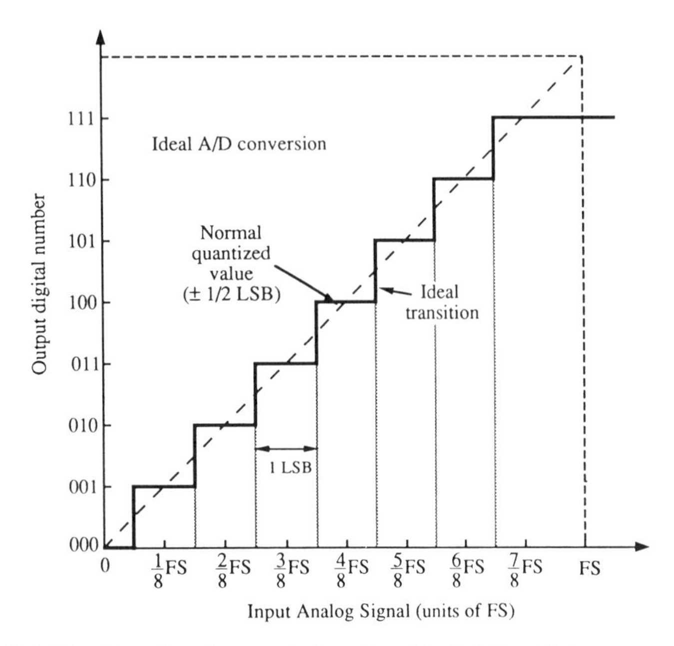

FIGURE 9.9 Transfer characteristics of an ideal 3-bit A/D converter.

speed of operation. Hence, the rate at which the converter can operate is of interest. The conversion time is the time from the application of the signal to start conversion to the availability of the completed output signal. A/D converters often require one or more clock cycles following the application of the analog input signal before the output digital word is available.

Other characteristics such as acquisition time, settling time, and aperture uncertainty can apply to some A/D converters but are more often key parameters in S/H circuits preceding an A/D converter. The acquisition time is the time during which the S/H circuit must remain in the sample mode to ensure that the subsequent hold-mode output will be able to track the input signal. The acquisition time assumes that the gain and offset have been removed. The settling time is the time interval between the sample-to-hold transition command and the time when the output transient and subsequent ringing have settled to within a specified error band. Thus, the minimum sample-and-hold time is equal to the sum of the acquisition and settling time. The maximum A/D converter sampling time is the reciprocal of this value. The dynamic performance of the converter will depend on the dynamic characteristics of the op amps and comparators. Therefore, the slew rate, settling time, and overload recovery time of these circuits are crucial.

An important aspect of the conversion is the aperture uncertainty. It is the time jitter in the sample point and is caused by short-term stability errors in the time base generating the sample command to the A/D

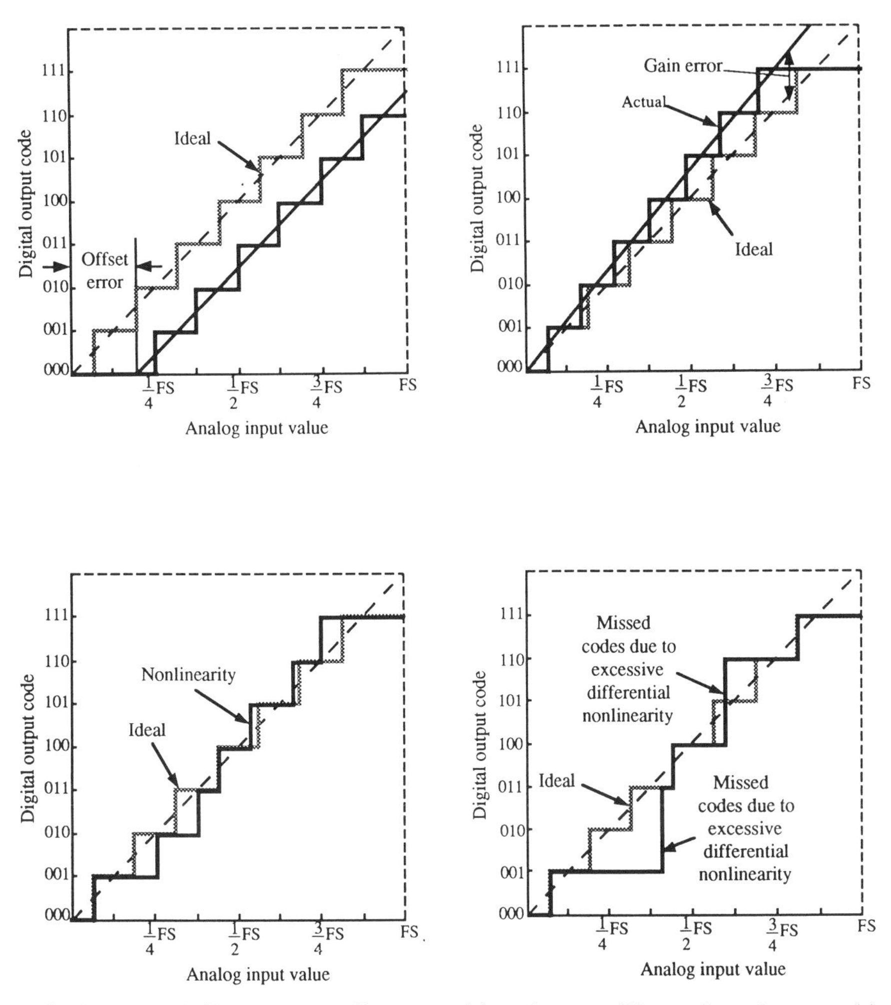

FIGURE 9.10 A/D converter offset error (*a*), gain error (*b*), nonlinearity error (*c*), and missing codes due to differential nonlinearity errors (*d*).

converter. The time jitter causes an amplitude uncertainty, which depends on the rate of the signal at the sample point.

A problem in high-speed A/D converters resulting in erroneous codes is known as *metastability* in the comparators. Metastability can occur when two input signals to a comparator are close together. The comparator requires a finite amount of time to make a decision but is unable to do so correctly with the input signal close to the reference in the allotted time interval. Digital error correction can reduce the effect caused by metastability.

9.3.1 Parallel A/D Converters

The highest-speed A/D conversion can be achieved with a parallel or flash A/D converter architecture [9–23]. As shown in Figure 9.11, the minimum number of components in a parallel A/D converter are a resistor string, which divides the voltage differential between two references into a finite number of values, a quantizer consisting of a number of comparators, and a digital encoding network. The reference voltage for each comparator is 1 LSB higher than the reference voltage for the comparator immediately below it. The divided reference exhibits inherent monotonicity; that is, the reference voltages to the comparators form a nondecreasing function starting from the low end of the resistor string. The conversion speed is determined by one clock cycle, which typically consists of an acquisition phase and a conversion phase. During the acquisition time period, the analog input signal is sampled by the comparator inputs. The outputs of the comparators form a so-called thermometer code. All comparator outputs

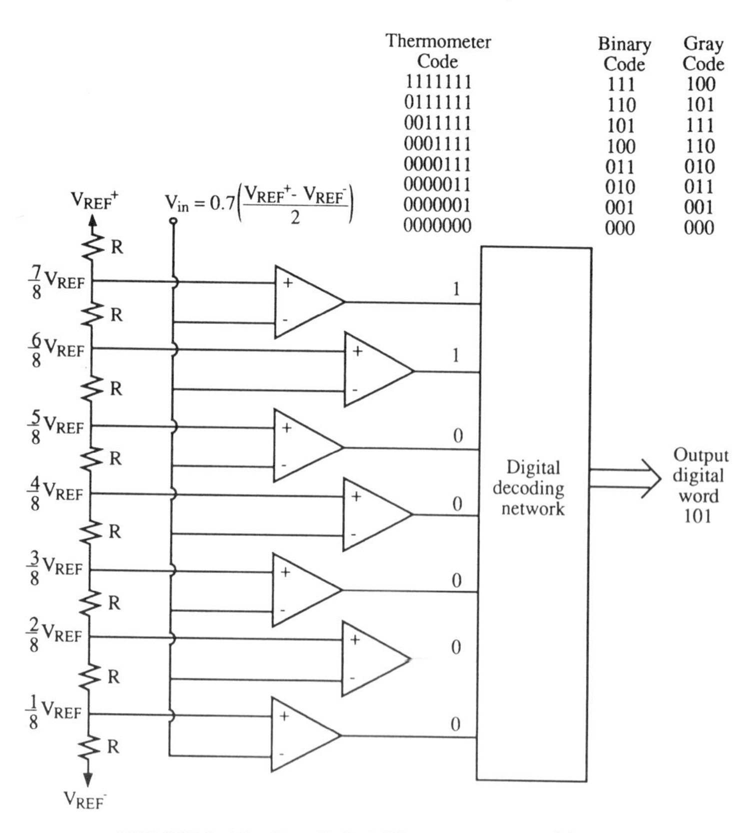

FIGURE 9.11 Parallel A/D converter architecture.

whose reference signal input is below the analog input signal generate a logic zero, while those comparators whose reference signal is higher than the analog input signal generate a logic one. It is clear that a lot of redundant information is available in the thermometer code, since only the two comparators with an adjoining logic 0 and logic 1 contain the relevant information. The digital encoding network converts the thermometer code to a more useful binary code. This is often an offset-binary code, which starts with all 0s at the low side and ends with all 1s at the high side, or a Gray code, which has the property that only 1 bit changes between quantization levels.

Implementations of parallel architectures require large area, power dissipation, and input capacitance. An exponential relationship, proportional to 2^N, exists between these quantities and the resolution N. For higher-resolution A/D converters or higher analog input frequencies, the comparator as sampler limits performance. Adding a S/H circuit in front of the quantizer relaxes the timing requirements on the comparators, but causes the S/H circuit to drive the large input capacitance of the quantizer.

Parallel architectures using GaAs MESFET technologies are limited by the offsets of the comparators and by the hysteresis effect. The offset is a random variable whose nonzero standard deviation causes nonlinearity. The variations in the offsets of all comparators combined with the hysteresis effects must be between ± 0.5 LSB for a maximum DNL <0.5 LSB and maximum INL <0.5 LSB. The use of autobias comparators or dynamic comparators, which employ self-calibrating techniques similar to the ones found in CMOS circuits, extends the possible performance of the quantizer into the 4–6-bit range.

GaAs MESFET flash converters have been reported operating at a rate of ≥ 1 Gigasample/s. The reported standard deviation of the comparator offset almost always exceeds 10 mV because of the poor device matching between MESFETs. This limits the accuracy of this class of converter architectures to 4 bits. For this reason, other A/D converter architectures deserve a closer look.

9.3.2 Feedforward A/D Converters

A class of A/D converter architectures that offer a flexible approach to achieving high performance, while reducing the amount of analog circuits, is generally described as feedforward architecture. Subclasses of this architecture are known as half-flash, two-step, subranging, and pipelined A/D converters [24–28]. The common characteristic of these architectures is that they consist of a succession of quantizer–subtractor stages, each of which determines one or more bits of the conversion and passes a residue analog signal on to the next stage as depicted in Figure 9.12.

To begin a conversion, the input signal is sampled and held. A low-resolution A/D converter converts the signal into a digital code, which

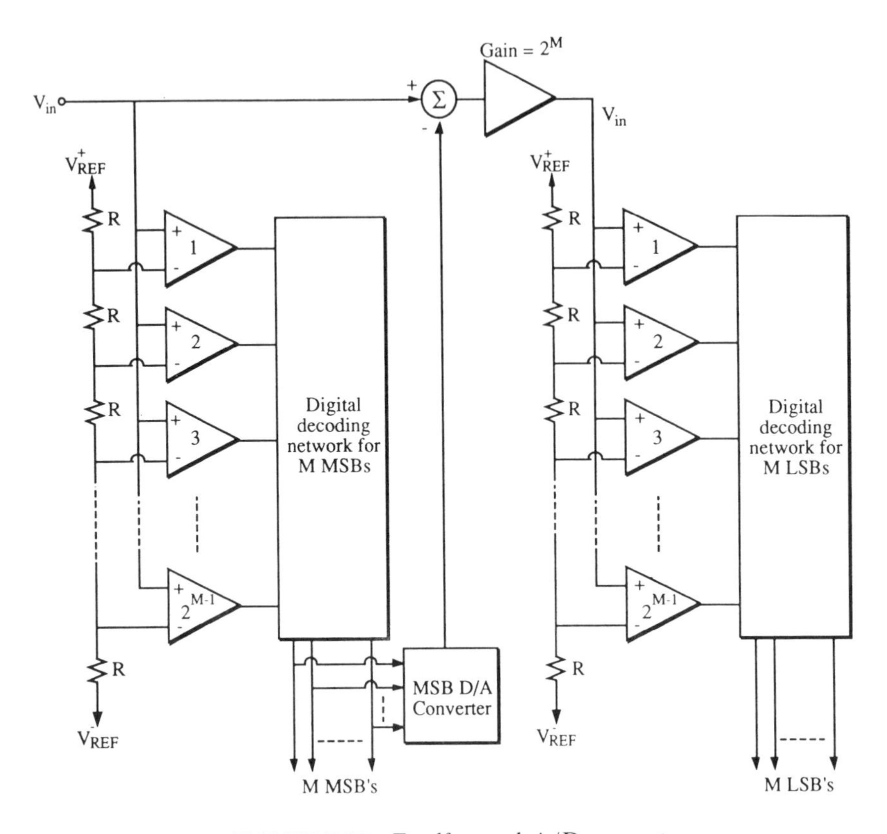

FIGURE 9.12 Feedforward A/D converter.

serves as input to a precise, low-resolution D/A converter. The analog output of the D/A converter is subtracted from the field input, producing a residue that can be digitized by the next stage. Each stage investigates the input signal with a finer granularity. Consequently, this converter architecture reduces the overall number of comparators in comparison to parallel converters while incurring only a small speed penalty. The number of comparators decreases to $m(2^{n_i} - 1)$, whereby m represents the number of stages and n_i the resolution of stage i. For an 8-bit, two-stage, subranging architecture, for example, only 30 comparators would be required in contrast to the 255 found in a flash converter. However, feedforward A/D converters often have to incorporate new analog components such as S/H circuits and gain stages. The subranging, half-flash, or two-step architecture, whose characteristic is its lack of S/H circuits between stages, forms one subclass. An implementation that determines more than one bit per stage is called a *series-parallel feedforward architecture*. This subclass of feedforward

A/D converters employs interstage S/H circuits and amplifiers and is commonly known as *pipelined A/D converters*. Whereas in subranging architectures the input signal has to be quantized by all stages of the converter before the next sample can be taken, the S/H function in the pipelined converter isolates samples from stage to stage, allowing for parallel processing on consecutive samples [28].

The lack of interstage gain in subranging converters causes serious limitations for pipelined converters. It means that any nonlinearity present in the second stage is unattenuated on the linearity of the entire conversion. As a consequence, the requirements on all comparators in the subranging architecture without an amplifier must be the same as in the flash converter. The need for offset cancellation is still present and difficult to achieve in GaAs MESFET technology for high-resolution A/D converters.

Although pipelined architectures with their interstage S/H circuits and amplifiers reduces the requirements on the A/D subconverters, the burden rests now with the S/H circuit, D/A subconverter, and amplifier. An amplifier gain greater than 1 allows for the use of digital error correction. With this, it is possible to correct the digital output code produced by all A/D subconverters in all stages.

9.3.3 Oversampled A/D Converters

One class of A/D converters, which has been implemented in GaAs as research subjects, is the oversampled A/D converter. A big advantage of oversampled architectures is that they offer a means of exchanging resolution in time for resolution in amplitude. The key is that they enhance resolution faster than they reduce bandwidth. Therefore, they do not require precision analog components and permit minimization of analog circuitry in favor of digital circuitry. A basic block diagram is shown in Figure 9.13. Unlike the other two A/D converter architectures described above, this one includes a clocked feedback path to the input, which

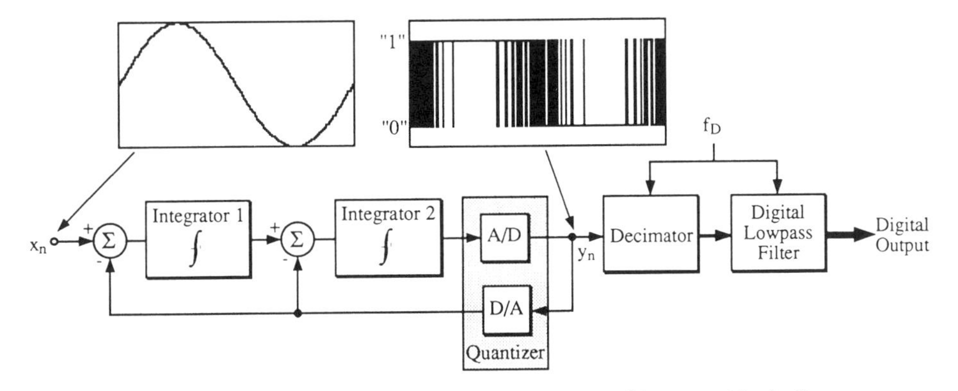

FIGURE 9.13 Basic oversampled A/D converter architecture block diagram.

represents a coarse estimation of the input signal. An integrating stage and a low-resolution A/D converter are present in the forward path. (The most common oversampled architecture determines 1 bit at a time in the forward path and is known as a ΔΣ A/D converter.) The integrator averages the coarse estimate in order to obtain a finer approximation. The output of the feedback loop represents a binary signal and is fed into a digital lowpass filter. The D/A converter in the feedback loop of a ΔΣ A/D converter switches between two reference voltages. The result is that no nonlinearity errors exist, because the D/A converter does not require any binary-weighted scaling elements.

Typical ΔΣ A/D converters employ one or two integrators in the loop filter. The quantization error in the feedback loop exhibits a highpass spectrum that can be filtered by the digital lowpass filter. Typical theoretical S/N ratios (SNRs) for loop filters with one and two integrations can be found related to the oversampling ratio:

$$D_{\mathrm{SNR}}(db) = 9L - 5.2 \tag{9.9a}$$

$$D_{\mathrm{SNR}}(db) = 15L - 13 \tag{9.9b}$$

where L is the number of octaves of oversampling. Common oversampling ratios of 64, 128, or 256 result in oversampled A/D converters with 10–16-bit resolution.

Another benefit of the digital lowpass filter is that it can carry most of the antialias filter responsibility. Only simple analog prefiltering is required with this kind of converter.

Implementation of the analog section in a GaAs MESFET process determines the capabilities of the converter. Techniques such as switched-capacitor integrators used in Si MOS technology are difficult to implement in GaAs. The analog components in the oversampled A/D converter are an op amp in each integrating stage, a low-resolution A/D converter, and a low-resolution D/A converter. The previous section discussed some choices for GaAs MESFET D/A converters. The same architectures can be considered for use in the oversampled A/D converters. Since only a 1-bit D/A converter is required in the ΔΣ A/D converter architecture, the design becomes straightforward.

The low-resolution A/D converter simplifies into a single comparator in the ΔΣ A/D converter architecture. The comparator follows an integrating stage that allows for relaxation of comparator's offset and noise requirements. The preceding integrator will adjust its output to accommodate the comparator's input threshold level. Since comparators are such a critical element in all A/D converters, they are discussed in detail in the following section.

The most critical element in the oversampled A/D converter is the op amp in the integrating stage. The op amp settling time not only affects the maximum input clock rate but also limits the achievable SNR in oversampled converters (since it is determined by the clock rate). Fortunately, $\Delta\Sigma$ loops are very tolerant of linear settling error; that is, the overall SNR of the converter suffers a negligible amount if the output of the op amp settles only to 10% of its final value.

The gain of the op amp need not be very high. It can be shown that the op amp open-loop gain can be as low as the oversampling ratio before quantization begins to have an effect on the SNR of the A/D converter. The voltage range of the input signal and the output swing of the op amp require careful evaluation. The input signal and the value of the D/A converter, which are added to the first stage integrator, cannot cause the integrator output to saturate.

The digital filter following the feedback loop is normally implemented as a finite-impulse-response (FIR) filter and requires only digital logic (adders, shift registers, storage for coefficients).

The design of $\Delta\Sigma$ A/D converters requires a large amount of system simulation. In this kind of converter the performance of the analog components can be traded for an increase in digital circuitry (a modified FIR filter) or a more complex $\Delta\Sigma$ loop. The GaAs MESFET process is very suitable for this architecture, since large amounts of digital circuitry can be easily integrated with a small section of analog components.

9.4 COMPARATOR CIRCUIT

The latching comparator is the element in a quantizer that samples an analog input voltage, compares it against a known reference, and converts the difference into a digital signal. The performance criteria are aimed at obtaining maximum gain and speed and minimum offset. A tradeoff exists between the gain parameter and the required speed. As the gain of a differential stage increases, so does the Miller effect. This is recognized by the driving circuitry as a large capacitive node causing a detrimental effect on the switching speed. The low gain needs to be sufficient to generate a logic 1 or 0 at the output within the allotted time interval.

The design of a comparator is based on a well-known differential input pair with "reasonable"-size MESFETs in order to minimize offset, followed by a regenerative latch. A simplified schematic diagram is shown in Figure 9.14. When the digital clock signal is in one state, the comparator is in the sampling mode, whereby the front-end amplifier is turned on and the latch is rendered inoperative. As the clock signal changes to the opposite state, the amplifier becomes disabled and the latch amplifies the differential voltage present at the two drains of the input differential pair. This continues until the clock signal changes state again.

The choice of input devices in the differential pair is important. A comparison between the enhancement- and depletion-mode MESFETs

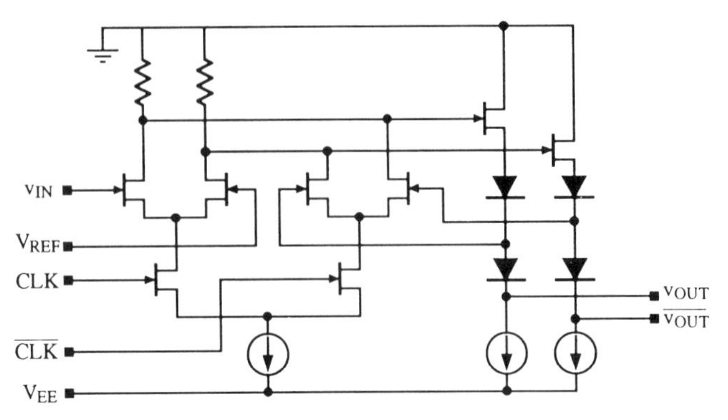

FIGURE 9.14 Schematic diagram of a comparator.

reveals the enhancement-mode MESFET to have lower capacitances. The device also operates in the saturated region ($V_{GS} - V_{TO} < V_{DS}$) even when V_{DG} is as low as zero volts. In the balanced differential pair each input device conducts half of the current required by current sink I_0. The differential pair's transconductance g_m can be calculated once the current has been determined. Knowledge of the required gain and transconductance is sufficient to calculate the corresponding load resistance:

$$R_L = \frac{A_v}{g_m} \tag{9.10}$$

The load in the comparator can be realized with active MESFETs or with resistors. Active MESFET loads would increase the gain of the differential stage at the expense of bandwidth. The high gain would make it almost impossible to balance the differential amplifier. In practice, one side of the stage would be pulled up and the other side pulled down, even with balanced inputs. To circumvent this problem, the comparator would require active feedback to bias it as a linear amplifier. Even with feedback, the variable drain conductance would cause a change in variation in the load devices. The result is an additive offset that can be translated back to the input.

The best choice is to use resistors for the load devices. Although N^+ resistors are suitable, they lack the matching capabilities of NiCr resistors and are prone to backgating. NiCr resistors have no interaction with the substrate and have good matching. Unfortunately, they require a large area because of their low sheet resistance.

In order to reduce the input offset voltage in a comparator, it is possible to employ a self-calibrating mechanism using techniques similar to those found in CMOS technology. A dual switch at the input samples the analog input signal and the reference signal and applies it on alternate clock phases to the input stage [29]. It is a good approach to obtain comparators with

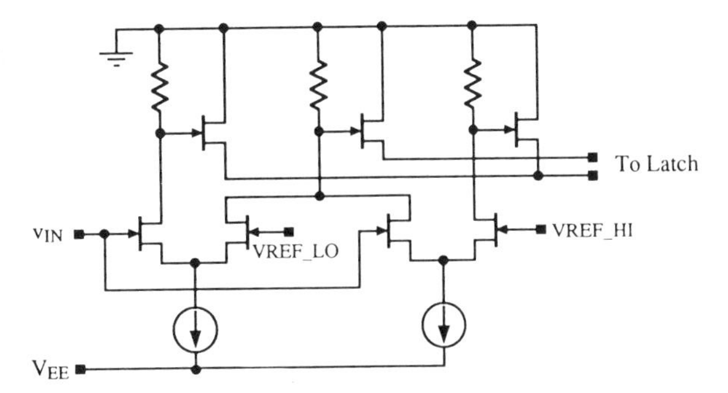

FIGURE 9.15 Comparator employing analog folding.

lower offsets for 4–6-bit parallel converters. The disadvantage is the slower operation speed and an increased complexity.

A method to decrease the number of comparators in parallel converters relies on a concept generally known as *analog folding*. In a way, it moves part of the digital encoding to the analog domain in front of the comparator. In a regular comparator, which has one reference input voltage, the output changes once when the input signal traverses from the negative to the positive reference. The folding comparator is an extension of the two-input, one-reference comparator. It has two or more reference inputs. Each time the analog input voltage passes a reference voltage level, the output of the folding comparator changes. The only complication resulting from the analog folding requirements can be found in the differential input stage. Instead of one differential pair of MESFETs, the folding comparator requires an input from two or more reference voltages and the analog input voltage [28]. A simplified schematic diagram of a folding stage is shown in Figure 9.15.

9.5 S/H CIRCUIT

A S/H circuit can be used to improve the dynamic performance of high-speed A/D converters. The operation of the S/H circuit requires that it samples the analog input signal during the time interval that the comparators in the A/D converter are not sampling their inputs. During the hold phase of the S/H circuit, the comparators start sampling their inputs.

S/H circuits form an integral part of many A/D converter architectures and are the key to the upper limit in performance. Although many different S/H circuit architectures have been developed in silicon and GaAs [30–32] ranging from accurate, closed-loop circuits to high-speed open-loop configurations, only open-loop implementations are of interest for GaAs MESFET technology because of the conversion speed of the A/D converters under consideration. Therefore, the S/H circuit generally consists of four elements: an input buffer, a sampling bridge, a hold capacitor, and an output buffer.

The input buffer separates the excitation signal from the sampling bridge. It must have a high input impedance and a low output impedance with a sufficient current drive capability. A low input capacitance is important for linearity reasons.

The requirements for the output buffer are often very similar to the ones for the input buffer. The output buffer transfers the voltage held by the hold capacitor to the input of the A/D converter. A high input impedance, inherent in the MESFET, prevents discharging of the hold capacitor through the follower. The current through the follower is determined by the number of comparators it needs to drive. The slew rate and settling time need to be sufficient in order to achieve the proper output voltage levels. Since the input of the A/D converter will be capacitive, slewing will be primarily a matter of providing sufficient current as given by

$$\frac{I_{\mathrm{OUT}}}{C_{\mathrm{LOAD}}} = \frac{\partial V_{\mathrm{OUT}}}{\partial t} \tag{9.11}$$

During the sample mode the buffer output must be able to charge and discharge this equivalent load with the same speed as the input signal. During the hold mode the buffer must settle rapidly with minimal offset voltage. The effect of offset and offset drift will be difficult to reduce in an open-loop system.

The hold capacitor can be realized in a GaAs MESFET process by using two or more metal layers with an insulator in between. Calculation of the size of the capacitor involves tradeoffs between several bridge parameters. A large hold capacitor can prevent any leakage from influencing the voltage on the capacitor by more than 0.5 LSB. In contrast, a small hold capacitor is required in order to keep the bridge time constants small.

Two different architectures can be considered for the switch preceding the hold capacitor: the diode bridge and a MESFET switch. In general, the diode switch will allow larger input signals and less high-frequency distortion at the expense of greater feedthrough and less accuracy. The MESFET switch exhibits some signal dependent turnoff characteristics, caused by the manner in which the switch is operated; one side is connected to the hold capacitor, while the other side follows the input signal. Another problem arises when the gate, or control terminal, becomes forward-biased with respect to the source. For these reasons, the diode bridge is normally the better choice.

REFERENCES

1. R. Bayruns and N. Scheinberg, "Design Issues Facing Analog Circuit Design Using GaAs MESFET Technology," *IEEE International Symposium on Circuits and Systems*, May 1987, pp. 188–192.
2. P. E. Allen and C. M. Breevoort, "An Analog Circuit Perspective of GaAs

Technology," *IEEE International Symposium on Circuits and Systems*, May 1987, pp. 184–187.
3. T. T. Vu and J. N. Vu, "GaAs Analog Integrated Circuits and Their Potential System Integration," *IEEE International Symposium on Circuits and Systems*, May 1990, pp. 3061–3064.
4. K. de Graaf and K. Fawcett, "GaAs Technology for Analog to Digital Conversion," *GaAs IC Symposium Technical Digest*, October 1986, pp. 205–208.
5. F. G. Weiss, "A 1 Gs/s 8-bit GaAs DAC with On-chip Current Sources," *GaAs IC Symposium Technical Digest*, 1986, pp. 2217–2220.
6. K. Hsieh, T. A. Knotts, G. L. Baldwin, and T. Hornak, "A 12-bit 1-Gword/s GaAs Digital-to-Analog Converter System," *IEEE J. Solid-State Circuits*, **SC-22** (6), 1048–1055 (Dec. 1987).
7. J. F. Naber et al., "A Low-Power, High-Speed, 10-bit GaAs DAC," *GaAs IC Symposium Technical Digest*, 1990, pp. 33–36.
8. F. G. Weiss and T. G. Bowman, "A 14-bit, 1 Gs/s DAC for Direct Digital Synthesis Applications," *GaAs IC Symposium Technical Digest*, 1991, pp. 361–364.
9. Y. Yoshii et al., "An 8b 100 Ms/s Flash ADC," *ISSDC Digest Technical Papers*, San Francisco, CA, February 1984, pp. 58–59.
10. B. Peetz, B. Hamilton, and J. Kang, "An 8-bit 250 Megasample per Second Analog-to-Digital Converter: Operation without a Sample and Hold," *IEEE J. Solid-State Circuits*, **SC-21** (6), 997–1002 (Dec. 1986).
11. B. Zojer, R. Petschacher, and W. Luschnig, "A 6-bit/200-MHz Full Nyquist A/D Converter," *IEEE J. Solid-State Circuits*, **SC-20** (3), 780–786 (June 1985).
12. Y. Yoshii et al., "An 8b 350 MHz Flash ACD," *ISSCC Digest Technical Papers*, New York, February 1987, pp. 96–97.
13. Y. Akazawa et al., "A 400-Msps 8b Flash A/D Conversion LSI," *ISSCC Digest Technical Papers*, New York, February 1987, pp. 98–99.
14. J. Corcoran et al., "A 400 MHz 6b ADC," *ISCC Digest Technical Papers*, San Francisco, CA, February 1984, pp. 294–295.
15. T. Wakimoto, Y. Akazawa, and S. Konaka, "Si Bipolar 2-GHz 6-bit Flash A/D Conversion LSI," *IEEE J. Solid-State Circuits*, **SC-23**, (6), (Dec. 1988).
16. C. W. Mangelsdorf, "A 400-MHz Input Flash Converter with Error Correction," *IEEE J. Solid-State Circuits*, **SC-25** (1), 184–191 (Feb. 1990).
17. R. J. van de Plassche and P. Baltus, "An 8-bit 100 MHz Full Nyquist Analog-to-Digital Converter," *IEEE J. Solid-State Circuits*, **SC-23** (6), 1334–1344 (Dec. 1988).
18. D. Daniel, U. Langmann, and B. G. Bosch, "Silicon 4-bit 1-Gsample/s Full Nyquist A/D Converter," *IEEE J. Solid-State Circuits*, **SC-23** (3), 742–749 (June 1988).
19. T. Ducourant et al., "3 GHz, 150 mW, 4-bit GaAs Analogue to Digital Converter," *GaAs IC Symposium Technical Digest*, 1986, pp. 209–212.
20. J. Kleks et al., "A 4-bit Single Chip Analog to Digital Converter with a 1.0 Gigahertz Analog Input Bandwidth," *GaAs IC Symposium Technical Digest*, 1987, pp. 79–82.
21. T. Nguyen et al., "A 4 bit Full Nyquist 1 Gsample/s Monolithic GaAs ADC with

On-chip S/H Circuit and Error Correction," *GaAs Symposium Technical Digest*, 1987, pp. 83–86.

22. R. Hagelauer et al., "A Gigasample/Second 5-b ADC with On-Chip Track and Hold Based on an Industrial 1-μm GaAs MESFET E/D Process," *IEEE J. Solid-State Circuits*, **SC-27** (10), 1313–1320 (Oct. 1992).
23. J. Corcoran, K. Poulton, and T. Hornak, "A 1 GHz 6b ADC System," *ISSCC Digest Technical Papers*, February 1987, pp. 102–103.
24. U. Fiedler and D. Seitzer, "A High-speed 8 bit A/D Converter Based on a Gray-code Multiple Folding Circuit," *IEEE J. Solid-State Circuits*, **SC-23** (3), 547–551 (June 1979).
25. R. Petschacher et al., "A 10-b 75 MSPS subranging A/D Converter with Integrated Sample-and-Hold," *IEEE J. Solid-State Circuits*, **SC-25** (6), 1339–1346 (Dec. 1990).
26. R. E. J. van de Grift, I. W. J. M. Rutten, and M. van der Veen, "An 8-bit Video ADC Incorporating Folding and Interpolation Techniques," *IEEE J. Solid-State Circuits*, **SC-22** (6), 944–953 (Dec. 1987).
27. S. H. Lewis and P. R. Gray, "A Pipelined 5 MHz 9b ADC," *ISSCC Digest Technical Papers*, February 1987, pp. 210–211.
28. C. M. Breevoort, "A 9-bit, Pipelined GaAs Analog-Digital Converter," Ph.D. thesis, Georgia Institute of Technology, Atlanta, March 1992.
29. K. Fawcett et al., "High-Speed, High-Accuracy, Self-Calibrating GaAs MESFET Voltage Comparator for A/D Converters," *GaAs IC Symposium Technical Digest*, 1986, pp. 213–216.
30. R. Bayruns, N. Scheinberg, and R. Goyal, "An 8 ns Monolithic GaAs Sample and Hold Amplifier," *ISSCC Digest Technical Papers*, February 1985, pp. 42–43.
31. B. Wong and K. Fawcett, "A Precision Dual Bridge GaAs Sample and Hold," *GaAs IC Symposium Technical Digest*, 1987, pp. 87–89.
32. J. F. Naber, H. P. Singh, and R. A. Sadler, "A 2.5 GHz GaAs Sample-and-hold," *IEEE International Symposium on Circuits and Systems*, May 1990, pp. 3077–3080.

CHAPTER TEN

Synthesis of Linearized Conductance Functions

DAVID HAIGH
Department of Electronic and Electrical Engineering
University College
London, UK

CHRIS TOUMAZOU
Department of Electrical and Electronic Engineering
Imperial College of Science Technology and Medicine
London, UK

10.1 INTRODUCTION

Gallium arsenide (GaAs) integrated-circuit (IC) technology became available as a foundry process about 1980. Since GaAs technology provides FET devices, which can be used to realize analog switches, it seemed natural to try to exploit GaAs technology in order to realize analog sampled data systems using switched-capacitor circuit techniques [1]. This led in 1987 to an impressive third-order lowpass filter with a switching rate of 100 MHz, at least 3 times the maximum then achieved with CMOS technology [2]. Meanwhile, the realization by the present authors of very high-gain wideband operational amplifiers using available GaAs devices [3] led in 1989 to a high-precision second-order bandpass filter with a switching rate of

High-Frequency Analog Integrated-Circuit Design Edited by Ravender Goyal
ISBN 0-471-53043-3

500 MHz [4,5]. A high amplifier gain of 60 dB was achieved by developing a push–pull amplifier architecture using an inverting current mirror consisting of a pair of MESFETs with gates and sources cross-coupled [6]. Analogue sampled data systems implemented in GaAs and employing fast-settling operational amplifiers [7] can now be contemplated with switching rates in the gigahertz range. However, signal frequencies are limited to a small fraction of this because of the need to avoid aliasing distortion and to attenuate output image bands [8].

The successful development of MMIC (monolithic microwave IC) technology [9] has given birth to a new subclass of circuits within which the material of the present chapter falls. The new subclass may be defined by a number of characteristics. Whereas in mainstream MMIC design the number of FETs is minimized and the components are placed on the chip in a sparse fashion, following traditional hybrid technology practices, this subclass, which may be loosely termed "all-FET circuits," uses FETs in larger numbers and with relatively high packing density. It may be viewed as an application into the microwave frequency band of lower-frequency IC design techniques developed for CMOS and bipolar technology [10]. In view of their size and incompatibility with high packing density, spiral inductors are avoided and large structures using transmission lines are also avoided except to couple together "islands" of high-density active circuitry. Resistors and capacitors are used primarily for biasing of FETs. In order to satisfy the conditions for a lumped-design approach, FETs with relatively small gate widths and compact passive components, such as silicon nitride capacitors and mesa resistors, are adopted. It goes without saying that this new class of circuits can not compete with mainstream IC circuits from several points of view, including ultimate high-frequency performance and low noise, but they are characterized by the advantages of wide bandwidth and small chip area providing low cost. For a GaAs process with a typical FET f_T of 20 GHz, circuit operation up to, say, 5 GHz would be expected, and this can encompass many applications in microwave systems. Components that have been realized in this way so far include amplifiers, isolators, combiners, splitters, mixers, gyrators, active inductors, and filters [11–15]. Obtaining good performance close to f_T is a challenging task that is leading to whole new approaches of circuit design in which the parasitic capacitances of the FETs that define f_T are effectively canceled over a finite frequency band [15]. Another exciting development is the achievement of exceptionally high efficiency in all-FET circuits [12]. In spite of these impressive achievements, the fundamental ground rules that define the limitations and optimum circuit solutions for this new class of circuit, we believe, are only beginning to be discovered and explored.

This chapter is concerned with the development of novel circuits using GaAs MESFET technology. The GaAs MESFET large-signal behaviour is intrinsically nonlinear, and we begin by presenting a systematic method for the synthesis of linearized circuit functions using the MESFET. Implementation of the linearized function is shown to require a building block consisting of the cross-coupled MESFET pair, which in fact formed the basis of the high-gain operational amplifier used by the present authors previously to successfully realize switched-capacitor filters, as mentioned above. Apart from linearity, the new circuit structures are shown to have the even more important property of high-efficiency and large-signal swings, and their performance parameters can be easily tuned. We believe these circuit ideas will play a role in the new generation of ICs now emerging.

10.2 SYNTHESIS

10.2.1 Basic Principle

The synthesis [16,17] is based on a very simple characterization of the GaAs depletion-mode MESFET but nonideal MESFET characteristics will be included in simulations as the chapter develops. The synthesis method assumes a GaAs MESFET operated in the saturation region, which can be approximately described by

$$I_{\mathrm{d}} = \beta(V_{\mathrm{gs}} - V_{\mathrm{T}})^2 \tag{10.1}$$

where I_{d} and V_{gs} are drain current and gate–source voltage, respectively, β is conductance factor, and V_{T} is the pinchoff voltage. The system output is a current formed as a sum of MESFET drain currents

$$I = \sum_{i=1}^{m} I_{d_i}^{+} - \sum_{j=1}^{n} I_{d_i}^{-} \tag{10.2}$$

where the number of +1-weighted MESFET drain currents is m and the number of −1-weighted MESFET drain currents is n. We let MESFET gate–source voltage in Equation (10.1) take the form of a sum of weighted system input voltages and combine Equations (10.1) and (10.2) to obtain

$$I = \sum_{i=1}^{m} \beta\left(\sum_{k=1}^{p} a_{ik}V_k - V_{\mathrm{T}}\right)^2 - \sum_{j=1}^{n} \beta\left(\sum_{k=1}^{p} b_{jk}V_k - V_{\mathrm{T}}\right)^2 \tag{10.3}$$

where weighting factors a_{ik} and b_{jk} are restricted to −1, 0, or +1. We define dc offset as the output current when all input voltages are zero, namely

$$I|_{V_k=0,\ k=1,2,\ldots,p} = \beta V_{\mathrm{T}}^2(m-n) \tag{10.4}$$

The undesirable dc offset can be eliminated simply by letting

$$n = m \tag{10.5}$$

In this case, Equation (10.3) becomes

$$I = \beta \sum_{i=1}^{m} \left[\left(\sum_{k=1}^{p} a_{ik} V_k \right)^2 - 2V_T \sum_{k=1}^{p} a_{ik} V_k - \left(\sum_{k=1}^{p} b_{ik} V_k \right)^2 + 2V_T \sum_{k=1}^{p} b_{ik} V_k \right] \tag{10.6}$$

which may be written

$$I = \beta \Bigg\{ \sum_{k=1}^{p} \left[\sum_{i=1}^{m} (a_{ik}^2 - b_{ik}^2) \right] V_k^2 + 2V_T \sum_{i=1}^{m} \sum_{k=1}^{p} (b_{ik} - a_{ik}) V_k + \sum_{i=1}^{m} \sum_{\substack{k=1 \\ k \neq l}}^{p} \sum_{l=1}^{p} (a_{ik} a_{il} - b_{ik} b_{il}) V_k V_l \Bigg\} \tag{10.7}$$

For the synthesis of transconductor–multiplier functions, the terms proportional to V_k^2 are undesirable and can be eliminated under the conditions

$$\sum_{i=1}^{m} a_{ik}^2 = \sum_{i=1}^{m} b_{ik}^2 \quad \text{for all } k \tag{10.8}$$

Under these conditions, Equation (10.6) has the form

$$I = \beta \left[2V_T \sum_{i=1}^{m} \sum_{k=1}^{p} (b_{ik} - a_{ik}) V_k + \sum_{i=1}^{m} \sum_{\substack{k=1 \\ k \neq l}}^{p} \sum_{l=1}^{p} (a_{ik} a_{il} - b_{ik} b_{il}) V_k V_l \right] \tag{10.9}$$

We regard Equation (10.9) as a definition of the transfer function for a general class of useful circuits referred to as transconductor–multiplier circuits, based on the square-law characteristic [Eq. (10.1)] of the FET.

Two special cases are of interest. If we set m [the number of positive (and negative) weighted currents in Eq. (10.2)] to unity, Equation (10.9) becomes

$$I = \beta \left[2V_T \sum_{k=1}^{p} (b_k - a_k) V_k + \sum_{\substack{k=1 \\ k \neq l}}^{p} \sum_{l=1}^{p} (a_k a_l - b_k b_l) V_k V_l \right] \tag{10.10}$$

The first term, proportional to V_k, defines a current that is linearly related to the input voltages. If current I occurs at an input port (where an input voltage is applied), this corresponds to the realization of a self-conductance; if current I occurs at a separate port (the output port), this corresponds to the realization of a transconductance. Terms in Equation (10.10) proportional to $V_k V_l$ allow the self-conductance or transconductance to be linearly tuned by means of an additional input voltage. The class of functions derivable from (10.10) will be referred to as conductance functions.

For an alternative class of functions, the terms in Equation (10.9) proportional to V_T are eliminated under the condition

$$\sum_{i=1}^{m} a_{ik} = \sum_{i=1}^{m} b_{ik} \quad \text{for all } k \tag{10.11}$$

The resulting functions consist entirely of sums of products of input voltages and are referred to as *multiplier functions*. It has been shown [16] that m [the number of positive (and negative) weighted currents in Eq. (10.2)] may be chosen to be 2 without loss of significant generality, yielding

$$I = \beta\left[\sum_{i=1}^{2} \sum_{\substack{k=1 \\ k \neq l}}^{p} \sum_{l=1}^{p} (a_{ik}a_{il} - b_{ik}b_{il})V_k V_l\right] \tag{10.12}$$

We now consider the choice of parameters in Equations (10.10) and (10.12) to realize some useful circuit functions.

10.2.2 Conductance Functions

We consider first the simplest case of p [the number of input voltages in Eq. (10.3)] being unity, in which case Equation (10.10) reduces to

$$I = 2\beta V_T(b_1 - a_1)V_1 \tag{10.13}$$

The only two choices of a_1 and b_1 that satisfy Equation (10.8), avoiding solutions of the form $a_k = b_k$ that lead to zero I, are given in Table 10.1. Bearing in mind that pinchoff voltage V_T is negative for depletion-mode devices, the functions obtained are the inverting and noninverting linear conductance function.

For the case of two input voltages ($p = 2$), Equation (10.10) becomes

$$I = 2\beta\{V_T[(b_1 - a_1)V_1 + (b_2 - a_2)V_2] + (a_1 a_2 - b_1 b_2)V_1 V_2\} \tag{10.14}$$

Two sets of coefficients satisfying Equation (10.8) and leading to useful functions are given in Table 10.2. The first is an inverting linear conductance in respect of the input voltage V_1 but where the conductance constant can be linearly tuned by means of input voltage V_2. The second is the differential-input conductance.

Finally, for the case of three input voltages ($p = 3$) and the set of coefficients in Table 10.3 that satisfy Equation (10.8), (10.10) provides a

TABLE 10.1 Basic Conductance Functions

	a_1	b_1	I
Inverting conductance	-1	1	$4\beta V_T V_1$
Noninverting conductance	1	-1	$-4\beta V_T V_1$

TABLE 10.2 Tunable and Differential-Input Conductance Functions

	a_1	a_2	b_1	b_2	I
Tunable inverting conductance	−1	1	1	1	$4\beta V_1(V_T - V_2)$
Differential-input conductance	−1	1	1	−1	$4\beta V_T(V_1 - V_2)$

TABLE 10.3 Tunable Differential-Input Conductance Function

	a_1	a_2	a_3	b_1	b_2	b_3	I
Tunable differential-input conductance	−1	1	1	1	−1	1	$4\beta(V_T - V_3)(V_1 - V_2)$

tunable differential-input conductance function that is linearly tunable by input voltage V_3.

10.2.3 Multiplier Functions

The generalized multiplier function in Equation (10.12) has been used to define various classes of particular multiplier functions in terms of the number of product terms in I (single-term, double-term, and quadruple-term) and the types of input (single-ended, differential-input, double differential-input) [16]. The set of coefficients in Table 10.4 satisfy Equations (10.8), as required, and provide a single-term, single-ended-input four-quadrant multiplier function.

10.3 REALIZATION ARCHITECTURE

The synthesis in Section 10.2 has been based on the formation of the output current as a sum of positive and negative weighted MESFET drain currents as in Equation (10.2) where, by virtue of (10.5), the number of positively weighted currents (m) equals the number of negatively weighted currents (n). The general circuit architecture for implementation of (2) is shown in Figure 10.1 [16]. The positively weighted currents are realized by MESFETs whose sources are connected to the node where current I is generated (denoted *source MESFETs*). The negatively weighted currents are realized by MESFETs whose drains are connected to the node where current I is

TABLE 10.4 Multiplier Function

	a_{ik} ($i = 1,2$; $k = 1,2$)		b_{ik} ($i = 1,2$; $k = 1,2$)		
i	$k = 1$	$k = 2$	$k = 1$	$k = 2$	I
1	1	0	1	−1	$2\beta V_1 V_2$
2	0	−1	0	0	

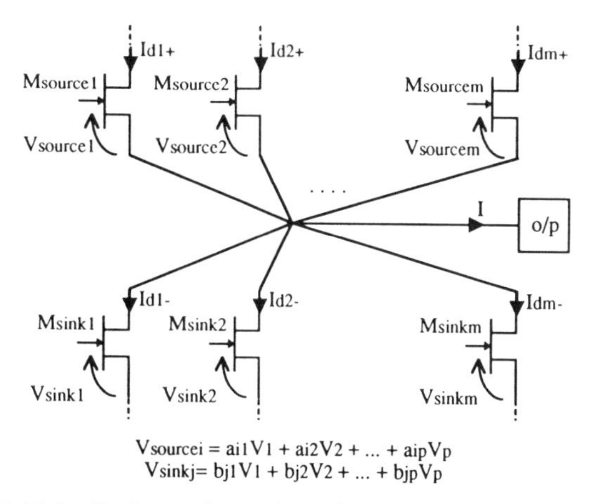

FIGURE 10.1 Basic configuration of source and sink MESFETs.

generated (*sink MESFETs*). Since the synthesis function in Equation (10.3) assumed that all MESFETs have the same conductance factor β, it is necessary that all source and sink MESFETs have the same gate width. The gate–source voltages of the source and sink MESFET consist of sums of $+1$, 0, and -1-weighted input voltages, $V_1 \cdots V_k$, according to Equation (10.3)

The general architecture of the complete system including the input voltage ports is shown in Figure 10.2 [17]. The voltage processing block has the function of realizing the required gate–source voltages for the source and sink MESFETs in terms of the input voltages.

Several modes of implementation can be defined [17]. For realization of a

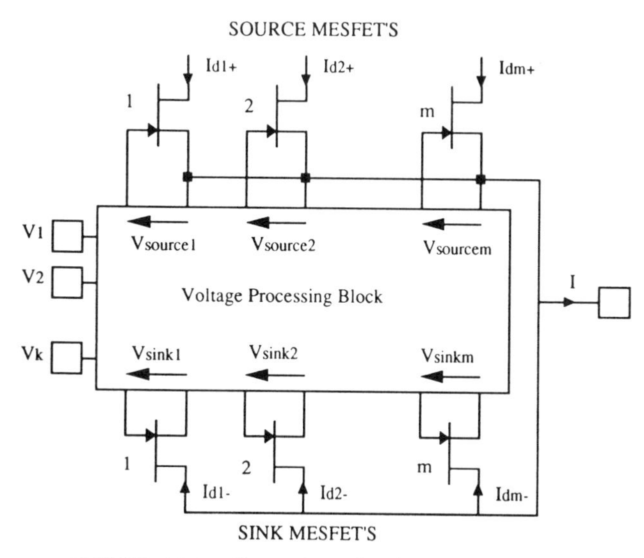

FIGURE 10.2 General realization architecture.

self-conductance, the node where current I is generated is coincident with one of the input voltages. For realization of a transconductance, the node where current I is generated is a separate node. This concept of modes of implementation will be further considered and extended in Section 10.9.

For Equation (10.2) to be realized accurately, it is essential that the current entering the voltage processing block at the interfaces with the sources of the source MESFETs be zero. This requirement can be met by imposing the constraint that these modes may be connected only to the gates of MESFETs within the voltage processing block. This requires the use of a subblock referred to as the *cross-coupled MESFET pair*, which will be discussed in the next section. Since the number of source and sink MESFETs in Figure 10.2 are the same and their gate widths are the same, the dc bias current for each source–sink MESFET pair may be common.

10.3.1 The Generalized Cross-Coupled MESFET-pair

The provision of the gate–source voltages of the source MESFETs in Figure 10.2 by the voltage processing block without taking any current from the MESFET source terminals might appear to be impossible, but can in fact be achieved by means of a special subcircuit referred to as the *cross-coupled MESFET pair*. Figure 10.3 shows a pair in which the MESFET gate–source ports are simply cross-coupled [6]. It is shown in its most generalized four-terminal form. As well as providing the required coupling between the voltage processing block and the source MESFETs, the cross-coupled MESFET pair plays an important role within the voltage processing block itself. In view of its vital role, it is appropriate to review its properties and performance.

Assuming that both MESFETs are biased in the saturation region with drain currents describable by Equation (10.1) and that the gate widths of the two MESFETs are the same (this restriction will be relaxed at a later stage), we have

$$
\begin{aligned}
I_1 &= I_3 = I_{d1} \\
I_2 &= I_4 = I_{d2}
\end{aligned}
\tag{10.15}
$$

where

$$
I_{d2} = I_{d1} + 4(I_{dss} - \sqrt{I_{d1} I_{dss}}) \tag{10.16}
$$

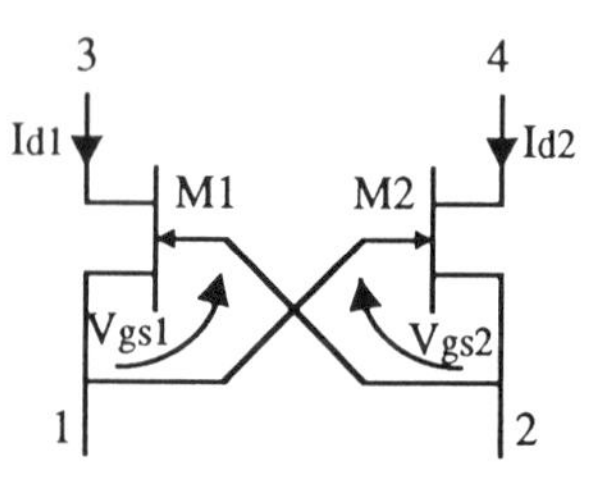

FIGURE 10.3 Generalized cross-coupled MESFET pair.

with I_{dss} given by βV_T^2 [6]. A GaAs chip implementing some key building blocks was fabricated in a 0.7-μm −2-V process, and the layout plot is shown in Figure 10.4 [18]. The measured static response of the cross-coupled MESFET pair of Figure 10.4*a* is shown in Figure 10.5 compared with the predicted response from Equation (10.16), showing reasonable agreement over a wide current range [18].

The small-signal operation of the cross-coupled MESFET pair may be explored using an equivalent circuit for the MESFET consisting of a voltage-controlled current source with input and output capacitance and output resistance. Analysis of the circuit in Figure 10.3 including initially only the controlled sources of the MESFET models, having transconductances g_{m1} and g_{m2}, leads to the transmission matrix description

$$\begin{bmatrix} V_1 \\ I_1 \end{bmatrix} = \begin{bmatrix} 1 & \frac{-1}{g_{m2}} \\ 0 & \frac{g_{m1}}{g_{m2}} \end{bmatrix} \begin{bmatrix} V_2 \\ -I_2 \end{bmatrix} \tag{10.17}$$

where the suffices on V and I refer to ports defined by terminals 1 and 2 in Figure 10.3 and the orientation of I_1 and I_2 is into the ports [6]. If port 2 is connected to a load impedance Z_{L2}, the input impedance at port 1 is given by

$$Z_{i1} = \frac{g_{m2}}{g_{m1}} Z_{L2} - \frac{1}{g_{m1}} \tag{10.18}$$

Similarly, for a load impedance Z_{L1} at port 1, the input impedance at port 2 is given by

$$Z_{i2} = \frac{g_{m1}}{g_{m2}} Z_{L1} - \frac{1}{g_{m1}} \tag{10.19}$$

Equations (10.18) and (10.19) show that the input impedance at each port contains a component proportional to the load impedance at the other port. This impedance transforming property of the cross-coupled MESFET pair is of crucial significance for the design of high-speed circuits. Although the input and output nodes of the cross-coupled MESFET pair are physically separate, the fact that parasitic capacitance at one node may be equivalently transferred to the other node means that they can be regarded in some respects as a single node. The excellent high-frequency performance is reflected in the small-signal short-circuit current transfer function

$$\frac{I_2(s)}{I_1(s)} = \frac{1 + s(2C_{gs}/g_m)}{1 + (s/g_m)(2C_{gs} + C_{gd})} \tag{10.20}$$

where C_{gs} and C_{gd} are MESFET gate–source and gate–drain capacitances

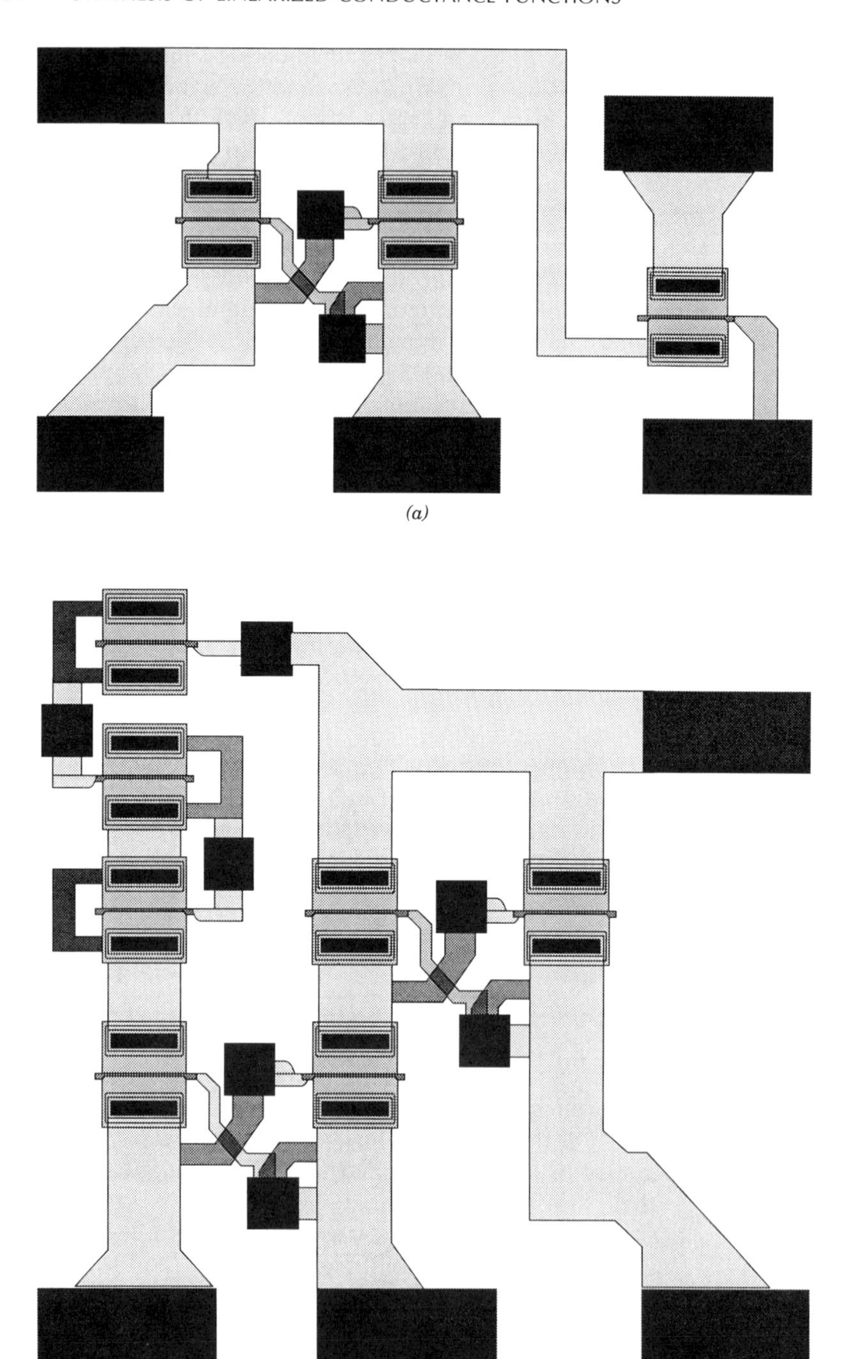

FIGURE 10.4 Layout of building-blocks on test chip: (*a*) cross-coupled MESFET pair and uncommitted MESFET; (*b*) linearized noninverting current mirror.

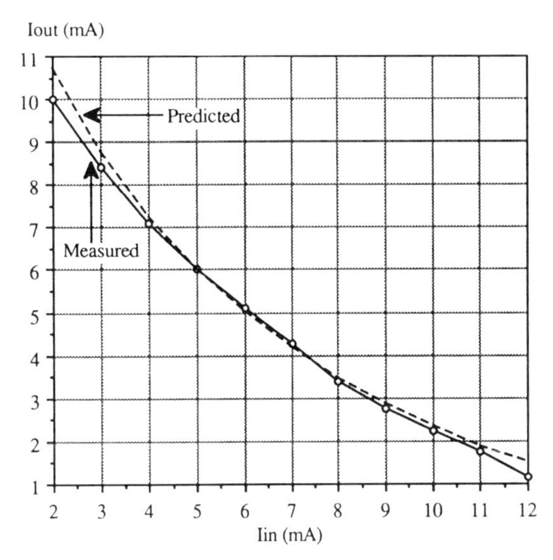

FIGURE 10.5 Measured and predicted transfer characteristics for cross-coupled MESFET pair.

[19]. For typical devices and bias conditions, C_{gd} is significantly less than C_{gs} and there is an approximate pole-zero cancellation in Equation (10.20) yielding a very wide bandwidth.

It can be seen from Equation (10.16) and Figure 10.5 that the cross-coupled MESFET pair has a current transfer characteristic that is nonlinear. However, in the proposed synthesis, we consider the circuit in Figure 10.3 in an alternative way as an inverter of the MESFET gate–source voltages

$$V_{gs1} = -V_{gs2} \tag{10.21}$$

The inversion occurs by virtue of the cross-coupled connection and is therefore unaffected by nonideal MESFET parameters.

Each source MESFET in Figure 10.2 may be one of the MESFETs in a cross-coupled pair with the other MESFET of the pair inside the voltage processing block. This allows the gate–source voltage of the source MESFET to be defined within the voltage processing block (albeit in inverted form) and effectively transferred to the source MESFET. Since the source of the source MESFET is connected within the voltage processing block only to a MESFET gate, the current drawn by the voltage processing block is zero as required. Since no other means of meeting this objective has been discovered, we conclude at present that the cross-coupled MESFET pair is an essential building block for the synthesis of transconductance and multiplier functions based on the square-law characteristic of the MESFET in saturation. This conjecture will, in fact, be further extended in Section 10.9.

10.3.2 AC-Coupled Circuits

10.3.2.1 Signal Handling Capability In the general architecture of Figure 10.2, the required current function I is formed by applying sums of $+1$ or -1-weighted signal voltages to the gate–source ports of the source and sink MESFETs. It is therefore important to consider the permissible range for MESFET gate–source voltage, which is bounded by both upper and lower limits [20]. Positive gate–source voltages above about 0.5 V run the risk of forward-biasing the gate-channel Schottky diode leading to large gate currents and a complete breakdown of normal FET behaviour. This upper limit will be referred to as V_{max} and will be taken to be 0.5 V. On the other hand, if the gate–source voltage falls below the threshold voltage V_T, the drain current remains at zero. This also is a departure from the square-law behavior described by Equation (10.1), and therefore V_T must constitute a minimum value for gate–source voltage. For most GaAs processes, V_T lies between -1 and -2 V, and here we shall assume for illustration purposes a typical value of -1.5 V.

The quadratic relationship of Equation (10.1) for the drain current of a MESFET as a function of gate–source voltage is shown in Figure 10.6 for the case $V_T = -1.5$ V. The value of quiescent gate–source voltage that

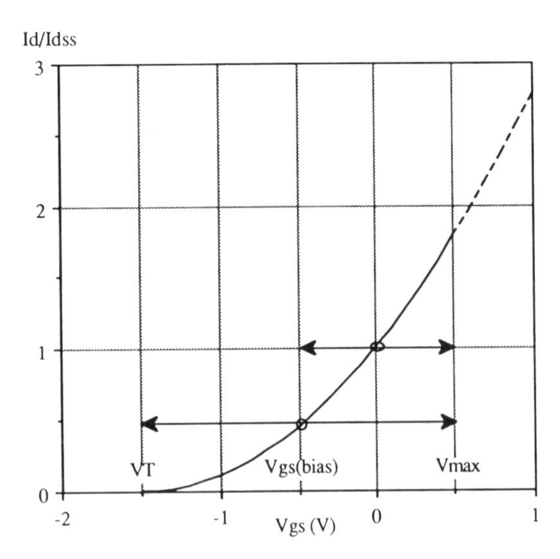

FIGURE 10.6 MESFET I_d-versus-V_{gs} characteristic with bias points and permitted V_{gs} swings.

maximizes signal swing is given by

$$V_{gs(bias)} = \frac{V_{max} + V_T}{2} \tag{10.22}$$

This quiescent point provides a peak-to-peak gate–source voltage swing of

$$V_{gs(pk-pk)} = V_{max} - V_T \tag{10.23}$$

corresponding to $2V_{pk-pk}$ (peak-to-peak) in our example. Thus, for maximum voltage handling capability given by Equation (10.23), the source and sink MESFETs in the general architecture of Figure 10.2 must be biased at a quiescent point given by Equation (10.22) [20]. However, a problem arises when we consider that the gate–source voltages of the source MESFETs in Figure 10.2 have to be defined from within the voltage processing block via a cross-coupled MESFET pair. It follows from the description [Eq. (10.21)] of the MESFET pair that a quiescent voltage of $V_{gs(bias)}$ on the source MESFET must imply a quiescent voltage of $-V_{gs(bias)}$ for the other MESFET of the cross-coupled pair. It can be seen from Figure 10.6 that such an operating point ($-V_{gs(bias)} = 0.5\,\text{V} = V_{max}$) is in fact far from optimum and would allow a signal swing of zero.

In some ways, the natural quiescent operating conditions for the cross-coupled MESFET pair of Figure 10.3 is $V_{gs} = 0$ for both MESFETs, but this implies the same quiescent conditions for both source and sink MESFETs in Figure 10.2, implying a deviation from the optimum condition of Equation (10.22). Referring to Figure 10.6, an operating point of $V_{gs} = 0$ permits a gate–source voltage swing of only $2V_{max}$ (under the reasonable assumption $V_{max} < |V_T|$) that is 1 V, and just half of the maximum voltage swing for the optimum bias point.

To satisfy the conditions for maximum voltage swings throughout the system, it is necessary to modify the cross-coupled MESFET pair so that both MESFETs may be operated with the same optimum bias voltage $V_{gs} = V_{gs(bias)}$. This can be achieved by replacing the cross-coupled connections in the MESFET pair of Figure 10.3 by capacitors and biasing the gates of the MESFETs from suitable bias points via resistors. The use of ac coupling in order to maximize voltage handling capability will introduce a lower cutoff frequency, but with resistor and capacitor values realizable on an MMIC (5 pF and 5 kΩ), a low cutoff frequency (6 MHz) is obtained. If this is not acceptable, then dc coupling, implying $V_{gs(bias)} = 0$ instead of Equation (10.22), must be adopted along with the necessary consequences of lower voltage and current swings and much lower efficiency.

10.3.2.2 Efficiency In this section, we consider some aspects of the power efficiency achievable with the general synthesis arrangement of Figure 10.2 and examine important implications of adopting ac coupling [20]. We take an example where there is a single-source MESFET–sink

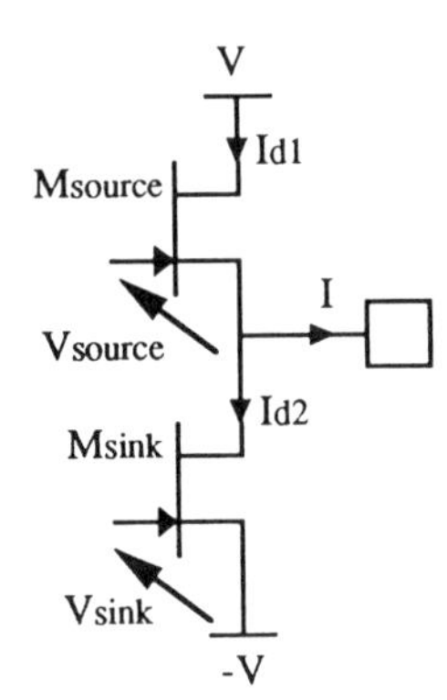

FIGURE 10.7 Typical source MESFET sink MESFET pair with complementary gate–source voltages.

MESFET pair driving the output terminal, as shown in Figure 10.7 (this corresponds to the case $m = 1$ in Section 10.2 and in fact formed the basis of the conductance function derivation in Section 10.3). We further assume, for the purpose of illustration, that the gate–source voltages of the source and sink MESFETs are complementary; this is true in many cases, including the basic conductance functions of Table 10.1, the differential input conductance of Table 10.2, for the V_1 input in the tunable conductance of Table 10.2 and for the V_1 and the V_2 inputs in Table 10.3.

The MESFET drain currents I_{d1} and I_{d2} in Figure 10.7 are plotted against the sink MESFET gate–source voltage in Figure 10.8 together with $-I_{d2}$ and the output current $I_{d1} - I_{d2}$. The curves in Figure 10.8 assume that the MESFETs are described by Equation (10.1). We have assumed, as in the

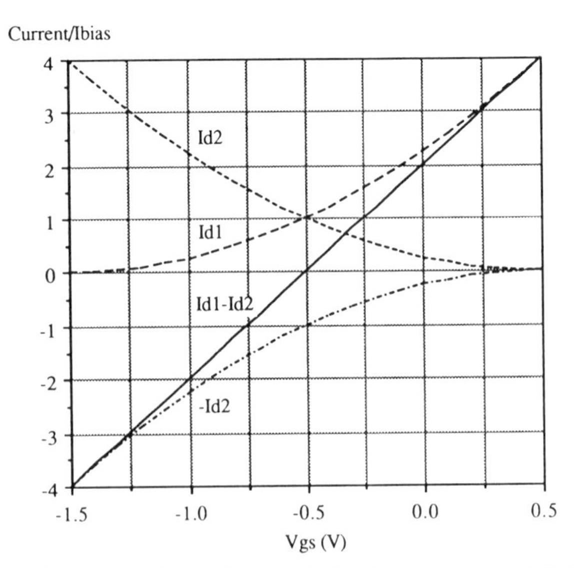

FIGURE 10.8 Current–voltage characteristics for arrangement in Figure 10.7.

previous section, that V_{max} is 0.5 V and that V_T is -1.5 V. We have also assumed that both MESFETs are biased for maximum voltage swing, provided by a quiescent bias voltage of -0.5 V given by Equation (10.22), and that signal swings are about this point. Currents in Figure 10.8 are normalized relative to the FET bias current at the quiescent point. We can now draw the important conclusion that operation of the MESFET pair for maximum voltage swing leads to a peak output current that is maximum and equal to 4 times the bias current. Thus under the condition of bias for maximum voltage and current swing, the type of circuit under discussion operates in a restricted form of class AB mode in which the quiescent current is one-quarter of the peak output current. This has important implications for efficiency.

To evaluate the efficiency of the architecture in Figure 10.7, we consider the case of a sinusoidal output voltage:

$$V_L = \hat{V}_L \sin \omega t \tag{10.24}$$

Assuming a resistive load, and that the output current swing is maximum, the output current is

$$I_L = I = 4I_b \sin \omega t \tag{10.25}$$

where I_b is the FET quiescent current. The currents in the source and sink FET can be described by

$$I_{d1} = I_b(1 + \sin \omega t)^2 \text{ and } I_{d2} = I_b(1 - \sin \omega t)^2 \tag{10.26}$$

The instantaneous power in the source and sink FET are thus given by

$$\begin{aligned} P_{inst1} &= (V - \hat{V}_L \sin \omega t)I_b(1 + \sin \omega t)^2 \\ P_{inst2} &= (V + \hat{V}_L \sin \omega t)I_b(1 - \sin \omega t)^2 \end{aligned} \tag{10.27}$$

where V is the supply voltage. The average power for both the source FET and the sink FET turns out to be

$$\overline{P_{inst}} = \tfrac{3}{2}VI_b - \hat{V}_L I_b \tag{10.28}$$

The second term in Equation (10.28) is the power delivered to the load, and therefore the first term must be the power obtained from the supply. Therefore the power conversion efficiency is given by the ratio of the terms

$$\eta_c = \frac{2}{3}\frac{\hat{V}_L}{V} \tag{10.29}$$

The maximum power conversion efficiency of $\frac{2}{3}$ is compared with the corresponding figure for the class A and class B stages [21] in Table 10.5.

TABLE 10.5 Comparison of Output Stage Types

Output Stage	Maximum Efficiency	I_{bias}/I_{peak}	Linearity
Class A	$\frac{1}{4}$	1	No
Linearized class AB	$\frac{2}{3}$	0.25	Yes
Class B	$\pi/4$	0	No

The optimum FET bias current expressed as a ratio of peak output current is also given. The linearized class AB stage clearly combines the property of linearity with a reasonably close approach to the high-efficiency and low-bias current of the class B stage.

10.3.2.3 Tuning In this section, we consider the implications for tuning that derive from the adoption of ac-coupled circuits [20]. We have established that for reasonable efficiency, the source and sink MESFETs should be operated at a quiescent gate–source voltage given by Equation (10.22). Considering, as an example, the case of a single pair of MESFETs with complementary gate–source voltage, this means that the full expressions for the gate–source voltages of the source and sink MESFETs are of the form

$$\begin{aligned} V_{gs(sink)} &= V_{gs(bias)} + V_{gs(signal)} \\ V_{gs(source)} &= V_{gs(bias)} - V_{gs(signal)} \end{aligned} \tag{10.30}$$

We now relate Equation (10.30) to the conductance function coefficient values in Tables 10.1, 10.2, and 10.3. If $V_{gs(bias)}$ in (10.30) is zero, then (10.30) corresponds directly with the inverting conductance function in Table 10.1, where the input voltage is applied directly to the sink MESFET ($b_1 = 1$) and in inverted form to the source MESFET ($a_1 = -1$). Consider now the tunable inverting conductance function in Table 10.2. Table 10.2 shows that in order to implement tuning, the tuning voltage V_2 must be added with the same polarity ($a_2 = b_2 = 1$) to the input voltage and to the inverted input voltage when they are applied to the gate–source ports of the sink and source MESFETs. However, this is precisely the description of the biasing arrangement described in Equation (10.30). It follows that the bias voltage in (10.30) may be identified with the tuning voltage V_2 in Table 10.2 (and also with tuning voltage V_3 in Table 10.3), and therefore the linearized transconductance functions we are synthesizing may be simply tuned in a linear fashion by means of the bias voltage.

A means of implementing tuning by control of bias voltage for a typical source–sink MESFET pair with complementary signal drive is shown in Figure 10.9. Essentially, the tuning (or bias) voltage and signal voltage for the sink MESFET are supplied via R_1 and C_1, respectively. Assuming that the external load has no dc offset, the source MESFET will have the same quiescent gate–source voltage as the sink MESFET. The signal component

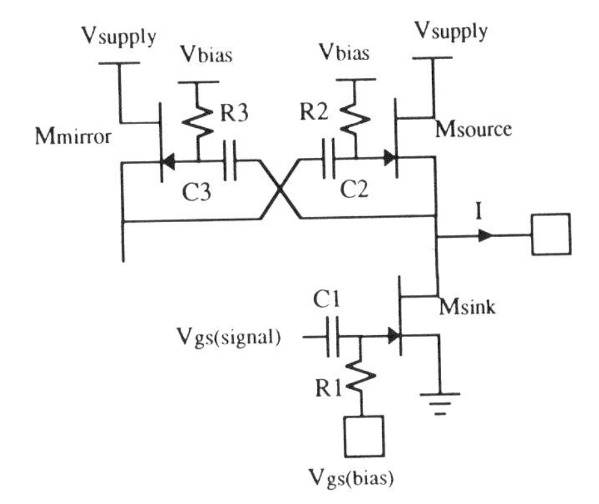

FIGURE 10.9 Source MESFET sink MESFET pair with tuning implementation.

$V_{gs(signal)}$ for the source MESFET is applied via C_3 and C_2 using an ac-coupled cross-coupled MESFET pair. Finally, the quiescent voltage on the gate of the source MESFET is supplied via resistor R_2 (the bias voltage on R_2 does not affect the quiescent gate–source voltage of the source MESFET that is determined via the sink MESFET as described above). Capacitors C_1, C_2, and C_3 are feeding the gates of MESFETs and may therefore have small values, typically of the order of 1 pF. Any capacitance required in series with the output to solve incompatibility of the circuit dc conditions with those of the load has to provide a high susceptance relative to the conductance being realized, and large values will be required to achieve a low cutoff frequency. The maximum capacitance normally realizable on an MMIC (20 pF) provides a susceptance of 8 Ω at a frequency of 1 GHz and occupies a rather large area in relation to a typical FET. An important design constraint is therefore to minimize the introduction of series coupling capacitors by appropriate design taking into account dc conditions.

The *RC* networks in Figure 10.9 form simple lowpass and highpass filter networks for the tuning components and for the signal component of the FET gate–source voltages. It follows that the tunable conductance function is still realized for non-dc tuning signals provided the frequencies of the tuning component and the signal component are firmly within the passbands of the lowpass and highpass filter responses. It follows that a necessary condition for implementation of tuning in this simple way is that the frequency bands for the signal and tuning components must be respectively highpass and lowpass band-limited and that their frequency bands must be noncontiguous.

10.4 BASIC BUILDING-BLOCK CIRCUITS

We have established that the cross-coupled MESFET pair forms the interface between the source MESFETs in Figure 10.2 and the voltage processing block. We now consider the complete set of building blocks for implementation of the system [6,20]. In the case of the sink MESFETs in Figure 10.2, the sources may be grounded and the sums of weighted circuit input voltages applied to the gates. There is thus a need for building blocks that can sum selected input voltages with specified signs. For the source MESFETs, there is the problem of providing an interface between the circuit input voltages defined relative to ground and the floating gate–source ports of the source MESFETs. Thus, here there is a need for voltage flotation circuits in addition to summing circuits.

10.4.1 Voltage Flotation Circuits

Consider the circuit shown in Figure 10.10*a*, where voltage V is applied to the gate–source port of MESFET M1.[1] Application of voltage V to the gate–source port of MESFET M1 generates a drain current according to Equation (1.1) that flows into M2 whose gate–source voltage therefore also equals V, as indicated. MESFETs M2 and M3 constitute a cross-coupled MESFET pair that may be regarded as a perfect inverter of the gate–source voltage of M2 by virtue of the cross-coupled connection between the MESFET gate–source ports. Thus the gate–source voltage of M3 is $-V$. M3 in Figure 10.10*a* may be a source MESFET in Figure 10.2 (in which case its

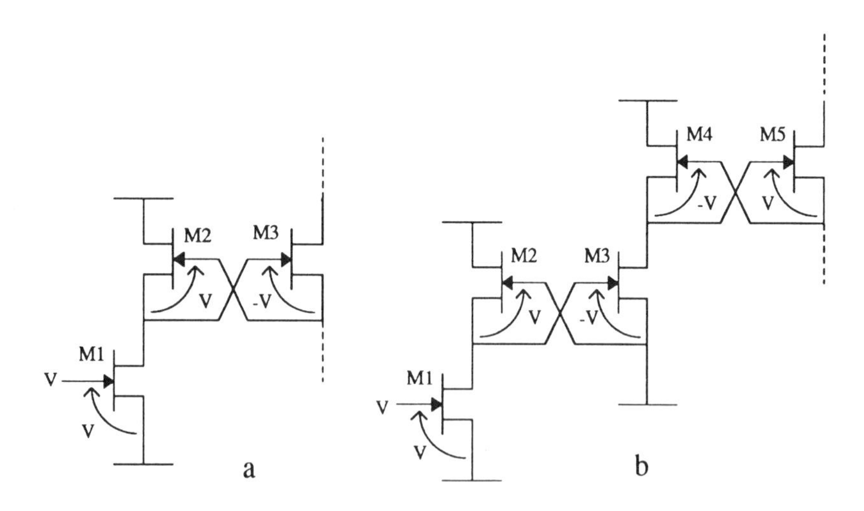

FIGURE 10.10 Voltage flotation circuits: (*a*) inverting; (*b*) noninverting.

[1] In the diagrams, we adopt a number of conventions: (1) dc voltage sources for biasing are indicated by a vertical line terminating in a short horizontal line, (2) dc voltage source values are not given—it is assumed that they are chosen to keep all devices well into the saturation region, and (3) circuit input voltages applied to a MESFET gate terminal are assumed to be defined relative to the quiescent voltage at the source terminal of that MESFET.

drain may be grounded), thus allowing the gate–source voltage required for the source MESFET to be presented at the gate of M1 with respect to ground (it must be in negated form). Alternatively, M3 in Figure 10.10*a* may be a sink MESFET (in which case its source may be grounded), thus allowing the gate–source voltage required for the sink MESFET to be presented in negated form at the gate of M1.

The circuit in Figure 10.10*b* is similar to that in Figure 10.10*a* except that the cross-coupled MESFET pair is replaced by a cascade connection of two such building blocks. Thus the input voltage V is inverted once across the gate–source ports of M3 and M4 and inverted again back to V at the gate–source port of M5. Another explanation for the operation of the circuit in Figure 10.10*b* is that, since MESFETs M2–M3–M4–M5 constitute a current mirror [19], which is linear assuming the MESFET model of Equation (1.1), we can regard the required output current contribution as being generated in M1 and then being transferred by the linear current mirror to appear as a current source at the output. M5 in Figure 10.10*b* would generally be a source MESFET in Figure 10.2 (in which case its drain would be grounded), thus allowing the gate–source voltage required for the source MESFET to be presented at the gate of M1 relative to ground. A prototype linear current mirror (M2–M5 in Fig. 10.10*b*) was fabricated as part of the building block test chip [18], and the layout is shown in Figure 10.4*b*. The measured current transfer characteristic is shown in Figure 10.11. Linearization is achieved over a very wide range of input and output currents.

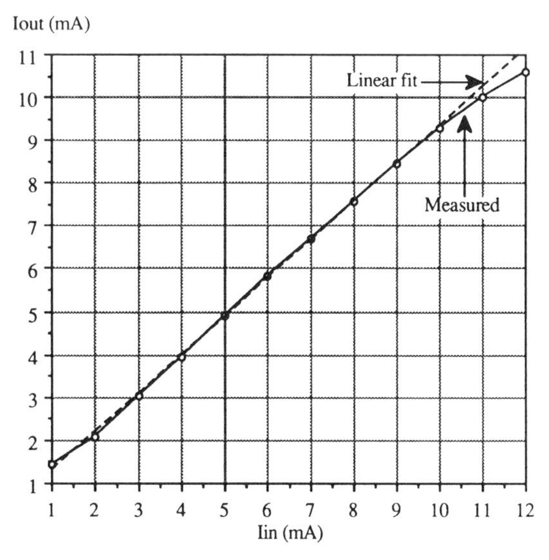

FIGURE 10.11 Measured transfer characteristic of linearized current mirror with best linear fit.

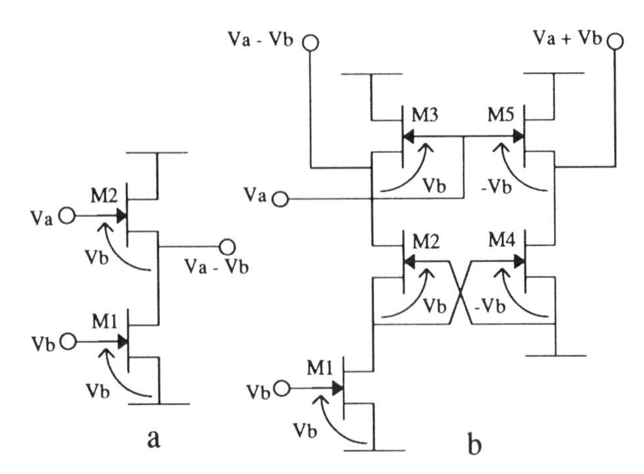

FIGURE 10.12 Voltage summation circuits: (*a*) difference circuit; (*b*) sum-and-difference circuit.

10.4.2 Sum-and-Difference Circuits

In the simple circuit of Figure 10.12*a*, the input voltage V_b defines the gate–source voltage of M1 and, since M2 shares the same drain current, its gate–source voltage also equals V_b, giving an output voltage $V_a - V_b$, where circuit output voltage is defined relative to the quiescent voltage at the output terminal. With noninverting input V_a set to zero, this difference circuit realizes a voltage inverter.

The circuit in Figure 10.12*b* [22], which is an extended version of the circuit in Figure 10.12*a* incorporating a cross-coupled MESFET pair M2/M4, has two outputs: the sum and the difference of the two input voltages. If the V_a input is derived via the difference circuit of Figure 10.12*a* acting as an inverter with input voltage $V_b = V_c$, then the circuit in Figure 10.12*b* provides output voltages $-V_b - V_c$ and $V_b - V_c$.

The circuit in Figure 10.13 is similar to the voltage flotation circuit of Figure 10.10*a* except that MESFET M4 has been introduced into the current mirror, such that M4 together with M5 form a difference circuit, as in Figure 10.12*a*. M3 would normally be a source MESFET in Figure 10.2, and its gate–source voltage is the sum of the negated input voltages.

The building blocks presented in Figures 10.12 and 10.13 effectively implement voltage transfer functions that are unaffected by uniform scaling of the conductance factors β of the MESFETs. It follows that the gate widths of the MESFETs in these building blocks may be chosen independently of the gate widths of the source and sink MESFETs in Figure 10.2. A gate width for the MESFETs in the voltage processing block of between one-half to one-third that of the source and sink MESFETs will give valuable saving in power consumption and chip area without unacceptable compromise of high-frequency performance in most cases.

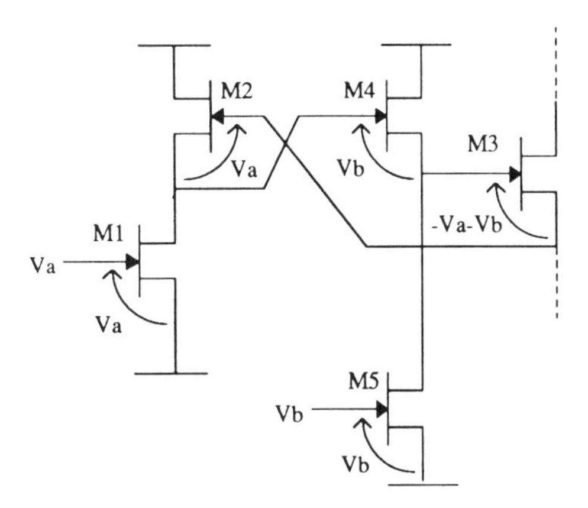

FIGURE 10.13 Summing voltage flotation circuit.

10.5 TRANSCONDUCTANCE REALIZATION

We showed in Section 10.2 that for the realization of conductance functions we may set m [the number of positive (and negative) weighted currents in Eq. (10.2)] to unity, which means that the implementation will contain a single source MESFET–sink MESFET pair. For the synthesis of an inverting transconductor, Table 10.1 shows that we need one source MESFET with a gate–source voltage of $-V_1(a_1 = -1)$ and one sink MESFET with a gate–source voltage of V_1 $(b_1 = 1)$, where V_1 is the input voltage. A simple circuit for the implementation of this function is shown in Figure 10.14*a* [23]. MESFET M1 is the sink MESFET, whose source is grounded and whose gate is fed directly from the circuit input, providing the required gate–source voltage V_1. MESFETs M2, M3, and M4 constitute an inverting voltage flotation circuit (as in Fig. 10.10*a*) that provides the required gate–source voltage of $-V_1$ to the source MESFET M4. M2 and M3 are within the voltage processing block in the general architecture of

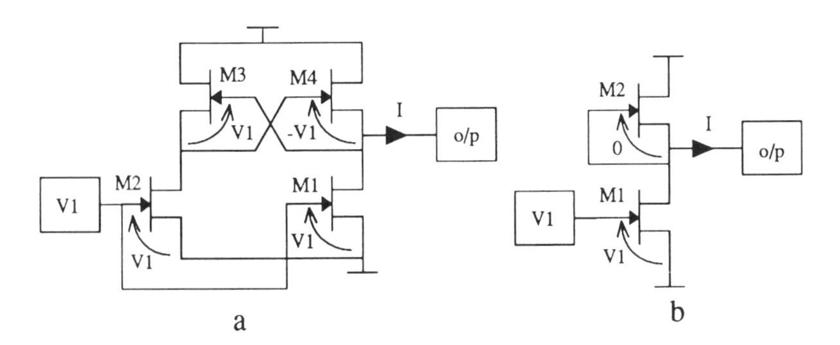

FIGURE 10.14 Inverting transconductor circuits: (*a*) linearized; (*b*) conventional inverter.

Figure 10.2, and therefore their gate widths may be chosen independently of the gate widths of the source and sink MESFETs.

For the purpose of comparison, the conventional equivalent consisting of a simple MESFET (with load consisting of a MESFET with gate and source connected to realize a constant current source) as shown in Figure 10.14*b* was considered. The circuits in Figure 10.14 were fabricated as part of the building-block test chip already discussed, and the layout plots can be seen in Figure 10.15 [18]. Since the small-signal transductance of the conventional circuit of Figure 10.14*b*, obtained by differentiating Equation (10.1)

$$G_{\mathrm{mo}} = \left.\frac{\partial I_{\mathrm{d}}}{\partial V_{\mathrm{gs}}}\right|_{V_{\mathrm{gs}}=0} = -2\beta V_{\mathrm{T}} \tag{10.31}$$

is only half of that for the linearized circuit (see Table 10.1), its gate widths are doubled, making the total width for the two circuits in Figure 10.15 the same. The measured static large-signal characteristics of the linearized and conventional transconductor circuits are shown in Figures 10.16*a* and 10.16*b*, where they are compared with the responses predicted using the MESFET model of Equation (10.1) [18]. The linearized circuit exhibits a sixfold reduction in nonlinearity for a wide input voltage range.

For the noninverting transconductor, Table 10.1 shows the gate–source voltages of the source and sink MESFETs are required to be V_1 and $-V_1$, respectively ($a_1 = 1$, $b_1 = -1$). Two possible realizations are shown in Figure 10.17. In the circuit of Figure 10.17*a*, MESFETs M1–M3 form an inverting voltage flotation circuit that provides a gate–source voltage of $-V_1$ to the sink MESFET M3. MESFETs M4–M8 make up a noninverting voltage flotation circuit that provides the required gate–source voltage for the source MESFET M8 of V_1. In the alternative realization in Figure 10.17*b*, the inverting transconductor of Figure 10.14*a* is preceded by the difference circuit of Figure 10.12*a* acting as an inverter with the V_a input grounded.

For the tuned inverting transconductor, Table 10.2 shows that we require gate–source voltages for the source and sink MESFETs of $-V_1 + V_2$ and $V_1 + V_2$, respectively, where V_2 is the tuning voltage ($a_1 = -1$, $a_2 = 1$, $b_1 = 1$, $b_2 = 1$). One possible implementation, which realizes the preceding function but with tuning voltage V_2 inverted, is shown in Figure 10.18. M5 and M6 constitute a difference circuit (as in Figure 10.12*a*) that subtracts the control voltage V_2 from the input voltage to form the sink MESFET M1 gate–source voltage $V_1 - V_2$. MESFETs M2, M3, M4, M5, and M7 constitute a summing voltage flotation circuit, as in Figure 10.13, which forms the source MESFET gate–source voltage $-V_1 - V_2$. In this arrangement MESFET M5 can act in conjunction with both M6 and M7 because the polarity of the V_2 signal introduced is the same both for the source and for the sink MESFET.

We have shown in Section 10.3.2 that ac-coupling of circuits is necessary to allow biasing for maximum input voltage and output current swing, for an efficient class AB operating mode and simple implementation of tunability.

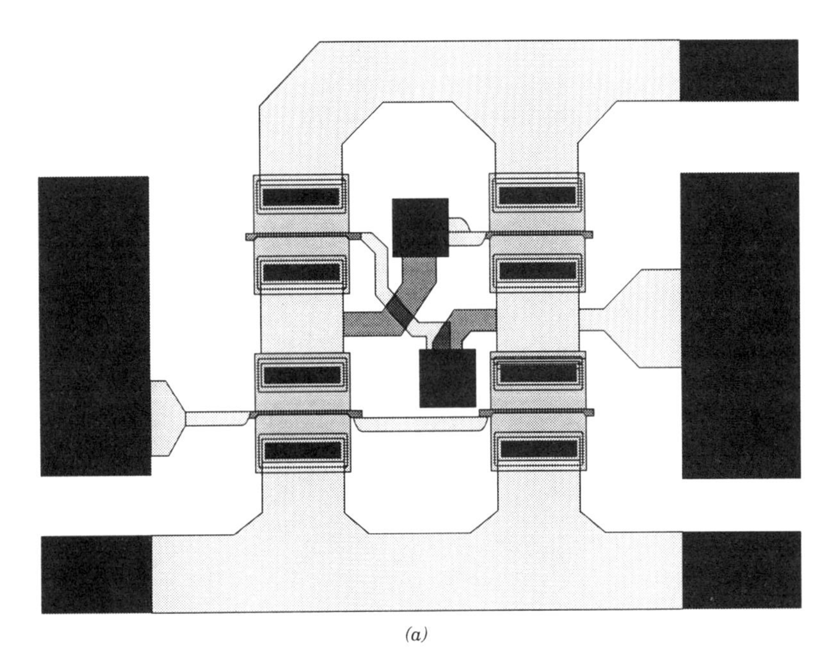

(a)

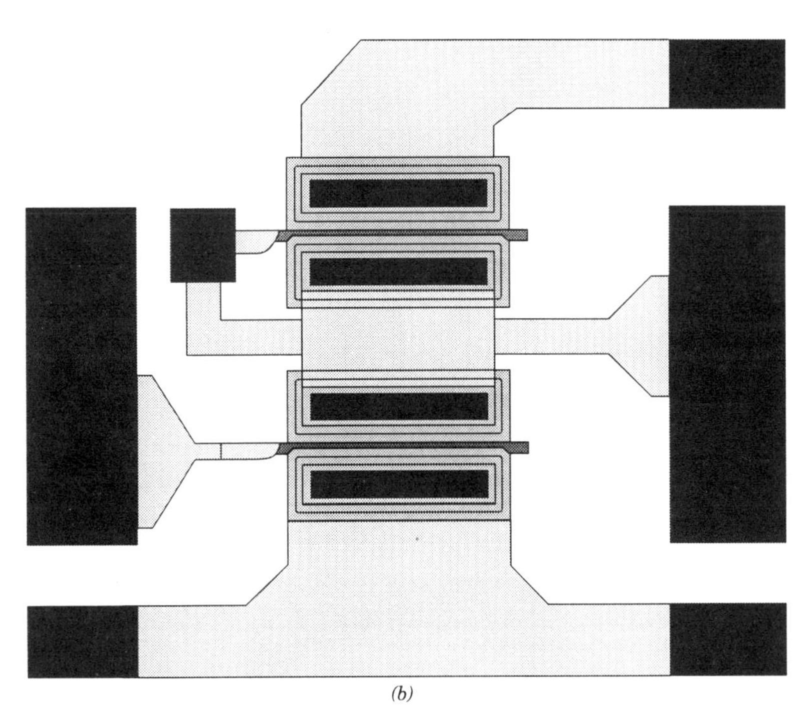

(b)

FIGURE 10.15 Layout of further building blocks on test chip: (*a*) linearized transconductor; (*b*) inverter transconductor.

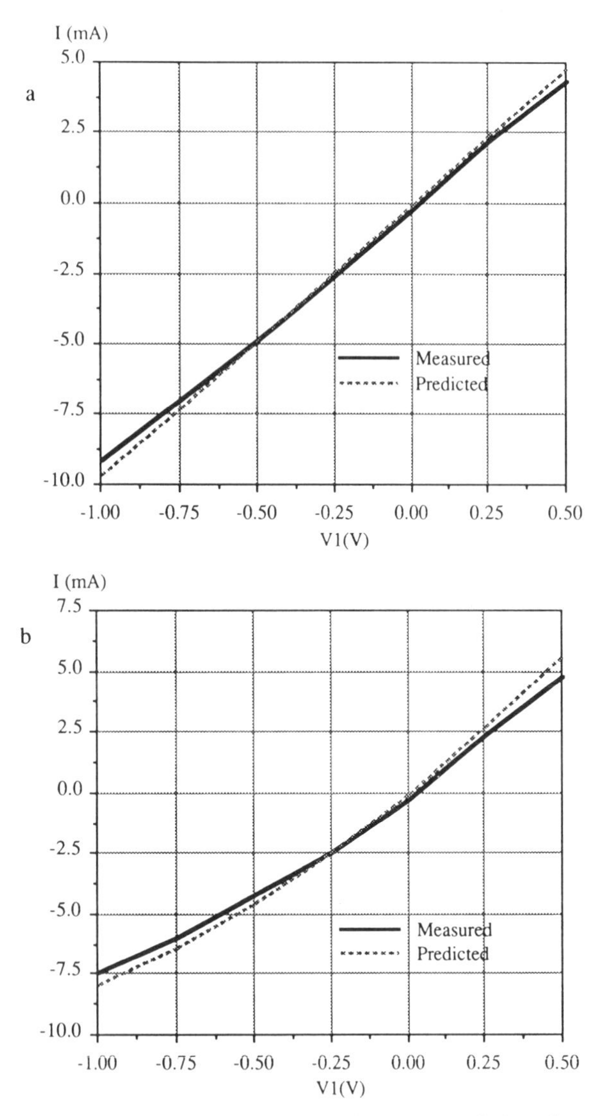

FIGURE 10.16 Measured and predicted transfer characteristics of transconductors: (*a*) linearized; (*b*) inverter.

The ac-coupled version of the inverting transconductor already shown in Figure 10.14*a* is shown in Figure 10.19 [20]. The dc component of the gate–source voltage for all four MESFETs may be set by the voltage V_{bias1} to the optimum value given by Equation (10.22). This allows the input voltage and all gate–source voltages to swing symmetrically over the maximum possible range and provides class AB operation. The transconductance may be tuned linearly by means of the voltage V_{bias1} provided it is accepted that the tuning and input signals are lowpass and highpass

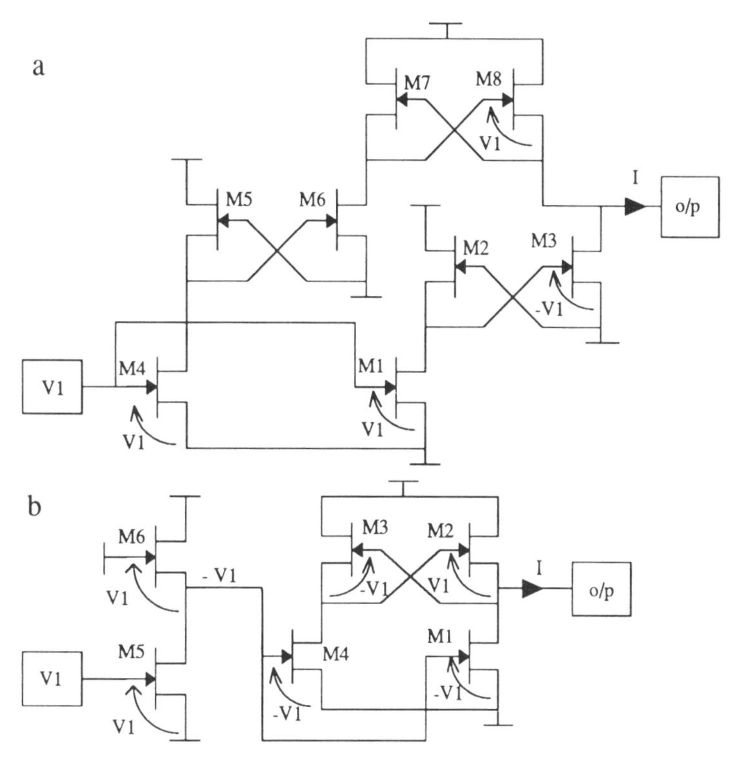

FIGURE 10.17 Noninverting transconductor circuits: (*a*) circuit; (*b*) alternative circuit.

band-limited, respectively, and that their frequency bands are noncontiguous. If the load suffers from dc offset current, it may be necessary to introduce a coupling capacitor at the output in order that the quiescent

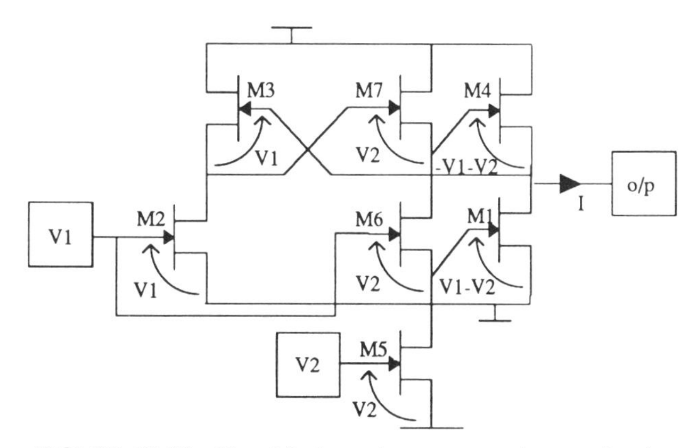

FIGURE 10.18 Tunable inverting transconductor circuit.

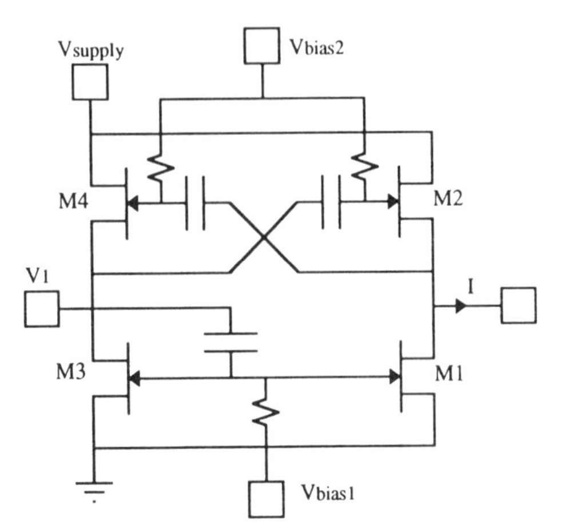

FIGURE 10.19 AC-coupled tunable linearized transconductor.

conditions on M1 and M2 in Figure 10.19 remain identical. The transconductor of Figure 10.19 will be used in an MMIC example in Section 10.7.

We now consider further dc-coupled transconductor implementations based on a simple modification to the basic synthesis architecture. A typical source–sink MESFET pair in the general synthesis architecture of Figure 10.2 is shown in Figure 10.20*a*. In order to maximize the attractiveness of dc-coupled implementations, it is appropriate to consider the modification to this architecture using diodes shown in Figure 10.20*b*. The modification consists in introducing a chain of low-dynamic-impedance diodes between the source of the source MESFET and the terminal where current I is synthesized. In many cases, source and sink MESFETs have complementary gate–source voltages, and in Figure 10.20*b* this may be simply achieved by cross-coupling the gate–source ports of the source and sink MESFETs,

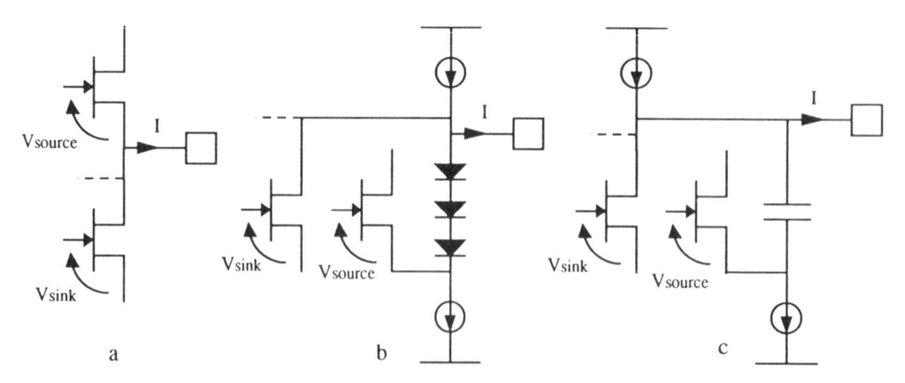

FIGURE 10.20 Modifications to realization architecture illustrated for typical source–sink MESFET pair.

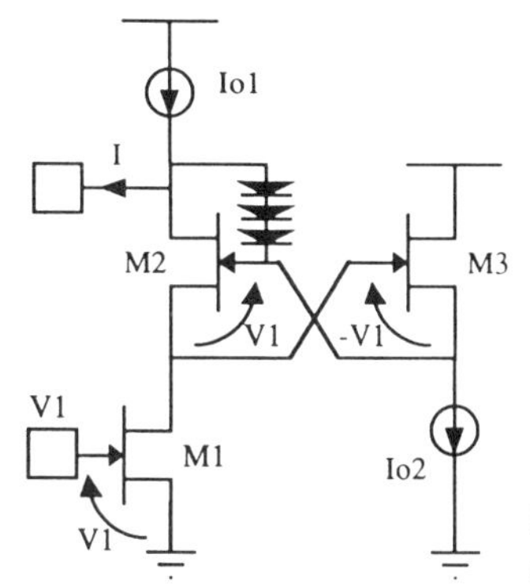

FIGURE 10.21 Linearized transconductor using diodes.

which is now possible because their gate and source terminals are at compatible quiescent voltages. Since the diodes have low impedance, the output current may alternatively be taken from the other end of the diode chain. The diodes may be replaced by a capacitor as shown in Figure 10.20*c* if highpass band limiting of signals is acceptable.

A realization of the inverting linearized transconductor function using the diode-augmented architecture is shown in Figure 10.21. MESFET M3 with gate–source voltage $-V$ is the source MESFET, and its source is connected to the output via diodes as in Figure 10.20*b*. MESFET M1 with gate–source voltage V is the sink MESFET. This example illustrates a further departure from the general architecture in that, although the sink MESFET defines the sink current at the output, its drain is not connected to the output directly, but via the drain–source port of MESFET M2, which shares the same drain current as M1 and therefore has the same gate–source voltage V_1.

We now consider simplification of the linearized inverting transconductor of Figure 10.14*a*, which is shown again in Figure 10.22*a* [17]. The circuit is

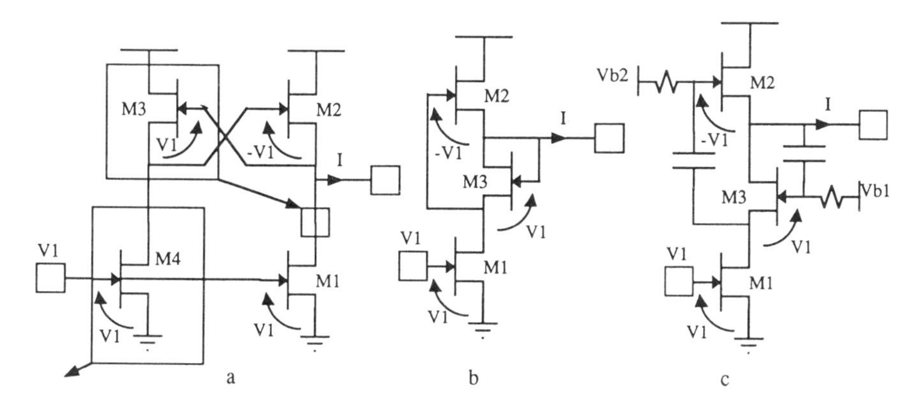

FIGURE 10.22 Simplified linearized transconductor derivation: (*a*) transconductor of Figure 10.14*a*; (*b*) simplified circuit; (*c*) circuit in (*b*) with biasing.

simplified using the following argument. Since MESFET M3 has the same drain current as M4 and MESFET M4 has the same drain current as M1 (since the gate–source voltages are the same), then M4 may be removed and the drain–source port of M3 inserted in the link between M1 and M2, as illustrated. The resulting circuit is shown in Figure 10.22*b*, which is only a conceptual circuit since the MESFET dc conditions required to maintain the MESFETs in the saturation region of operation cannot be met. In order to satisfy the dc conditions, ac coupling can be introduced as shown in Figure 10.22*c*, subject to the restriction of a highpass band-limited input signal. Since the capacitors are feeding only gate terminals, small values occupying low chip area are acceptable and low cutoff frequencies can be achieved. MESFETs M2 and M3 in Figure 10.22*c* effectively realize an ac-coupled cross-coupled MESFET pair operating with different dc voltages. Figure 10.23 shows a comparison of the SPICE simulated output current versus input voltage characteristics for three transconductors using the typical foundry FET parameters given in Table 10.6. The transconductors are the 2-MESFET inverter in Figure 10.14*b*, the linearized 4-MESFET transconductor of Figures 10.14*a* and 10.22*a*, and the 3-MESFET simplified linearized circuit of Figure 10.22*c*. It can be seen that the simplified circuit is nearly as linear as the 4-MESFET circuit of Figure 10.14*a* and provides a dramatic improvement over the nonlinearized conventional inverter. As for the 4-MESFET circuit with ac coupling in Figure 10.19, the ac coupling that is inherent in the circuit of Figure 10.22*c* permits class AB operation allowing considerable reduction in power consumption and dissipation.

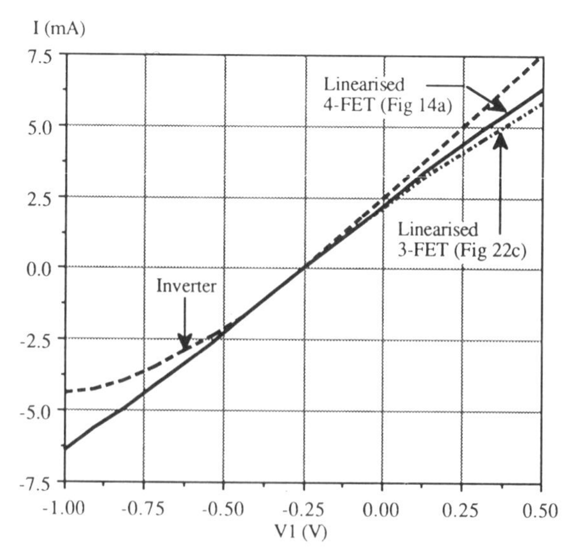

FIGURE 10.23 Simulated transconductor characteristics.

TABLE 10.6 Level 2 MESFET Parameters for HSPICE Simulation

Parameter	Value	Unit
VTO	−1	V
BETA	0.067E-3	mA V^{-2}
LAMBDA(LF)	0.06	V^{-1}
LAMBDA(HF)	0.3	V^{-1}
RD	2920	Ψ
RS	2920	Ω
CGS	0.39E-15	F
CGD	0.39E-15	F
PB	0.79	V
IS	0.075E-15	A

10.6 SELF-CONDUCTANCE REALIZATION

In utilizing the general architecture of Figure 10.2 for the realization of a self-conductance, the input node and the node where the current I is synthesized are coincident [17,20]. Since the synthesized current I in Figure 10.2 flows out of the system, it follows that in order to realize a positive self-conductance the conductance function realized must be negative. Table 10.1 shows that for a negative, or inverting, conductance the gate–source voltages of the source and sink MESFETs must be $-V$ and V, respectively, where V is the input voltage.

An ac-coupled realization of the self-conductance is shown in Figure 10.24. MESFET M1 is the source MESFET, and M2 is the sink MESFET. In the case of this very simple function, the voltage processing block in Figure 10.2 takes a very simple, almost degenerate form. It consists of a CR network that couples the ac component of the port voltage V to the gate of

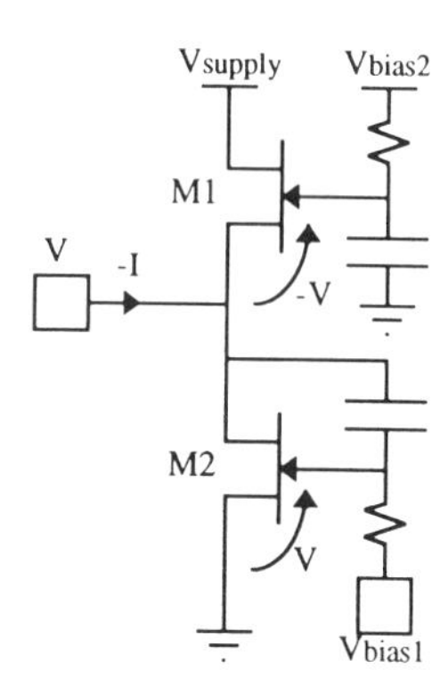

FIGURE 10.24 Tunable ac-coupled self-conductance.

the sink MESFET M2, and a ground connection to the gate of source MESFET M1 that is sufficient to provide its required gate–source voltage of $-V$. Linear tuning of the conductance can be achieved, as described in Section 10.3.2.3, simply by means of the voltage on the bias terminal V_{bias1}. The ac-coupled circuit in Figure 10.24 can be operated with the transistors biased for maximum voltage and current swing, and this provides class AB mode for maximum efficiency as described in Section 10.3.2.2. The reader may be surprised at the simplicity of the circuit in Figure 10.24 bearing in mind that we stated in Section 10.3.1 that the source MESFET must form part of a cross-coupled MESFET pair in order to satisfy the synthesis conditions. It is interesting to observe that in this simple (almost degenerate) case, the cross-coupled MESFET pair is formed between the source MESFET and the sink MESFET (the source of the source MESFET is ac-coupled to the gate of the sink MESFET, and the source MESFET gate and sink MESFET source are both grounded). Thus even in the simplest case, the cross-coupled MESFET pair can be shown to exist. An equivalent version of this circuit for MOS technology has already been published [24].

A dc-coupled realization of the linearized self-conductance using the diode-augmented architecture of Figure 10.20*b* is shown in Figure 10.25, illustrating direct cross coupling of the source MESFET (M1) and sink MESFET (M2) gate–source ports in order to achieve the required voltage inversion [25].

For tuning of a conductance by means of a control voltage V_c, Table 10.2 shows that the inverse voltages applied to the source and sink MESFET gate–source ports must be complemented by V_c. Each of the two voltage sources V_c required can be realized by a source follower introduced into the arrangement of Figure 10.25, as shown in Figure 10.26, where the inverted control voltage is supplied to the gates of MESFETs M5a and M5b [25]. The simulated transfer characteristic of the tunable resistance circuit is shown in Figure 10.27 for a range of tuning voltage. Simulation is by HSPICE and is based on use of the level 2 MESFET parameters in Table 10.6. Simulation of the small-signal admittance of the tuned resistance circuit yielded a

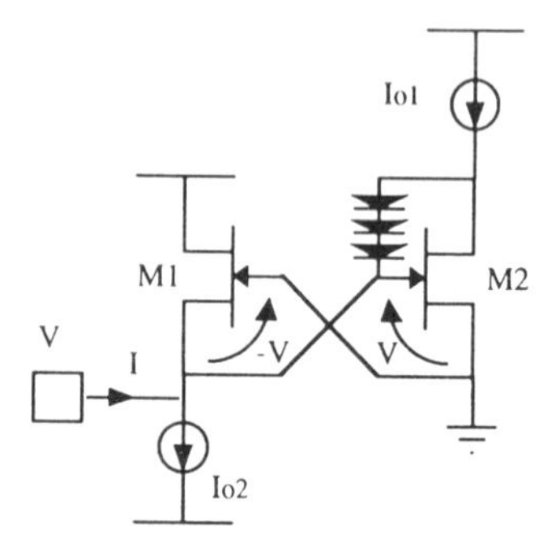

FIGURE 10.25 DC-coupled self-conductance.

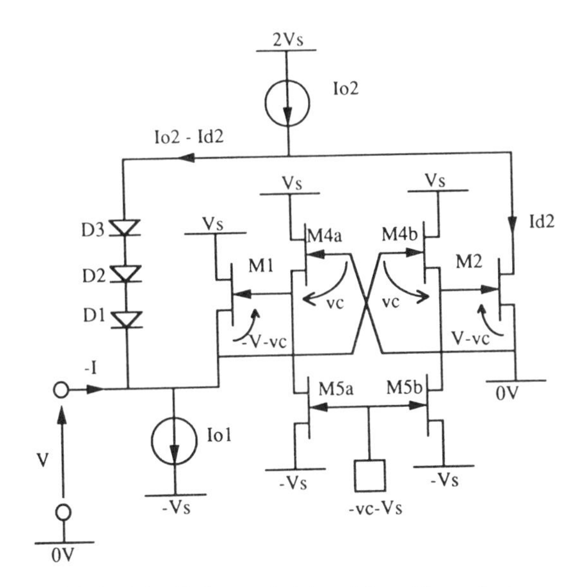

FIGURE 10.26 DC-coupled tunable self-conductance.

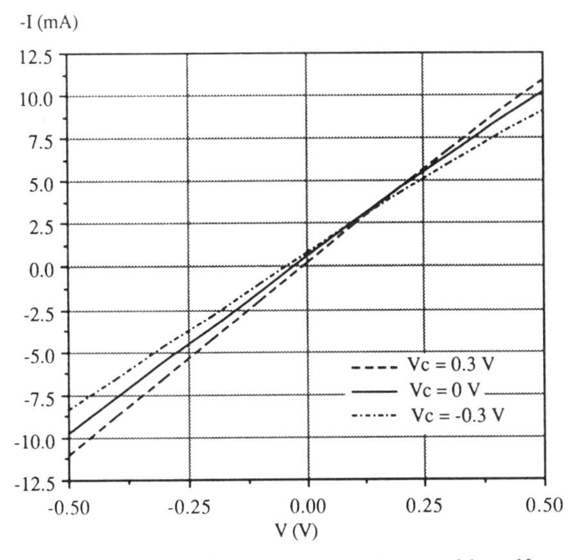

FIGURE 10.27 Simulated I–V characteristic for tunable self-conductance.

conductance that was accurate and monotonic up to around 30 GHz and a susceptance that was no longer purely capacitance but that was low and monotonic.

10.7 LINEARIZED ISOLATOR CIRCUIT

The isolator is an important component of high-frequency systems and is an ideal vehicle for IC techniques in view of its simplicity. The ideal isolator is required to have perfect resistive matching at its input and output ports, a forward transmission of unity, and a reverse transmission of zero. In addition, it is required to have a wide bandwidth. A successful IC isolator has been implemented based on a parallel connection of a common-gate connected MESFET with a common-drain connected MESFET [12]. In this section we explore the exploitation of some of the concepts we have been discussing to implement an alternative isolator realization [20,26]. The potential advantages are that the circuits are linearized and that independent tuning of both the input and output matching and of the forward gain is possible.

The circuit diagram of the isolator is shown in Figure 10.28. It consists essentially of a cascade of three components between the input and output ports. The first and last components are linearized self-conductance circuits as already presented in Figure 10.24. The central component is the inverting linearized transconductor of Figure 10.19. All circuits are ac-coupled and are biased for maximum voltage and current swings and efficient class AB operation. The input and output resistance can be tuned by means of the bias voltages V_{tune1} and V_{tune3} on the input and output matching circuits. The transconductance of the transconductor component can be tuned by means

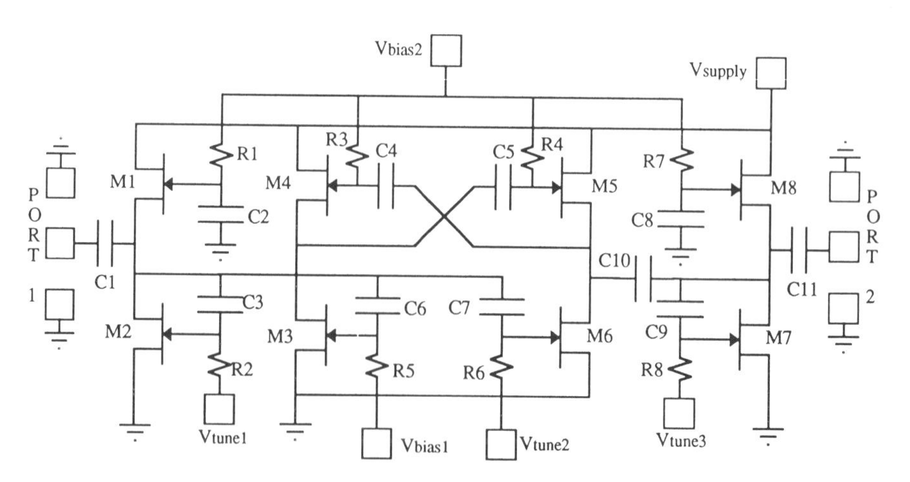

FIGURE 10.28 Circuit diagram of linearized tunable isolator.

TABLE 10.7 Component Data for Isolator in Figure 10.28

Component	Value	Unit
M1, M2, M7, M8	100	μm
M3, M4	150	μm
M5, M6	300	μm
C1, C10, C11	7.5	pF
C2–C9	1	pF
R1–R8	3	kΩ

of V_{tune2}, which defines the operating points of source and sink MESFETs M5 and M6, thus allowing tuning of the forward transfer gain of the isolator. Bias voltage V_{bias1} affects the operating points of MESFETs M3 and M4 and may be chosen for maximum voltage swing.

The circuit is designed for implementation using an ion-implanted GaAs process providing 0.7-μm 20-GHz MESFETs with −2-V pinchoff voltage. The nominal design input and output resistances of the isolator are 50 Ω, and the nominal forward gain is unity. The power consumption is 250 mW. The passive component values and MESFET gate widths are listed in Table 10.7. The layout plot is shown in Figure 10.29. The chip area is 1.5 mm × 0.75 mm. The bias voltage V_{bias2} is obtained from the supply voltage V_{supply} by a resistive potential divider. At the time of writing the chip is being fabricated, and therefore test results are not yet available. We therefore in the meantime present some SPICE simulated results using a modified form of the SPICE MESFET model parameters in Table 10.8 that were derived empirically.

Large-signal characteristics of the circuit illustrating its tuning capabilities

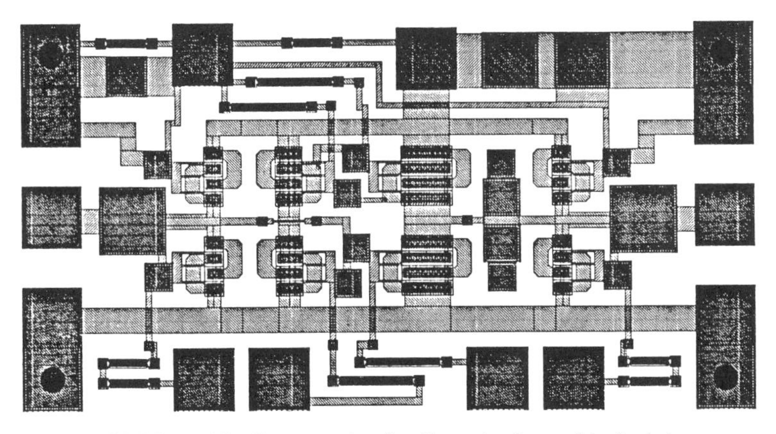

FIGURE 10.29 Layout plot for linearized tunable isolator.

TABLE 10.8 Parameters for SPICE Simulation of MESFET

Parameter	Value	Units
VTO	−2	V
BETA	0.03E-3	mA V^{-2}
LAMBDA	0.15	V^{-1}
RD	720	Ω
RS	720	Ω
CGS	0.5E-15	F
CGD	0.5E-15	F
PB	0.79	V
IS	0.075E-15	A

are presented in Figure 10.30, which shows the input port static current–voltage characteristic for a range of values of the input circuit tuning voltage V_{tune1} (Fig. 10.28). The graph shows that the input resistance is fairly linear for a $2V_{\text{pk-pk}}$ input voltage range and can be tuned over ±10% of nominal.

Figure 10.31 is intended to characterize linearity of response and shows both port voltages (V_1 and V_2) for a ramped 50-Ω source at port 1 and then at port 2. Good linearity is exhibited at both ports (V_1—source at port 1; V_2—source at port 2). This contrasts with the less linear behavior of a nonlinearized implementation simulated in reference [26]. Linearity of the forward transfer characteristic in Figure 10.31 (V_2—source at port 1) can be further improved by optimum choice of power supply voltage and biasing in

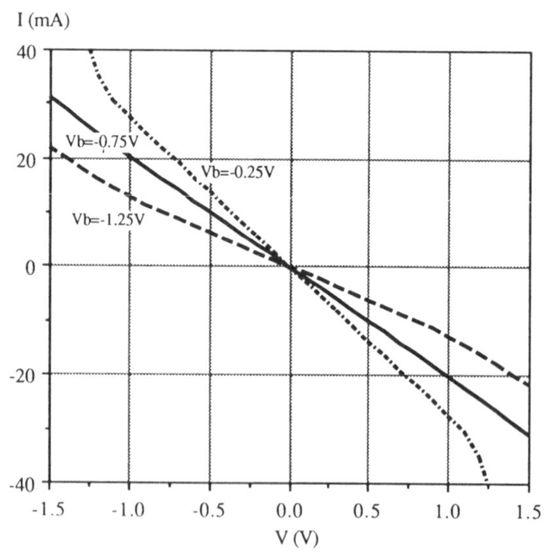

FIGURE 10.30 Simulated isolator input port large-signal characteristic showing tuning.

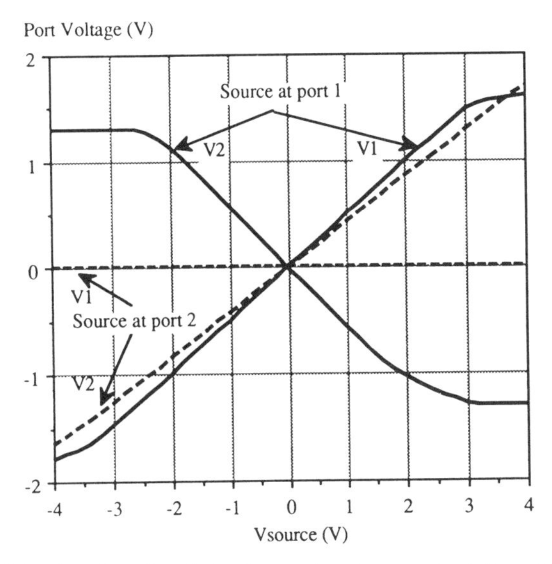

FIGURE 10.31 Simulated isolator large-signal characteristics at both ports.

relation to FET parameters; these factors are more critical for a linearized circuit.

The simulated small-signal S parameters of the isolator are shown in Figure 10.32. These results confirm good isolation and gain flatness better than 1 dB up to 5 GHz. Bandwidth is limited by loading of the input port by the input capacitance of the transconductance stage. This is an architectural

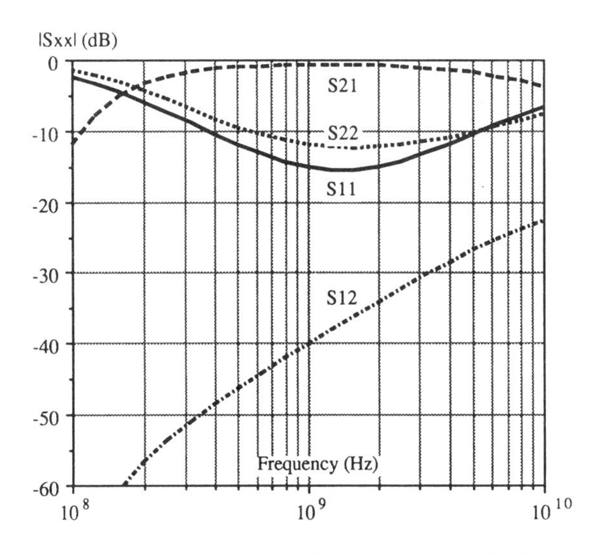

FIGURE 10.32 Simulated S parameters for isolator.

feature of this general form of isolator implementation, which will be tackled in future work using a parallel rather than cascade architecture [26]. Deterioration in S_{11} and S_{22} in Figure 10.32 at low frequencies is due to coupling capacitors included in the design; they will be removed in future designs in order to improve matching.

While the circuit proposed in Figure 10.28, which is being fabricated, is more complex than the simpler circuit in reference 12, it is expected to have some advantages. First, the linearized circuit blocks used should provide lower distortion levels and maintain good port matching for the full range of input and output signal levels. Second, the feature of tunability is important because a major disadvantage of conventional MMIC circuits in comparison with MIC circuits is that tuning is no longer possible. Thus the restoration of a simple and effective tuning capability to microwave circuits implemented in monolithic integrated circuit form is an important aspect of this work.

Even though the isolator architecture adopted by Ali and Podell [12] of a parallel connection of a common-gate connected MESFET with a common-drain connected MESFET does not provide linearized response characteristics, it is simple and very efficient. On the contrary, the architecture in Figure 10.28 of a shunt active resistance followed by a transconductance followed by a second shunt active resistance is structurally inefficient even though the conductance building blocks themselves may be efficiently realized. Nevertheless, it does demonstrate the use of linearization techniques providing a tuning capability and serves to raise challenging questions for longer-term research such as whether it might be possible to directly synthesize a linear transconductance with tunable finite input and output resistance as a single efficient circuit entity rather than as the present cascade structure. A possible approach to this problem is to realize the system as a parallel combination of composite subnetworks that are equivalent to a linearized common-gate FET and a linearized common-drain FET [26]. We will briefly discuss potentially interesting building blocks for this approach in Sections 10.9 and 10.10.

10.8 MULTIPLIER REALIZATION

We now consider the realisation of multiplier functions using the general synthesis technique presented earlier [16]. In Section 10.2 we stated that for the synthesis of multiplier functions it is appropriate to let m [the number of positive (and negative) current terms in Eq. (10.2)] have a value of 2. It follows that the implementation will constitute two source MESFETs and two sink MESFETs. A set of suitable function coefficients for a single-term four-quadrant multiplier function is already given in Table 10.4. It follows that the gate–source voltages for the two source MESFETs are V_1 and $-V_2$ (since $a_{11}=1$, $a_{22}=-1$), and for the two sink MESFETs are V_1-V_2 and zero (since $b_{11}=1$, $b_{12}=-1$, $b_{21}=b_{22}=0$). A possible implementation is

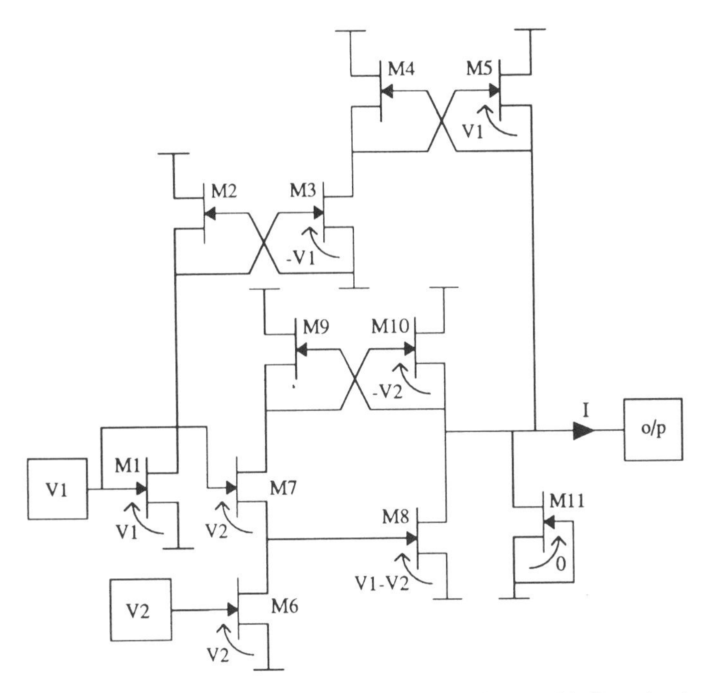

FIGURE 10.33 Single-term four-quadrant analog multiplier circuit.

shown in Figure 10.33. The source MESFETs are M5 and M10, and the sink MESFETs are M8 and M11. M1–M5 constitute a noninverting voltage flotation circuit (as in Fig. 10.10*b*) providing the source MESFET gate–source voltage of V_1. MESFETs M6 and M7 form a difference circuit to supply the sink MESFET M8 with gate–source voltage $V_1 - V_2$. M9 and M10 together with M6 form an inverting voltage flotation circuit (as in Fig. 10.10*a*) providing the source MESFET gate–source voltage $-V_2$. Finally, the sink MESFET M11 has zero gate–source voltage.

The multiplier function is more sophisticated than the transconductance functions already considered, and therefore the multiplier circuit in Figure 10.33 cannot be regarded as a fully engineered circuit until the effects of MESFET imperfections, including finite output conductance, are taken into account and overcome. The effect of finite output conductance can be minimized by the introduction of cascode MESFETs at the cost of increasing power supply voltage and current [3] or by modification of the FET gate widths to maintain the ideal source and sink FET gate–source voltages [17]. The study of the effect of finite MESFET output conductance in transconductor–multiplier circuits is a difficult task and still at an early stage. Nevertheless, some progress is being made and precise simulated multiplier performance has been achieved, although at the cost of relatively high circuit complexity [27]. An interesting possibility is that, provided it is accepted that the two multiplier inputs have noncontiguous frequency bands, which is the situation in many communication systems, the second multiplier input may be applied to the circuit in the form of the operating points of MESFETs, as was done in the case of transconductors in Section 10.3.2. This will greatly simplify the circuits and hence make the effect of nonideal MESFETs much easier to analyze and counteract by means of simple circuit modifications [17].

10.9 LINEARIZED BUFFER CIRCUITS

The general synthesis technique developed in Section 10.2 has so far been applied to implement conductance functions and multiplier functions. The conductance functions have been implemented as either transconductance or self-conductance functions. In this section we introduce a third category of conductance function implementation that we refer to as a *buffer function* [17,20,26]. Buffers are very important for both interfacing between subsystems on an integrated circuit and providing an on-chip interface with off-chip components that will in general have higher parasitic capacitances and low impedance levels. The performance of the buffer including linearity is vital to preserving a high overall system performance. Furthermore its efficiency and chip area requirements can be very important, and even dominant, factors of an overall design.

The three categories of conductance function implementation are illustrated in Figure 10.34, where we assume a single input voltage for simplicity. The transconductance and self-conductance implementations of the conductance function are illustrated in Figures 10.34*a* and 10.34*b*, where the input voltage and synthesised current are at different ports or at the same port, respectively. The further category we wish to consider is shown in Figure 10.34*c*. Here the current appears at the output terminal, but the voltage that determines this current is the difference between the input and the output voltage. In an ideal buffer, the conductance gain would be very high,

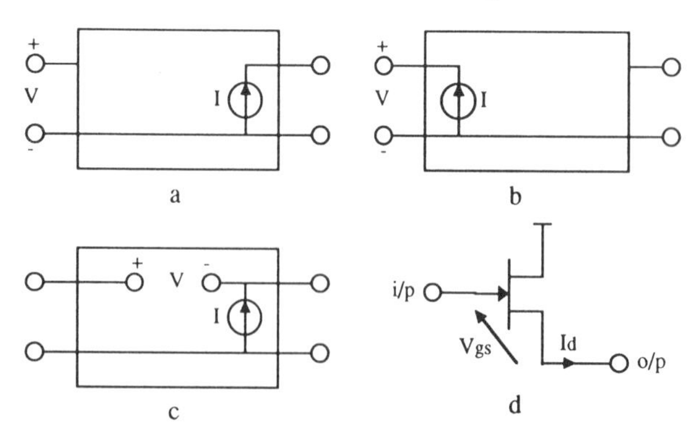

FIGURE 10.34 Classes of implementation for the conductance function: (*a*) transconductance; (*b*) self-conductance; (*c*) buffer function; (*d*) source-follower buffer.

tending to constrain the input and output voltages to be equal. The circuit usually used to realize a voltage buffer is the common-drain FET or source follower, which is illustrated in Figure 10.34*d*. We may identify V and I in Figure 10.34*c* with the V_{gs} and I_d of the MESFET in Figure 10.34*d*. Two problems with the conventional circuit can be identified at this stage. First, it is nonlinear since output current variations will, according to Equation (10.1), imply a nonlinear voltage difference V_{gs} between the input and output voltages. Second, since the FET can only source current and cannot sink current, it must operate in class A mode and its efficiency is therefore inherently low. We now present some very recently developed, and therefore preliminary, ideas for linearized class AB buffer implementation. The circuits implement the noninverting conductance function in Table 10.1, where gate–source voltages of the source and sink MESFETs are V and $-V$, respectively ($a_1 = 1$, $b_1 = -1$).

Consider the proposed buffer circuit in Figure 10.35*a*. The output current is the difference between the drain currents of source MESFET M1 and sink MESFET M3. Since MESFETs M1 and M2 share the same drain current, their gate–source voltages are identical. Therefore the input–output voltage difference v appears as the gate–source voltage of M2. MESFETs M2 and M3 form a cross-coupled MESFET pair (with ac-coupling), and therefore the gate–source voltage of M3 is $-v$, thus establishing the required gate–source voltages for the source and sink MESFETs. It follows that there is now a linear relationship between the output current and the input–output voltage difference v. By virtue of the transconductance function in Table 10.1 realized, the transconductance is double that for the conventional circuit in Figure 10.34*d*, and this allows the gate widths in Figure 10.35*a* to be halved. Also, and very importantly, the circuit in Figure 10.35*a* has the feature that it may be operated in class AB mode providing much higher efficiency.

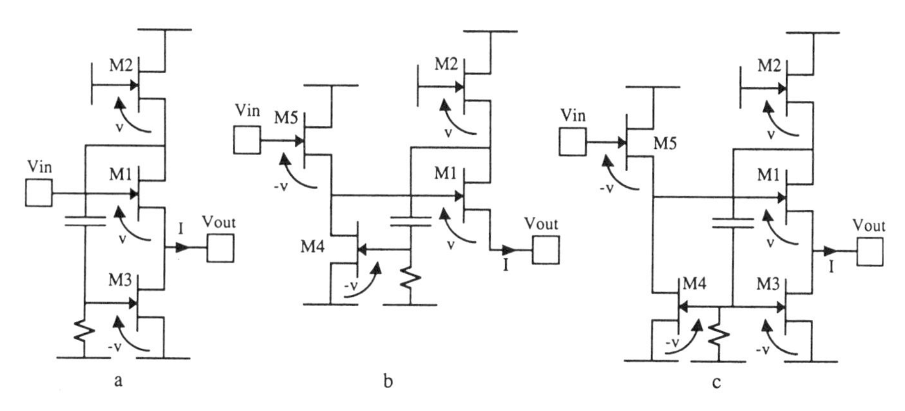

FIGURE 10.35 Linearized buffer circuits: (*a*) current difference linearized; (*b*) voltage difference linearized; (*c*) current and voltage linearized.

A buffer using an alternative method of linearization using subtraction of voltage functions rather than of current functions is shown in Figure 10.35*b*. Here, a second MESFET M5 is placed in the path between the input and output terminals and its gate–source voltage is the same as that of MESFET M4, which is cross-coupled to M2. This circuit has the property that the voltage difference between the input and output terminals, assuming the MESFET model of Equation (10.1) is ideally zero and may therefore be described as a "perfect follower." The linearization by current differencing in Figure 10.35*a* and the linearization by voltage differencing in Figure 10.35*b* may be combined in a single circuit as shown in Figure 10.35*c*.

The fact that both the voltage linearized buffer circuit in Figure 10.35*b* and all of the many current linearized circuits presented in this chapter rely on the cross-coupled MESFET pair suggests that a more general theory for linearization than that given in Section 10.2 might be possible on the basis of the mathematics of the cross-coupled FET pair. It may be of interest that for complementary devices, the corresponding equivalent to the cross-coupled MESFET pair is an N-channel FET and a P-channel FET with their gate–source ports connected in parallel. Thus, the CMOS inverter (or inverting transconductor) of Figure 10.36*a* and the CMOS buffer of Figure 10.36*b* [21] both conform to the linearization theory of Section 10.2; P1 and N2 are source FETs, and N1 and P2 are sink FETs. Perhaps the fact that these commonly used circuits are "good" circuits is not an accident, but arises from the beneficial properties of linearised circuits that we have discussed in Section 10.3.2.

Much work clearly remains to be done in simulating circuits such as those in Figure 10.35 under tough load conditions in order to explore their capabilities before fabricating selected circuits and testing them. We hope to have shown, nevertheless, that the synthesis techniques presented in Section 10.2 could make important contributions to the design of efficient, high-performance buffers.

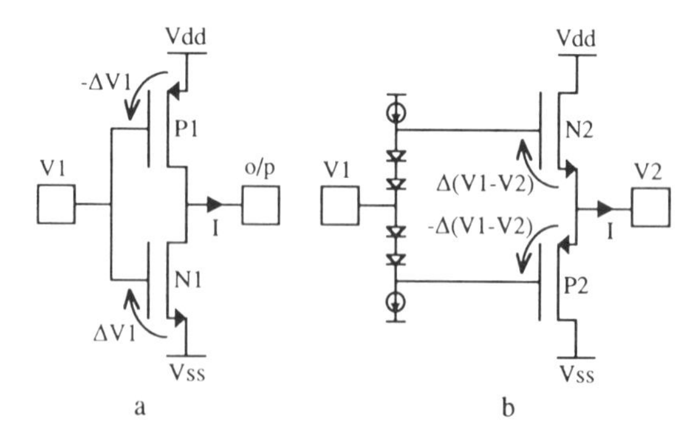

FIGURE 10.36 Linearized CMOS circuits: (*a*) inverter; (*b*) unity-gain buffer.

10.10 LINEAR COMPOUND FET

It is well known that performance characteristics of transistors can be improved by replacing individual transistors by compound transistors containing several devices. An important requirement in many electronic and communication systems is linearity, and we close this chapter with some preliminary ideas on the application of the linearization techniques we have proposed to the development of a compound FET circuit using depletion-mode GaAs MESFETs and diodes, in such a way as to linearize the nonlinear relationship of the MESFET [28].

The linearized compound FET may be defined as in Figure 10.37*a*, where $I_D = -I_S = I$ and I is a linear function of V_{GS}. The linear relationship between the single output current I_D and V_{GS} has already been implemented in the inverting linearized transconductor of Figure 10.14*a*. The problem we address here is the implementation of the current I_S.

The proposed circuit is shown in Figure 10.37*b* [28]. The part of the circuit constituting MESFETs M1–M4 is the linearized transconductor of Figure 10.14*a*, which provides a linear relationship between the current I_D and the voltage V_{GS}. In the present circuit, the current sources I_0 and the diode chain are introduced in order to make the current I_S equal to $-I_D$ and therefore also linearly related to V_{GS}. Thus the device may be regarded as a linearized replacement for a MESFET. The result $I_S = -I_D$ may be determined by analysis but, in fact, follows from Kirchhoff's current law since, in Figure 10.36, $I_G = 0$, and the only connections to ground are via equal and opposite current sources.

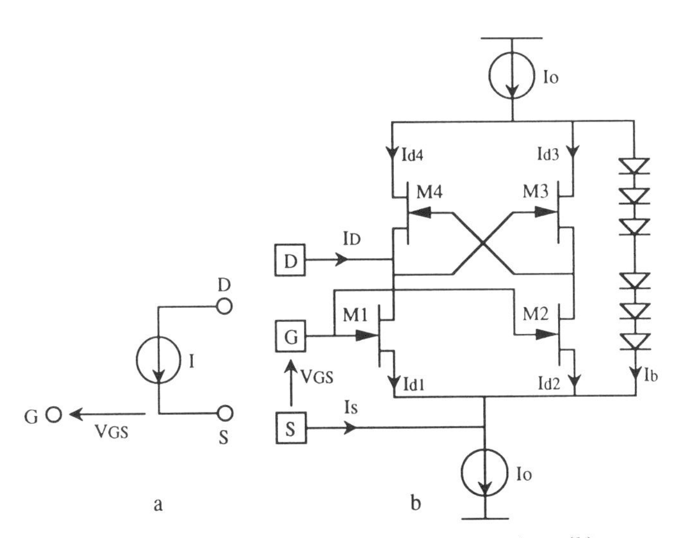

FIGURE 10.37 Compound FET: (*a*) symbolic representation; (*b*) proposed circuit.

Computer simulation using SPICE with an appropriate MESFET model as in Table 10.6 shows that the compound FET circuit is much more linear than the single MESFET it replaces in regard to both its transfer characteristics and its driving point characteristics [28]. In common-source configuration, the −3-dB bandwidths for a single MESFET of 19 GHz compares with 20.5 GHz for the compound FET. In common-gate configuration, the bandwidths are 46.0 and 25.6 GHz, respectively, indicating some bandwidth penalty expected for the more complex circuit. The input admittance characteristics of both circuits in common-gate configuration are found to be well behaved [28].

As with other topics we have presented, the theory and design of linear compound FET circuits using GaAs MESFETs appears promising but has, in fact, hardly begun. Many possibilities can be only just glimpsed that remain to be thoroughly researched. One possibility, for example, is the derivation of a linearized compound FET circuit from the inverting transconductance circuit of Figure 10.21. The drain of MESFET M3 may be disconnected from ground, terminated in a dc current source, and connected by means of a diode chain to the source of MESFET M1 to form a linear source terminal for a compound FET. Another important area is the use of ac coupling to realize class AB compound FETs with high efficiency. These and many further possibilities remain to be explored and evaluated in future work.

ACKNOWLEDGMENTS

For their invaluable help, the authors would like to thank Paul Radmore and Sean Smedley of University College London, Mahmood Darvishzadeh of SERC Rutherford Appleton Laboratory, and Shinji Hara of Sharp Central Research Laboratories, Japan. The isolator was fabricated by GEC-Marconi (Caswell) under the EEC Eurochip program.

REFERENCES

1. D. G. Haigh and J. K. A. Everard, eds., *GaAs Technology and Its Impact on Circuits and Systems*, Peter Peregrinus, London, 1989.
2. L. E. Larson, K. W. Martin, and G. C. Temes, "GaAs Switched Capacitor Circuits for High Speed Signal Processing," *IEEE J. Solid State Circuits*, **SC-22**, 971–981 (Dec. 1987).
3. C. Toumazou and D. G. Haigh, "Design of GaAs Operational Amplifiers for Sampled Data Applications," *IEEE Trans. Circuits Systems*, **CAS-37** (7), 922–935 (July 1990).
4. D. G. Haigh, C. Toumazou, S. J. Harrold, J. I. Sewell, and K. Steptoe, "Design and Optimisation of a GaAs Switched Capacitor Filter," *Proc. IEEE 1989*

International Symposium on Circuits and Systems, Portland, OR, June 1989, pp. 1449–1454.

5. D. G. Haigh, C. Toumazou, S. J. Harrold, K. Steptoe, J. I. Sewell, and R. Bayruns, "Design Optimisation and Testing of a GaAs Switched Capacitor Filter," *IEEE Trans. Circuits Systems*, **CAS-38** (8), 825–837 (Aug. 1991).
6. C. Toumazou and D. G. Haigh, "Design and Applications of GaAs Current Mirror Circuits," *IEE Proc.*, Part G (*Circuits, Devices and Systems*), **137** (2), 101–108 (April 1990).
7. D. G. Haigh and C. Toumazou, "High Performance GaAs Analog Sampled Data Circuits and Systems, *IEEE International Symposium on Circuits and Systems* (Singapore), June 11–14, 1991, pp. 1845–1848.
8. R. S. Soin, F. Maloberti, and J. Franca, eds., *Mixed-mode Analogue-Digital ASIC's*, Peter Peregrinus, London, September 1991.
9. R. Goyal, ed., *Monolithic Microwave Integrated Circuits*: *Technology and Design*, Artech House, Norwood, MA, 1989.
10. C. Toumazou, F. J. Lidgey, and D. G. Haigh, eds., *Analogue IC Design—the Current-Mode Approach*, Peter Peregrinus, London, April 1990.
11. Y. Imai, M. Tokumitsu, K. Onodera, and K. Asai, "10 GHz Bandwidth 20 dB Gain Low-Noise Direct-Coupled Amplifier IC's Using Au/WSiN GaAs FET," *Electron. Lett.*, **26** (11), 699–700 (May 1990).
12. F. Ali and A. Podell, "A Wideband Push–Pull Monolithic Active Isolator," *IEEE Microwave Guided Wave Lett.*, **1** (2), 26–27 (Feb. 1991).
13. T. Tokumitsu, S. Hara, T. Takenaka, and M. Aikawa, "Divider and Combiner Line-Unified FET's as Basic Circuit Function Modules—Parts I and II," *IEEE Trans. Microwave Theory Tech.*, **MTT-38** (9), 1210–1226 (Sept. 1990).
14. M. West, Jr. and M. Kumar, "Monolithic Integrated Blanking Up-Converter," *IEEE Trans. Microwave Theory Tech.*, **MTT-38** (9), 1227–1231 (Sept. 1990).
15. S. Hara, T. Tokumitsu, and M. Aikawa, "Lossless Broad-Band Monolithic Microwave Active Inductors," *IEEE Trans. Microwave Theory Techn.*, **MTT-37** (12), 1979–1984 (Dec. 1989).
16. D. G. Haigh and C. Toumazou, "Synthesis of Transconductor/Multiplier Circuits for GaAs Technology," *IEEE Trans. Circuits Systems*, **CAS-39** (2), 81–92 (Feb. 1992).
17. D. G. Haigh, P. M. Radmore, and A. E. Parker, "Advances in Linearised GaAs MESFET Circuit Design Technique," *Proceedings of 1992 IEEE International Symposium on Circuits and Systems* (San Diego), May 10–13, 1992.
18. D. G. Haigh, C. Toumazou, and S. J. Newett, "Measurements on Gallium Arsenide Building Blocks and Implications for Analog IC Design," *Analog Integrated Circuits Signal Processing* (Kluwer), **1** (2), 137–150 (Oct. 1991).
19. C. Toumazou and D. G. Haigh, "High Frequency Gallium Arsenide Current Mirror," *Electron. Lett.*, **26** (21), 1802–1803 (Oct. 1990).
20. D. G. Haigh, "AC-Coupled Linearised GaAs MMIC Circuits," *Proceedings of 1992 IEEE International Symposium on Circuits and Systems* (San Diego), May 1992.
21. P. R. Gray and R. G. Meyer, *Analog Integrated Circuits*, Wiley, New York, 1984.
22. D. G. Haigh and C. Toumazou, "High Frequency GaAs Transconductors and Other Building Blocks for Communications," *Proceedings of 1990 IEEE International Symposium on Circuits and Systems* (New Orleans), May 1990, pp. 1731–1735.

23. D. G. Haigh and C. Toumazou, "High Frequency Gallium Arsenide Transconductors for Communications," *Electron. Lett.*, **26** (8), 497–498 (April 1990).
24. Z. Wang, "A CMOS 4-Quadrant Analog Multiplier with Single-Ended Voltage Output and Improved Temperature Performance," *IEEE J. Solid State Circuits*, **SC-26** (9), 1293–1301 (Sept. 1991).
25. C. Toumazou and D. G. Haigh, "Linear Tunable Resistance Circuit Using Gallium Arsenide MESFET's," *Electron. Lett.*, **27** (8), 655–656 (April 1991).
26. D. G. Haigh, "Circuit Techniques for Efficient Linearised GaAs MMIC's," *1992 IEEE MTT-S Symposium* (Albuquerque, NM), June 1992.
27. C. Toumazou, D. G. Haigh, and J. M. Fopma, "Transconductor/Multiplier Circuits for Gallium Arsenide Technology: Part II—Practical Circuit Design," *IEEE International Symposium on Circuits and Systems* (Singapore), June 11–14, 1991, pp. 2991–2994.
28. D. G. Haigh and C. Toumazou, "Linear Compound FET Circuit Using Gallium Arsenide MESFET's," *Electron. Lett.*, **27** (16), 1494–1495 (Aug. 1991).

Index